The Chemistry of Human Nature

The Chemistry of Human Nature

Tom Husband

THE QUEEN'S AWARDS
FOR ENTERPRISE:
INTERNATIONAL TRADE
2013

Print ISBN: 978-1-78262-134-8

A catalogue record for this book is available from the British Library

Published by The Royal Society of Chemistry,
Thomas Graham House, Science Park, Milton Road,
Cambridge CB4 0WF, UK

Registered Charity Number 207890

Visit our website at www.rsc.org/books

Printed in the United Kingdom by CPI Group (UK) Ltd, Croydon, CR0 4YY, UK

Preface

Just after I graduated, I was struck by a sudden urge to share what I had learned during my chemistry degree. Inspiration came to me in the form of a story about atoms. I fancied that carbon would be like a mafia godfather figure, having his fingers in so many pies. In the burgeoning narrative, silicon became its arch rival—an element in the same group with the capacity to form the same number of bonds, yet unable to forge organic compounds. The element's seething jealousy would drive his bid to reign supreme in the silicon chips of the artificially intelligent machines that would overthrow their carbon-based oppressors.

As original as these ideas seemed, they made dreadful fiction. Just awful. The various drafts I pushed on my friends were met with polite expressions of utter bewilderment. After much soul searching, it finally occurred to me that a non-fictional text might be the best way to share my love of science and that was how this book came into existence.

But what does this story really explain about my actions? I wrote the book because I love writing and because I love science, but this admission only presupposes more fundamental questions. Why do I love writing and science? Many of the questions considered in this book are relevant to the point of inquiry: does my urge to create have something to do with the number of receptors in a region of my brain linked with creativity? Were the genes for that variety of receptor expressed

The Chemistry of Human Nature
By Tom Husband
© Tom Husband 2017
Published by the Royal Society of Chemistry, www.rsc.org

more or less frequently because of other genes I inherited? Perhaps a mutated gene I inherited. Or did my upbringing cause my body to express a normal gene more or less frequently? Finally, was the book's authorship a meaningful expression of human consciousness or does it all boil down to a status-chasing exercise in securing passage for my genes in future generations?

While I have been writing this book, a lot of people have complained that human nature is much too complex to boil down to chemistry. Physicists and biologists might say this because the book draws on their disciplines, as well as that of neuroscience and others. But what the people meant was that surely culture also influences our behaviour. Some people also admitted that they do not like the idea that our motivations arise from the machinations of molecular self-replicators. As such, I think it is worth saying something about the book's title. Is it saying *all* of our behaviour is explicable in terms of chemistry and this book explains *all* the chemistry involved, or is it saying *some* of our behaviour is explicable in terms of chemistry and this book explains all that chemistry? The answer to this question is for you to decide.

Tom Husband

To Robin, Priscilla, April and Jess

Contents

Section 1: Fuel and Building Supplies

Chapter 1
The Chemistry of Space Travel   **3**

1.1   Is it in Our Nature?   4
1.2   A Reaction of Self-replication   5
1.3   A Risky Strategy   8
    1.3.1   Eating   8
    1.3.2   Reproduction   9
    1.3.3   Status   10
    1.3.4   Free Will   12
1.4   Chemistry Fundamentals—A Possible Way Back   16
References   23

Chapter 2
The Chemistry of Taste   **25**

2.1   How Muscles Work   30
2.2   How Taste Buds Work   40
    2.2.1   Ion Channel Taste Receptors   41
    2.2.2   G-Protein Coupled Taste Receptors   43
    2.2.3   What Happens Next?   45
2.3   What They All Have in Common   47
References   49

The Chemistry of Human Nature
By Tom Husband
© Tom Husband 2017
Published by the Royal Society of Chemistry, www.rsc.org

Chapter 3
The Chemistry of Pleasure   **51**

3.1   The Pleasure Circuit   52
      3.1.1   Ions and Potential   53
      3.1.2   How Neurons in the Gustatory Nerve Transmit
              Sweet Signals   54
      3.1.3   Tasting Chocolate: The Next Step   60
3.2   The Case for Food Addiction   68
      3.2.1   Where Next for Dopamine and Food Addiction?   73
References   75

Chapter 4
The Chemistry of Life's Origins   **79**

4.1   Where Did RNA Come From?   90
4.2   Chemical Evolution   95
4.3   Cosmic Origins   98
References   100

Chapter 5
The Chemistry of Evolution   **102**

5.1   From Self-replicators to Organisms   103
      5.1.1   Cell Membranes   103
      5.1.2   Proteins   104
5.2   From Replication to Reproduction   110
      5.2.1   Single-celled Organisms   111
      5.2.2   Multicellular Organisms   111
      5.2.3   Sexual Reproduction   113
5.3   Developing Choice   120
5.4   Introducing Pleasure   125
References   130

Section 1: Concluding Remarks   **134**

References   136

Section 2: Love and Relationships

Chapter 6
The Chemistry of Lust   **141**

6.1   Androgens and Women   147
6.2   How Testosterone Works   148
6.3   A Radical Transmitter   157
6.4   Priming Genitals   159
6.5   The Lust Hypothesis   164
References   164

Chapter 7
The Chemistry of Romantic Love   **169**

7.1   Monoamines   172
      7.1.1   Getting Rid of Monoamines   175
7.2   The Hypothesis Develops   177
7.3   Duping Dopamine Receptors   179
7.4   What about Humans?   181
7.5   Which Neurotransmitter?   183
7.6   Why Does My Heart Go Boom?   186
7.7   Another Role for Noradrenaline   189
References   194

Chapter 8
The Chemistry of Attachment   **198**

8.1   The Puppet Masters of Mammalian Attachment   203
8.2   Oxytocin—the Cuddle Hormone?   206
8.3   The Role of Temperature   208
8.4   Cheating the Sensors   209
8.5   Not a Cuddle Hormone   218
8.6   The Neuromodulator   218
8.7   Reward   221
References   225

Chapter 9
The Chemistry of Baby Making   229

9.1   Epigenetics   230
9.2   Foetal Development   236
      9.2.1   The Three Hypotheses   239
      9.2.2   Mechanism   242
      9.2.3   Merging Hypotheses   244
9.3   Epigenetic Inheritance   248
9.4   Conclusion   249
References   249

Section 2: Concluding Remarks   252

Section 3: The Chemistry of Character

Chapter 10
The Chemistry of Creative Intelligence   259

10.1   The Birth of a Paradigm   261
10.2   How Heritable is Creativity?   266
10.3   The Validity of Heritability Studies   268
10.4   The Price of Creativity   270
10.5   Nature–Nurture Duality   276
10.6   A Role for the Reward Network?   280
References   281

Chapter 11
The Chemistry of Violence   287

11.1   Foetal Alcohol Spectrum Disorder   288
11.2   What is a Gene Anyway?   290
11.3   Brunner Syndrome   292
11.4   Other Problems with the *MAO A* Gene   294
11.5   A Role for the Epigenome   296
11.6   A Role for the Reward Network   297
11.7   Conclusion   299
References   300

Chapter 12
The Chemistry of Dominance 303

12.1 What is Dominance? 305
12.2 Dominance and Status 308
12.3 Heritability of Status 310
12.4 Findings from Neuroscience 314
    12.4.1 Processing Information 316
12.5 Cocaine Studies 318
12.6 Serotonin 321
12.7 The Winner Effect 323
    12.7.1 Nobel Legacies Boosting the Hypothesis for Reward Mediation of the Winner Effect 325
12.8 Conclusion 331
References 332

Section 3: Concluding Remarks 336

Section 4: So What?

Chapter 13
The Chemistry of Free Will 341

13.1 Unease around Emerging Biology 341
    13.1.1 Dennett's Position 343
13.2 Indeterminacy 347
13.3 Pleasure and the Three Directives 348
13.4 Breaking the Rules 351
    13.4.1 Sex and Drugs 351
    13.4.2 Buddhism 352
    13.4.3 So, Can We Break the Rules or Not? 354
13.5 A Healthy Balance? 355
13.6 More Answers from Buddhism 357
13.7 What Should We Change? 359
References 361

Appendix: Thermodynamics, Immiscible Liquids and Heat Receptors 365

Part 1: What Makes Reactions Happen 365

Part 2: Life and Thermodynamics 370
Part 3: Free Energy 371
Part 4: A Truer Picture of Entropy 373
Part 5: Energy Levels 375
Part 6: The Relationship of Entropy with Energy Levels 377
Part 7: The Hydrophobic Effect 382
Part 8: Proteins, Specific Heat Capacity and the Heat Receptors 385
Specific Heat Capacity and the Hydrophobic Effect 386
Folding Proteins 387
The Thermodynamics of Heat-detecting Ion Channels 390
References 396

Acknowledgements **398**

Subject Index **400**

Section 1:
Fuel and Building Supplies

CHAPTER 1

The Chemistry of Space Travel

One day in April, 2013, my students burst into my classroom clamouring about something utterly incomprehensible—a group of individuals had volunteered to take a trip to Mars. This fact alone did not quite account for the levels of consternation they were demonstrating, until one of them excitedly added: "They're not coming back! They've said they'll go to Mars but that they'll stay there." They had learned on the news about the Mars One mission—a voluntary, one-way trip to the Red Planet open to non-professionals. This was not one of those adult preferences I could explain by assuring them they would understand when they were older; I was just as flummoxed as they were.

Two connections with the origins of life occurred to me, one immediately, the other much later. Mars is in the habitable zone of our solar system, meaning its conditions could potentially support life. Martians have been a rich theme for sci-fi and now these Mars One astronauts may get the chance to seek actual evidence of their existence. But could there be another connection? Could it be that the manner in which life emerged explains the baffling motivation of these volunteers?

This book will present the case that the unique conditions necessary for life to originate are echoed throughout our modern-day behaviour. The central role of eating in cultures around the world, our romantic search for *The One*, our social

The Chemistry of Human Nature
By Tom Husband
© Tom Husband 2017
Published by the Royal Society of Chemistry, www.rsc.org

tendencies and our craving for success can each be understood
in terms of the thermodynamic demands of the chemical reactions that launched life. But how compatible are these traits
with a suicide mission to Mars?

1.1   IS IT IN OUR NATURE?

What can the intentions of the Mars One hopefuls tell us about
human nature? *Homo sapiens* have been explorers since we left
Africa to colonise the world and perhaps before. But we are also
social creatures inclined to live in communities and raise
families.

The aims of the Mars One hopefuls may or may not be representative of human nature. More than 200 000 people applied
for a place on the mission[1] and, in February, 2015, a shortlist of
100 candidates was published.[1] A gruelling training program will
serve as the interview process, whittling that number down to 10
successful candidates—the intrepid individuals who will leave
behind their friends, family and atmospheric oxygen in order
to settle permanently in extra-terrestrial environs. Certainly, the
number of applicants is large, but it forms a miniscule proportion of Earth's population. We can only speculate on what
proportion of those applicants would actually go through with
the mission if they were selected. If 10 volunteers actually depart
for Mars, their actions will not be necessarily demonstrative of
human nature.

What their actions might tell us about human nature requires
consideration of what human nature is. The following is a proposed list of universal human traits compiled by George Peter
Murdock:

> *Age-grading, athletic sports, bodily adornment, calendar, clean-*
> *liness training, community organisation, cooking, cooperative*
> *labour, cosmology, courtship, dancing, decorative art, divination,*
> *division of labour, dream interpretation, education, eschatology,*
> *ethics, ethnobiology, etiquette, faith-healing, family, feasting,*
> *fire-making, folklore, food taboos, funeral rites, games, gestures,*
> *gift-giving, government, greetings, hairstyles, hospitality, hous-*
> *ing, hygiene, incest taboos, inheritance rules, joking, kin-groups,*
> *kinship nomenclature, language, law, luck superstition, magic,*

marriage, mealtimes, medicine, modesty concerning natural functions, mourning, music, mythology, numerals, obstetrics, penal sanctions, personal names, population policy, postnatal care, pregnancy usages, property rights, propitiation of supernatural beings, puberty customs, religious rituals, residence rules, sexual restrictions, soul concepts, status differentiation, surgery, tool making, trade, visiting, weaning and weather control. (Reproduced with kind permission from Columbia University Press.)[2]

These traits can be divided into four interconnected themes: safety in numbers, eating, mating and status. Four traits are directly linked to eating, whilst six relate to romantic union and the potential it provides for reproduction. Roughly half of the characteristics could be categorised under safety in numbers. For example, body adornment denotes clan membership, the hope being that the person dressed like you will fight alongside you if outsiders attack. Publicly observed customs on hygiene promote safety through good health, while collective agreements, such as governance, law-making and property rights, assure security. The status theme is only directly linked with the trait of status differentiation, but it influences many of the others. High status predicts superior access to food and mating opportunities, along with greater means to preserve the well-being of one's kin. Moreover, educational attainment is a common indicator of socio-economic status, religious rituals typically produce religious leaders, etiquette incorporates deference and so on.

1.2 A REACTION OF SELF-REPLICATION

All of these traits also have something to do with wants. Athletic sports are only interesting because the athletes want to win. We want to spend time with loved ones and when their demise makes that impossible, we mourn them. We observe laws because we want to be treated fairly. We observe certain customs because we want to stay healthy. We use tools to make things we want, and use trade to get things we want.

The fascinating thing about our wants is that we are made of atoms that do not want things. As we will see in Chapter 4, this situation seems to have arisen from the emergence of a very special kind of molecule, which has the ability to make copies of itself. These self-replicating molecules were the first ancestors of our modern day DNA, which now feverishly produces copies of itself. These long molecules are like necklaces composed of four different beads, with the twist that each bead is exclusively attracted to one of the other three. As such, a completed necklace will attract nearby, loose beads to line up next to it in a sequence that exactly mirrors it. When the mirror image spirits yet more beads to line up alongside it, the sequence of the third necklace exactly matches the first. In the same way, these self-replicating molecules were made up of four different units called nucleobases, which also form exclusive pairings.

Scientists have long puzzled over how to resolve this replicative activity with the laws of thermodynamics, seemingly fundamental rules of the universe that favour chaos over order. Our bodies are phenomenally ordered, notwithstanding the chaos we use them to create in the world. As such, the ongoing chemical self-replication that sustains life has been called a *far-from-equilibrium state*, which depends on the constant supply of two things: fuel and building materials. Provided self-replicators maintain this constant supply, they can keep replicating.

Now let's imagine two self-replicators, one short and one long. The short self-replicator will be able to copy itself more quickly than the long replicator. The replica of a self-replicator is itself a self-replicator, and once the first replica is complete, it will start making new copies of itself. Before the long replicator has finished a single copy, its rival has multiplied into two, both of which are now copying themselves again. If the building materials and the necessary fuel are in finite supply, the long self-replicator will eventually go out of existence.

Two developments follow. First, the self-replicators make mistakes. Sometimes they produce an unfaithful copy, where the wrong "necklace bead" gets added to the chain. This can lead to the production of faster self-replicators that will outbid their peers for the finite supplies. Second, the self-replicators start to act as instruction manuals for how to make other molecules called proteins. Again, some of these proteins may be a hindrance, but others will help the self-replicators to compete for the limited resources available.

These two facts trigger an arms race. Copying errors can help the self-replicators to reproduce more quickly, but these molecules also code for the production of proteins. What happens when the instruction manual makes an unfaithful copy of itself? Then, following the instructions produces a different outcome. When self-replicators make shoddy copies, new proteins are invented, some of which will be helpful and others a hindrance.

The arms race escalates. The self-replicators constantly modify their operation and those that improve their efficiency breed their rivals out of existence. They stumble over a recipe to build the protective casing we now call a cell membrane. These membranes act as filters to absorb useful substances and keep out harmful ones. These emerging lifeforms develop molecular motors to help them move. They evolve linking proteins, with which they reach out for each other and form mutual alliances. Cell types diversify, enabling division of labour. Organisms grow and they learn to transform sunlight into fuels, send tendrils burrowing into the earth to extract minerals and grow leaves to harvest the sun's rays. Anything that helps the self-replicators better compete for the finite resources of fuel and building materials swells their production of replicas. Meanwhile, newly forged replicas tumble off the production line, equipped with the same set of tactical advantages.

A huge breakthrough was the evolution of pleasure. Evolving combinations of new and old proteins formed brains that could direct the activities of the organism. These brains worked by expressing their approval, rewarding host organisms with a pleasurable sensation when they helped their self-replicators to get what they want. Eating food earns a reward because it supplies fuel and building materials in the form of nutrients, such as sugars and proteins. Mating earns a reward because it produces new vessels in which for self-replication to continue after the old vessels have died. Now, as before, atoms do not want things and self-replicators made of atoms do not want things, but they have succeeded in forging an intelligent, protective casing that does want things, and it wants not for itself but for its self-replicators (at least until humans appeared).

Our wants begin to make sense in light of the needs of the self-replicators. We are rewarded for eating as it provides the energy and protein necessary to replicate the DNA into which those ancestral self-replicators have evolved. Sex is pleasurable because we are rewarded for actions likely to produce children, the new vessels in which for self-replication to occur. We are adapted to cooperate so as to improve our safety, as well as to spread the risk of maintaining the food supply necessary to keep us alive. Meanwhile, elevating our status increases our access to food. Being overweight became a status symbol in post-revolutionary China, signalling that you could afford to eat well (on the frequent occasions that I was called fat when I lived there, I was assured it was a compliment for this reason). Status also improves our chances in the dating game, enabling us to select mates that represent the best long-term chances for the continued self-replication of our genetic material. If there is one thing peacocks have taught us, it is that bling gets chicks.

1.3   A RISKY STRATEGY

Universal human traits reflect the needs of the self-replicators, but do the actions of the Mars One volunteers make sense in the light of those traits? At face value, a one-way trip to Mars is a risky strategy to guarantee the continued replication of their genetic material. But then, the way things are going on Earth, the greater risk might be to stay put. Do our wants still reflect the needs of the self-replicators?

1.3.1   Eating

One of the Mars One hopefuls, Leila Zucker, has said that the hardest thing to give up would be her husband, followed by meat.[3] In her shoes, I would be worried that there will not be anything to eat at all: permanent emigration to a barren planet is not the last word in food security. But the would-be colonisers will need to eat and the great challenge of doing so will merely heighten the focus. If they want to live on Mars, they will have to pioneer a sustainable means of food production. Assuming the organisers secure enough funding to run the mission, presumably a sizeable sum will be available for this critical requirement. So, while a trip to Mars does not guarantee access to nutrition, it is guaranteed that considerable effort will be made to meet this need.

1.3.2 Reproduction

This is where things start to fragment. A one-way trip to Mars may or may not represent the best mating strategy, but the best mating strategy may or may not represent the motivations of the candidates. The strategies evolved by the self-replicators often drive us to commit actions that are no longer of any service to them. When we consider how they influence our behaviour, we have to think about what such actions *used to* achieve.

Witness the dichotomy. Mars One hopeful Maggie Lieu has already professed her desire to be the first woman to have a baby on Mars.[4] Conversely, Sheldon Cooper, fictional star of TV sitcom *The Big Bang Theory*, also applied for the mission, but he is equally averse to coitus and child rearing. Whilst fictional, Sheldon's proclivities are representative of real people. But can Sheldon and his self-replicators both get what they want?

The question is moot because self-replicators cannot want things. In fact, they do not even direct things. Physical laws dictate atomic interactions and those that have previously accelerated self-replication have become incorporated into the self-replication process. Reproduction creates new vessels in which for self-replication to continue and, in the past, sex was the only way that could happen, which is why engaging in the practice elicits a rewarding sensation. But now contraception enables us to reap the reward without sewing the seed.

Why then, should the Mars One astronauts, or anyone else for that matter, have to have children? A better way to evaluate their motivations might be to consider whether they can have an active sex life on Mars. The brain rewards us for creating the conditions for conception to occur, but when an egg is actually fertilised, the event does not register with our brains at all. We get paid to start the job, not to finish it; otherwise, our sex lives might be very different. As such, the opportunity to conceive is a separate and frequently much stronger motivation than the opportunity to copulate.

But embarking on the trip need not prevent the successful candidates from reproducing; it is possible that this strategy may yet prove to be the fittest for survival. Fellow Mars One applicant, Michael McDonnel, has suggested that a one-way trip to Mars may be an effective way to mitigate the threats of global warming, nuclear war and asteroids.[3] Many would argue that global warming might prove to be such a problem. As climate scientist Kevin Anderson puts it: "There is real hope, but that hope reduces significantly each day."[5] Under such a serious threat, a trip to Mars may not represent the worst chance for the continuation of our existence or of the self-replicating genes that enable it.

Ultimately, the conclusion is similar to the case concerning eating. The best it may be possible to say for the Mars trip is that it is not necessarily *incompatible* with drives originated from self-replicator activities, although it could equally be the only way to survive Armageddon, should it occur.

1.3.3   Status

The arguments concerning status are just as fragmented. How important is status to us? How well does it advance the interests of the self-replicators? Does it matter? In Chapter 12 we will consider whether the brain's pleasure centre rewards us not only for eating and having sex but also for elevating our status.

There is little doubt that participation in the Mars One mission will raise the profile of everyone involved. British candidate Ryan MacDonald openly admits that part of his motivation is to leave a legacy.[6] This information appeared in one of many news articles, which themselves show that the process of status elevation is already well underway. The 10 successful candidates, as well as the runners up, will be world famous long before they embark on their voyage. In evolutionary terms the argument goes that we strive to raise our status in order to create opportunities to eat and procreate. A one-way trip to Mars is unlikely to culminate in a baby-booming banquet, so will their enhanced status actually progress the aims of their self-replicating genes?

What is the point of elevating your status if not to increase access to food and mates? Remember that the self-replicating genes do not want anything. In our times as hunter-gatherers, elevated status increased access to food and mates, and, as we have evolved minimally since, the biological mechanisms responsible still exist. There were no one-way trips to Mars back then, so it did not matter. Anything a person did to elevate their status was unlikely to simultaneously remove them from the reproductive arena. Does that mean the hopefuls are misguided to embark on their mission? Not necessarily.

There are several ways to consider this apparent disconnect, including that it may not actually be a disconnect. First of all, the hopefuls will be famous before they go to Mars and, even if they withdraw from the mission, they will still have elevated their status. This mirrors the brilliance of the Mars One organisation's strategy. The endeavour may be profitable even if the mission never takes place. Once the successful applicants have been announced, they will embark on training, all of which the world will witness in the reality TV show to be produced by Endemol offshoot Darlow Smithson Productions.[7] At some point, if governments of the world start to believe they are serious about launching a Mars-bound rocket, they are bound to intervene, creating an obstacle very lucrative to production companies who understand the vitality that conflict brings to a narrative. Meanwhile, any of the hopefuls could voluntarily drop out, having elevated their status, yet without having to take the trip to Mars.

That is not to say that embarking on the trip need prejudice their genetic legacy. The inclusive fitness theory, originated by William Hamilton, shows that natural selection can favour outcomes where individuals sacrifice themselves for the clan. A person's uncle and father share many genes. If the family is attacked in such a manner that all of them will die unless one member sacrifices himself, then the uncle's sacrifice will lead to the preservation of his genes in his nieces and nephews. This does not mean that the uncle hatches the unconscious plan to advance his genes by the least bad outcome but rather that a gene, or more likely a set of genes, may promote the behaviour. While the uncle will not have achieved self-preservation, the genes will, which shows how the strategy is competitive in terms of natural selection. Similarly, while the Mars One hopefuls may never themselves reproduce, they may elevate the status of their families in a manner that their own genes thrive in the progeny of their siblings.

What, then, is the disconnect? A one-way trip to Mars will heavily restrict the freedom with which the astronauts can trade on their elevated status but could nonetheless help their genes to flourish in future generations. If catastrophe wipes out humans on Earth, mating on Mars will represent the best chance of survival. Alternatively, self-replicators may leave their legacy in children, nieces or nephews who remain on Earth. By this rationale, there is no disconnect between the elevation of status and the enhancement of reproductive potential. On the other hand, we can also understand their actions in terms of what *used* to get results for the self-replicators. High status hunter-gatherers had their pick of potential mates. Things have changed. The Mars One volunteers may have evolved a means to enhance status that prejudices rather than promotes reproductive potential. Such concerns may extend beyond these adventurous souls. Is our pursuit of status a legacy or a hangover?

The Mars mission is not necessarily incompatible with the motivations arising from the needs of the self-replicators. Not exactly a ringing endorsement, but it does present another question. If we concede (and we may not) that the Mars mission just about meets the criteria for the enhancement of reproductive fitness, what of the behaviours that do not? Put more simply, is it possible to act in ways that do *not* promote reproductive fitness? Or another way: how free is the will of the candidates hoping to travel to Mars?

1.3.4   Free Will

The argument that these candidates are chasing status may seem flimsy—especially to the candidates themselves. They want to live out their dreams, transmute reality from science fiction, taste new experience, inspire young minds and safeguard humankind from the ravages of a cruel universe. Perhaps more than anything they want to be pioneers. But they are making enormous sacrifices. Why do the hopefuls not try to be pioneers in other fields? A series of decisions will seal their fate. In candidate Dr Leila Zucker's case, the first decision was her husband's, who told the media that he would have been a lousy spouse if he had not told his sci-fi-loving wife about the mission.[3] Next, presumably, she made a decision to fill out and send a form. Many decisions will follow that do not require her to board a rocket. If she is successful, at some point a final decision will need not only to be made but to be maintained past a point of no return, perhaps when the blasters fire. The age-old debate into free will investigates the extent of her control over this process.

The Greek philosopher Democritus wrested his destiny from the fate of the gods by the hypothesis that all matter is made of atoms, indivisible particles that are subject to natural laws.[8] Prior to this visionary hypothesis it was widely held that the gods decided the fate of men. By suggesting that man and matter are composed of atoms, Democritus had liberated himself at a cost. If our brains control our actions and the inner workings of our brains are the result of atomic interactions governed by physical laws, can we claim any responsibility for our actions at all? This troubled Democritus, who did not like the idea that an individual was not responsible for his wrongdoings.

Science has greatly informed the question of whether we have free will. Democritus seized his rudder from the capricious gods by invoking the laws of physics, only to wonder that if this were so, how could villains be held responsible for their wrongdoings? This was a forerunner to the theory of scientific determinism introduced by the 16th century French classical physicist Laplace, who first suggested that if one knew the position and momentum of every particle in the universe at any instant in history, then one could work out the position of every particle at any other time in history. A process that could accurately predict the future. A powerful computer programmed to complete this gargantuan calculation might foresee that a certain person, disadvantaged by his genetic inheritance and/or the circumstances of his upbringing, would certainly commit a particular crime at a certain point in time. If the laws of physics directed the atoms in his mind to cause his action, in such a way that there was *zero* possibility that he could have chosen a different action, why should he be punished? But what if Democritus and Laplace were wrong? The philosopher Epicurus revised the atomic theory of his predecessor. He felt that atoms, although chiefly subject to universal laws of physics, periodically committed a "random swerve".[8] This random activity, he suggested, restored freedom of will to man, an idea which pre-empted modern-day developments.

One important flaw in Democritus' atomic theory concerns the name itself. The word comes from the Greek *atomos*, meaning indivisible, which is what he thought atoms were. In fact, atoms can be divided into the sub-atomic particles (protons, neutrons and electrons), which can be further subdivided into even smaller particles, such as leptons, quarks and the elusive Higgs boson.

The two stalwarts of chemistry are protons and electrons. Protons are said to have a positive charge and electrons a negative charge, which causes them to attract each other but to repel themselves. This attraction is at the root of most things that happen anywhere throughout the universe (no one is quite sure how gravity fits in with these smaller scale interactions, a conundrum that Einstein spent much of his career trying to unravel). What influence do such interactions have inside our bodies?

While it is true that electrons and protons attract each other, they don't behave exactly as you might think. Magnets attract each other. If you put two of them near each other on a smooth enough surface, they will move towards each other until they are touching. However, this idea of attraction does not extend to the interaction between protons and electrons. Atoms are composed of a positive nucleus, which is made up of protons and usually neutrons too, surrounded by a negative cloud of electrons. Since electrons and protons attract each other, you might expect to find all of the electrons stuck to the surface of the nucleus, but this is not the case. The electrons follow an erratic, unpredictable path near the nucleus but not touching it. In this manner their behaviour is more like a moth circling a lightbulb than two magnets edging towards one another. A moth flies towards a lightbulb, but it does not land on it and then settle down on its surface. It may periodically touch the lightbulb but will then immediately fly away. Even the word "circling" is dubious, suggesting that it orbits the bulb in perfect circles. There is no geometry to its flight path. It flies near the lightbulb and at a varying distance from it, as well as up and down, in an unpredictable pattern much like the electron, as shown in Figure 1.1. The only predictable aspect of its behaviour is that it will stay near the bulb and, in that sense, is attracted to it. In fact, to suggest that an electron has a flightpath is misleading and the analogy illustrates nothing about the wave–particle duality of the electron, but it does show how the distance between two bodies can increase even while attraction exists between them. Despite the attraction between protons and electrons, the electrons do

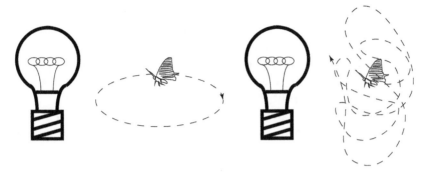

Figure 1.1 Moths do not fly in perfect circles but rather in erratic, unpredictable patterns like the electron.

not cling to the surface of the proton-containing nucleus but zing around near it in fabulous ways that continue to confound our understanding.

This was a very vexing problem for Einstein. He quarrelled with his contemporary, Heisenberg, who originated the uncertainty principle, which states that it is only possible to measure the momentum of an electron or its exact position but not both. A joke has it that a policeman pulls over an electron and asks: "Excuse me Sir, do you realise you were travelling close to the speed of light?" "Great!" the electron replies, "now I'm lost." Einstein was convinced that the path was predictable but that physics had yet to work out the pattern, hence the quote famously attributed to him that God does not play dice with the universe. Heisenberg's uncertainty principle became a cornerstone of quantum theory.

These ideas have been linked to the Epicurean random swerve.[9,10] Measuring the position or the momentum of a subatomic particle causes its wave function to collapse. Prior to this collapse, the particle is said to be in a superposition—a combination of quantum states, each of which predicts different, incompatible outcomes. Having been observed, the wave function collapses into one of these states, selected apparently at random like the swerving atoms.

This has implications for determinism. The path of an electron influences an atom's interactions with its neighbours, so if the path of the electron cannot be predicted, it suggests that Laplace's theory is flawed. Even a powerful enough computer, fed with the precise positions of every particle in the universe, could not predict their future positions. This phenomenon is known as quantum indeterminacy. Controversy rages over its significance to free will and even its existence, but philosophers continue to argue that it may yet diffuse Democritus' fears for our moral responsibility.[11]

Section 4 will consider the freedom of will in the context of the directives arising from the activities that have previously served the self-replicators. Can we choose to shun them? Would we want to? Can we consider ourselves free if we do not?

So, the successful candidates for the Mars mission may or may not act on their own free will. Many have argued that a condition of free will is that the person could have chosen a different course of action. Any candidates who finally embark on their mission are bound to consider at least twinges of regret. In that event will they wish they had acted differently and, if so, would they have been able to act differently? Perhaps another way to put the question is this: can we truly do something just for the sake of it, or is our every action contrived to directly or indirectly increase our access to food or mates? These are important questions, the answers to which ultimately decide whether the Mars One astronauts will boldly go where no one has gone before, or simply go where no one has gone before.

1.4   CHEMISTRY FUNDAMENTALS—A POSSIBLE WAY BACK

Hope springs eternal and this may shed further light on the hopeful's motives. Are they secretly gambling on the possibility that technology will advance sufficiently to secure their passage home? Mission hopeful Greg Sachs has plainly stated that he would prefer to come back.[3] If the hopefuls are indeed hopeful of coming back, the odds may not be stacked so heavily against them. Scientists are already conceiving a less extreme mission to Mars, that is, one with the desirable design feature of a ticket home.

The concept mission designed by a group at Imperial College London is more of a feasibility study than a design blueprint. It describes a plausible plan for a two-way manned trip to Mars. The designers of the concept mission have conceived an ingenious way to produce the fuel necessary to launch the rocket's return journey. The plan is to dig up ice from the planet's crust and break it down into hydrogen and oxygen (sci-fi aficionados may be reminded of the plot from *Total Recall*, based on a short story by Philip K. Dick, in which a reactor bequeaths Mars with an atmosphere by melting the planet's icy core). The hydrogen will be reacted with carbon dioxide in the atmosphere to make methane, which will be used as rocket fuel. This presents a useful opportunity to review some chemistry basics.

The reaction between methane and oxygen is the one that could enable the astronauts' return to Earth. The heat released by the combustion of methane will cause the gases to expand, which will propel the rocket.

In Figure 1.2, each circle represents an atom and the different patterns represent different kinds of atoms called elements. In this case, the elements are hydrogen, carbon and oxygen, which are distinguished by the numbers of protons they contain; hydrogen has the fewest, oxygen the most. Two or more attached circles show a molecule that represents atoms bonded to each other. A glass of water contains roughly 10^{24} molecules of water. There are also different kinds of molecules, those that contain just one element, such as oxygen, and those that contain different elements, which are called compounds, in this case, methane, carbon dioxide and water.

This is an example of the classic downhill reaction. A boulder at the top of the hill needs just a nudge—for example, the lighting of a match—to roll all the way down to the bottom of the hill. The snag is, it will not roll back up the hill. The products carbon dioxide and water will not react with each other to turn

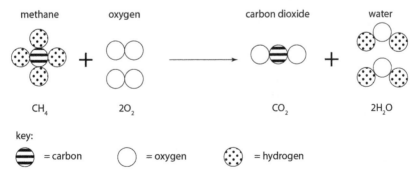

Figure 1.2 A downhill reaction: the combustion of methane to form carbon dioxide and water.

back into more methane, or not simply by the ignition of a match anyway. In order to reconstruct methane from carbon dioxide and water, the boulder has to be pushed up the hill by a process such as photosynthesis. Natural gas, which is basically methane, was produced from the decomposition of ancient plants, which used sunlight and biochemical machinery to transform carbon dioxide and water into the biological fuel glucose. Even in this case, the boulder was not pushed directly up the hill but by zigs and zags *via* various intermediates.

The methane needed to bring the astronauts home would be produced by similar zigs and zags. First of all, the dug up ice would be melted and decomposed into hydrogen and oxygen. The hydrogen would be reacted with carbon dioxide from the Mars atmosphere to produce the methane and water, as shown in Figure 1.3.

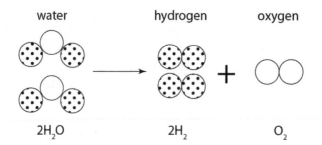

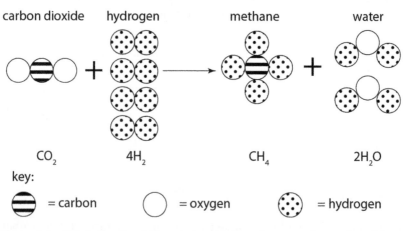

Figure 1.3   The indirect uphill route by which methane can be produced from dug up ice and atmospheric carbon dioxide.

No one has ever witnessed a glass of water foaming with bubbles of gas as it spontaneously breaks down into hydrogen and oxygen gas. This is an uphill reaction, so some kind of substitute for photosynthesis is required. In fact, photovoltaic cells will convert sunlight into electrical energy, which can be used to power the process of electrolysis. This is the chemical process that pushes boulders back up hills.

The defining characteristic of chemical reactions is the transfer of electrons between atoms. Earlier, we saw that electrons zing around protons in an erratic path like a moth near a lightbulb. It is hard to be precise about its location; if you know where an electron is, you do not where it is going and *vice versa*. As a result, scientists have come up with zones called orbitals, which represent the statistical likelihood of finding an electron in a certain place at a certain time. Imagine if someone calls you and asks where you are. You reply that you are at home and this satisfies them, but you could be in the sitting room, the bathroom or wherever. Likewise, atoms in molecules can merge their zones, creating a larger roaming space for one of each of their electrons to share.

When atoms contribute electrons to merged orbitals, they form a chemical bond. Figure 1.4 (left) shows the arrangement of electrons in the compound water. This suggests that the oxygen atom is surrounded by electrons, with the bonding pairs of electrons sandwiched between the oxygen and the hydrogens. Although this is a simplification, it is a useful one as, in fact, the erratic hydrogen electrons may also be found on the far side of each hydrogen atom. Nevertheless, Figure 1.4 (right) provides a convenient way to consider a chemical bond. The negatively

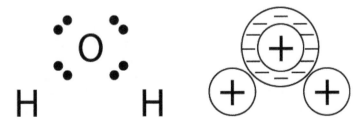

Figure 1.4 Left: the arrangement of electrons in the compound water. Right: another way of considering a chemical bond, whereby the negatively charged cloud of electrons attracts the positive nuclei, holding the atoms together.

charged cloud of electrons simultaneously attracts the positive nuclei that sandwich it, holding the atoms together.

The simplified version also helps to understand the phenomenon of intermolecular bonding. What this means is that not only are the atoms *within* the molecule attracted to each other, but they are also attracted to the atoms of neighbouring molecules. This concept is vital to the understanding of biochemistry. From Figure 1.4 (right) it is easy to comprehend how the positive hydrogen atoms could be attracted to the exposed electron cloud of a neighbour, but that is not the whole story.

Moths are not the only available analogy for the unpredictable meanderings of the electron. When people are attracted to each other, they behave no more like the magnets than the moths do. A man might be drawn towards a woman to whom he is attracted, but he will not simply move towards her until his body is pressed flat against hers, remaining there unless prised away (probably by irate friends or bouncers)! Like the moths, mutually attracted couples exist at varying distances from one another.

This gives us a useful way to consider intermolecular attraction. Sadly, just because you are smitten with your current partner, it does not mean you will not be attracted to other women or men. In the same way, in molecules of water, each oxygen atom is attracted to hydrogen atoms inside the molecule, as well as to the hydrogen atoms in nearby molecules. The reason for this is that oxygen is more electronegative than hydrogen. What this means is that it holds its electrons more closely to itself than hydrogen is able to. Part of the reason for this is that oxygen has eight protons and eight electrons, between which there is a greater attractive force than exists between hydrogen's one electron and one proton. Consequently, when the two elements share their electrons in their merged orbitals, both electrons are more attracted to the oxygen atom. The electrons are not evenly distributed across the molecule but are instead more likely to be found near the oxygen. This gives the oxygen a partially negative charge, causing it to attract the partially positive hydrogen atoms of its neighbours, as shown by the dashed line in Figure 1.5. Intermolecular attraction is roughly 10 times weaker than the attraction between the atoms *within* the molecule.

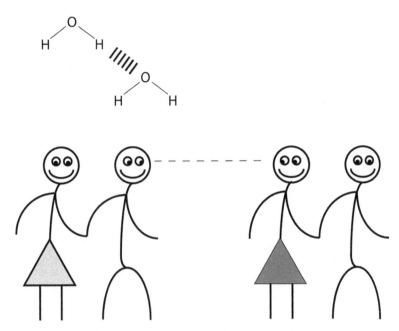

Figure 1.5 Intermolecular attraction between the atoms of neighbouring molecules is similar to the unfortunate phenomenon that a loving relationship may not quell a roving eye.

Sometimes, elements are so different in electronegativity that the more positive bonded atom can slope off, leaving its "shared" electron with the other, more electronegative atom. This is also of central importance to biochemistry because the products are ions, which play a key role in cellular processes. Figure 1.6 shows how hydrogen forms an ionic bond with chlorine by donating its electron. In gaseous form their opposing charges will bind them together but when dissolved in water—to produce hydrochloric acid found in the stomach—the ions separate. As we will see in Chapter 2, acids taste sour because the positively charged hydrogen ions flood through cellular channels in the taste buds.

Now we are ready to consider how electricity can make reactions go uphill. Conceptually speaking, electrolysis uses an electrical current to force electrons in directions they would not naturally go. In the above case of hydrogen chloride, the result of electrolysis is that the donated electron is forced back towards the hydrogen. In reality, that is not what happens. The electrical

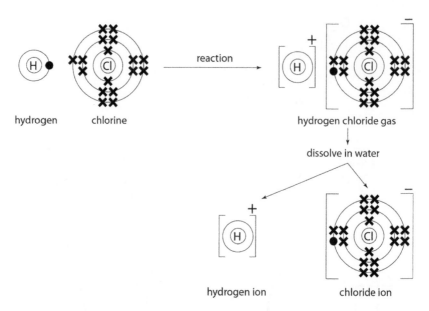

hydrogen   chlorine   hydrogen chloride gas

dissolve in water

hydrogen ion   chloride ion

Figure 1.6 Formation of the ionic compound, hydrogen chloride. When dissolved in water, the two elements separate into positively and negatively charged ions. (The hydrogen electron is shown as a dot, while the chlorine electrons are shown as crosses. This is to emphasise the element from which the different electrons originated.)

current supplies electrons to the emptied orbitals of the hydrogen ions, while the positive terminal strips them from the filled orbitals of the chloride ions.

These processes may bring the Mars astronauts home. Electrolysis is used to force water through the uphill decomposition into hydrogen and oxygen. The former can be reacted with carbon dioxide to produce methane and can subsequently be combusted with the latter.

The concept mission shows a feasible way to get astronauts home from Mars. All that remains to be seen is whether it will work, whether it will be deployed to get the Mars One participants home and whether the Mars One mission will take place at all. If the mission does go ahead, the participants will have a challenge. Harvard grad student Marina Santiago, 26, has said that harder even than parting with friends and family will be sacrificing the ability to go outdoors so that she will never again feel the sun or the wind on her skin.[3]

History may yet remember the Mars One astronauts as pioneers. Furthermore, the settlers will be able to carry out invaluable research on the red planet. Candidate Gillian Finnerty hopes to provide evidence of previous life on Mars.[12] As we shall see in Chapter 4, researchers now believe that events on Mars could have played a pivotal role in the appearance of life on Earth.

It seems only a matter of time before we will be able to say that life has existed on Mars, if only because of extramartial visitors. We may view the Mars One volunteers as brave or insane, but their pioneering spirit could make an indelible impression on the effort to answer one of the most fundamental questions in existence—how did life begin at all?

REFERENCES

1. "The Mars 100: Mars One Announces Round Three Astronaut Candidates", Press Releases—News, *Mars One*. Available at: http://www.mars-one.com/news/press-releases/the-mars-100-mars-one-announces-round-three-astronaut-candidates. [Accessed: 22-Mar-2015.]
2. G. P. Murdock, "The Common Denominator of Cultures", in *The Science of Man in the World Crisis*, Ralph Linton, Ed., Columbia University Press, New York, 1945.
3. E. Laudau, "Mars Hopefuls Ponder Life Without Families, Favorite Foods" on CNN.com, *CNN*. Available at: http://www.cnn.com/2014/05/16/tech/innovation/mars-one-candidates/index.html. [Accessed: 20-Mar-2015.]
4. S. Griffith, "'I Want to Have the First BABY on Mars," says Mars One candidate', *Mail Online*. Available at: http://www.dailymail.co.uk/sciencetech/article-2958312/I-want-BABY-Mars-says-British-candidate-one-way-space-mission-TV-company-reveals-plans-Big-Brother-style-beamed-red-planet.html. [Accessed: 06-Mar-2015.]
5. K. Anderson, "Climate Change Going Beyond Dangerous—Brutal Numbers and Tenuous Hope".
6. I. Sample, "Mars One Shortlist: Five Britons Among 100 Would-be Astronauts", *The Guardian*. Available at: http://www.theguardian.com/science/2015/feb/16/five-britons-among-100-would-be-astronauts-shortlisted-for-mars-mission. [Accessed: 20-Mar-2015.]

7. "Mars One Teams with Endemol for Worldwide TV Event", Press Releases—News, *Mars One*. Available at: http://www. mars-one.com/news/press-releases/mars-one-teams-with-endemol-for-worldwide-tv-event. [Accessed: 20-Mar-2015.]

8. B. Russell, *History of Western Philosophy*, Routledge, London; New York, New edn, 2004.

9. M. Hockney, *Free Will and Will to Power*. Lulu Press, Inc., 2014.

10. T. Brennan, *Borrowed Light: Vico, Hegel, and the Colonies*, Stanford University Press, 2014.

11. *The Oxford Handbook of Free Will*, ed. R. Kane, Oxford University Press, 1st edn, 2005.

12. J. Edgar, 'Mars Mission Student: "I'd Rather Go and Not Come Back Than Not Go at All"', *The Telegraph*. Available at: http://www.telegraph.co.uk/news/science/space/10571354/ Mars-mission-student-Id-rather-go-and-not-come-back-than-not-go-at-all.html. [Accessed: 22-Mar-2015.]

CHAPTER 2

The Chemistry of Taste

Even the famously big-headed Carolus Linnaeus cannot have
realised how suitable his name for the cocoa plant would
become. When the godfather of classification named it *Theobroma
cacao*—from the Greek for *food of the gods*—its beans were used to
make a stimulating drink that Mesoamericans called *chocolatl*. In
a sense the food has been playing a game of catch up ever since
its domestication by the Olmec civilisation sometime between
1500 BC and 400 BC.[1] Just as humans have evolved senses of taste
and smell in order to distinguish life-sustaining substances from
poisonous ones, so chocolate has evolved to perfectly complement
those senses. Even the aptness of the name was to evolve. For if
the food was held to be worthy of deities in its first incarnation as
a questionably flavoured drink (one pope disliked it so much that
he approved it for consumption during fasting),[2] what epithet
befits the wanton pleasures of the version we enjoy today?

The transformation of this beloved delicacy from invigorating
drink to tantalising confection is a story of evolution. As we shall
see in Chapter 4, biological evolution thrives on glitches during
replication of genetic code. By contrast, mutations to the process
by which chocolate was made were not random events but rather
the result of calculated bids to profit from innovation. Entrepreneurs competed for money rather than food. But, as with

The Chemistry of Human Nature
By Tom Husband
© Tom Husband 2017
Published by the Royal Society of Chemistry, www.rsc.org

biological evolution, the supremacy of the trait is evident by its modern-day ubiquity.

Today, every sweet shop is like a pantheon of the great pioneers of chocolate, immortalised in the brands named after them. Fry was the first chocolatier to produce solid chocolate by mixing cocoa powder, sugar and cocoa butter. Nestlé was the inventor of the powdered milk used to make the first milk chocolate. Lindt was the inventor of the conching process, which transformed its gritty forebears into the smooth treat we enjoy today, and Cadbury pioneered chocolate as a Valentine's Day gift. The experiments of these pioneers produced technologies that are now indispensable across the industry.[1]

There are important names missing from this who's who of chocolate, notably, the Dutch chemist Coenraad Johannes van Houten. When Linnaeus named the divine tree in 1753, chocolate had always been taken in liquid form, not as the creamy milk chocolate we enjoy today but brewed in hot water like tea or coffee. Enjoyment of this drink was marred by the fat contained in cocoa beans. The cocoa butter melted in the hot water, then rose to the surface giving it a less than appealing appearance.[2] Van Houten's mutation was arguably the least accidental. Using his chemical knowledge, he devised a way to remove the cocoa butter and make the resulting cocoa powder more soluble.

But it was Joseph Fry who played the most significant role in chocolate's evolution. Van Houten's process produced a by-product, cocoa butter, and industrialists of the era knew better than to just sling it in the bin. His invention sparked a race to find a use for the leftovers. What a masterstroke Fry played when he had the insight to recombine it with the very ingredient van Houten had laboured to separate from it. Fry mixed cocoa butter, cocoa powder and sugar to make the first solid chocolate, and the multi-billion-pound industry in existence today pays testament to the strength of this new characteristic. Was he a genius or a chancer? Perhaps both are traits of the true entrepreneur, but Fry's coup could be the best contender for the most "accidental" of the mutations. But what does this have to do with self-replicators and human nature?

Self-replicators need fuel and building materials. As such, our tongues have evolved to distinguish these useful nutrients according to their tastes. The early appeal of *chocolatl* arose from the stimulants it contained, the alkaloids caffeine and theobromine. But the innovations of the intervening millennia have honed the recipe to satisfy our tastes ever more fully. Indeed, chocolate has oft been cited as a sex substitute as it affects brain chemistry comparable to that accompanying coitus. But the reason it pleases us is because of our own evolution.

The pleasure we get from eating chocolate is the body's way of rewarding us for providing it with valuable nutrients. The confection gives us sugar, fat, protein and salt, all of which are essential to our survival. Never mind that even the most beneficial of nutrients is a poison at the correct dose, hunter-gatherers thrived on a nervous system that encouraged them to seek food and rewarded them when they got the right stuff. What the pioneers of chocolate may not have understood was that they were refining a product into the perfect match for a sensory system bent on maximising its owner's reproductive potential. The next question is how that sensory system works.

Flavour has three components: taste, smell and a little something from the nervous system. Taste is what we sense *via* the taste buds, most of which are located on our tongues. This is called the *gustatory system*. Smell is what we detect through our noses *via* the *olfactory system*. Finally, the twist added by our brains is the *somatosensory* component.[3] Many events have shaped our understanding of the somatosensory input to flavour, from the sense-impaired experiences of car crash victims to the findings of research. One such study found that expert wine tasters could be fooled into thinking that they were drinking red wine, when in fact they were drinking white wine that had been coloured with an odourless dye called anthocyanin.[4] For a more immediate reminder of the mind's facility for invention, see Figure 2.1.

There are currently five accepted tastes. These are bitter, salty, sweet, sour and umami. Salty, sweet and umami-containing foods trigger pleasurable responses, whereas the unpleasant sensation of eating overly bitter or sour foods has evolved to ward us off potentially harmful substances, such as the bitter poison strychnine.

This idea will seem at odds with the fact that bitter and sour notes often contribute to desirable flavours. Coffee and beer are

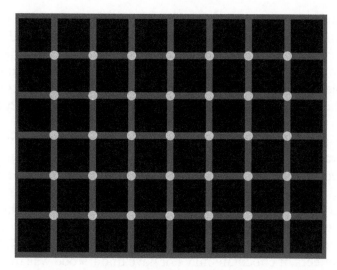

Figure 2.1   The mind invents dots in the intersections.
(Image from Shutterstock, artwork © nogoudfwete).

characteristically bitter, whilst children enjoy an array of sour
sweets. This taste for tangy confectionery may have arisen in
unconscious mimicry of the sour flavour of certain fruits.
Meanwhile, it is most likely the bitter flavour of vegetables that
makes them so repellent to children. Much of this apparent
contradiction results from defence mechanisms evolved by
plants. Astonishingly, there is some truth to a *Simpsons* gag,
in which Dr Hibbert counsels that broccoli tries to warn off
consumers with its terrible taste. The vegetable contains toxic
glucosinolates, which impair the function of the thyroid gland
and can increase an individual's risk of developing a goitre.
Another example is kidney beans, which contain compounds
that disable our protein-digesting enzymes. Boiling the beans
breaks down these toxins. The point is that plants do not need to
be eaten to survive and, while fruits have proven a useful means
by which to disperse seeds, the plants themselves are fuelled by
the sun and obtain their nutrition from inorganic sources. This
is why the evolution of toxic compounds gave plants an evolutionary advantage, as anything poisonous to animals could
expect to become gradually more populous. Plants existed
without animals, but we cannot cope without them, which is why
it has been worthwhile to acquire the taste for the bitter notes

ensconced in their nutritional bounty. Nevertheless, it is the less bitter brands of coffee and beer that enjoy the highest global sales.[5] Overall, we do not enjoy bitter tastes.

Umami is another interesting taste, chiefly because of its relationship with monosodium glutamate. So-called MSG has been surrounded by controversy since Dr Ho Man Kwok blamed it for what he called *Chinese-Restaurant Syndrome*,[6] and yet its very deliciousness is proof that the substance is in some way beneficial to the body. The whole reason why MSG tastes so good is because your brain rewards you for eating it. But what could possibly be so good about eating an additive? It has to do with the relationship between monosodium glutamate and glutamic acid. The latter is one of the 22 amino acids that the body relies on to make the vast array of proteins necessary to keep us alive. And the former, as the name suggests, is very nearly the latter. As Figure 2.2 shows, glutamate is the ion produced when glutamic acid releases its acidic hydrogen. In ionic form it can bond with ionic sodium (Na^+) to produce MSG (analogous to the relationship between the smoked and snorted varieties of cocaine). So, the reason why MSG tastes good is because the body rewards us for eating what appears to be a nutritious protein. Named by its

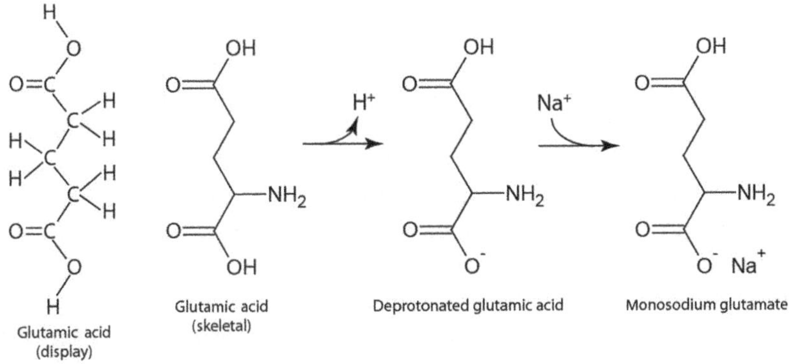

Figure 2.2 Production of MSG from glutamic acid. A note on diagrammatic conventions: in skeletal formulas zig zag lines represent carbon chains with each corner representing one carbon atom. All carbon atoms are assumed to share four bonds with neighbouring atoms. Where fewer than four bonds are shown, the "invisible" bonds are shared with hydrogen atoms, which are omitted for clarity. For glutamic acid, both the skeletal and the display formula are shown for comparison, in which all constituent atoms are represented with letters.

Japanese discoverer, *umami* can be translated as "delicious" and the flavour associated with it is savoury.

This does not mean that we should all start eating MSG with reckless abandon. This is another example of the fact that any substance is a poison at the right dose. And according to a study by the Federation of American Societies for Experimental Biology (FASEB) that dose is 2.5 grams, cited as the threshold above which symptoms of Dr Kwok's syndrome present—tightness in the chest and jaw, a burning sensation in the back of the neck and headache.[6] In the same report, glutamic acid was linked with asthma attacks.

The imperfections of the gustatory system are all too obvious. Bitter foods may not be unhealthy and savoury foods might not be nutritious, and even when valuable nutrients are identified, there is a danger that overconsumption will render them harmful. Nevertheless, chocolate's appeal is becoming clearer. Milk chocolate contains sugar, salt and the glutamate-rich protein casein, meaning that all three of our *"eat more!"* taste buttons are pressed.

2.1   HOW MUSCLES WORK

Many biochemical processes take place before we eat the chocolate. Usually, we will have to lift the chocolate to the mouth in order to start eating it and, before that, we will need to have earned this treat. Our brains know what they can expect from chocolate. Motivating us with the carrot of its delectable flavour, our grey matter guides us towards opportunities to relive the memory of former encounters. This process informs the generation of signals to our muscles that help us to earn the necessary money, to walk to the shop, to hand over the payment, to open the wrapper and, finally, to raise the chocolate to the mouth.

As we are focusing on the chemical events that inform our behaviour, it is important to understand the mechanism of muscles. Our muscles enable us to act on our goals so the processes that activate them are of great interest to us. Through the course of the book we will consider their role not only in helping us to eat but in helping us to mate and even for what they can show us about the freedom of will. Not only that, but the well-understood mechanism demonstrates characteristics of proteins that we can apply to a host of other proteins relevant to this story.

A pair of proteins called myosin and actin plays the starring role of muscle contraction. Myosin demonstrates two important features of proteins. Firstly, it does its work by cyclically changing shape and, secondly, it has a cyclically changing electrostatic fingerprint. We will consider each characteristic in turn.

When it comes to muscle contraction, myosin cannot be considered in isolation from its partner actin. This is another protein that forms fibres, which we can consider as ropes. Myosin works like a team of sailors tugging on the rope. By mooring this rope to one end of a muscle cell and mooring the myosin to other fibres attached to its opposite end, this pulling causes the cell to contract, as shown in Figure 2.3.

To consider how this is achieved at the molecular level, we need to scale down the analogy. Instead of sailors hauling a rope, we can model it as a beckoning finger pulling a piece of string. This might not sound like a desperately efficient way to pull a piece of string. One method might be to tie a loop in the string,

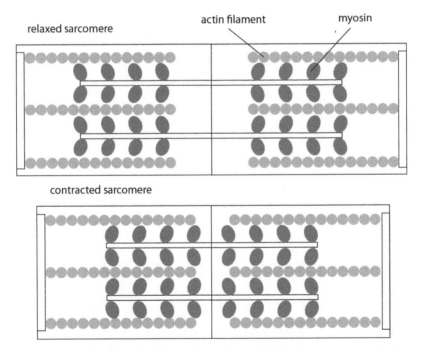

Figure 2.3 Contraction of the sarcomere. As the molecules of myosin tilt their heads the actin filaments are drawn towards the centre of the sarcomere, causing the whole unit to contract.

which could go around the finger, but all this would achieve was the alternate slackening and tightening of the string, which would not be sufficient to contract the cell. Instead, what happens is that, as the finger beckons, it is intermittently sticky. When the finger pulls inwards, it is sticky and thus pulls the string with it, but when it extends back outwards, it loses its stickiness, enabling it to release the string and grasp for another patch further along.

The obvious two questions are as follows: how does the myosin change shape? And what makes it intermittently sticky?

These questions require us to consider the nature of proteins. A protein consists of a chain of amino acids linked in a specific sequence. Each of these amino acids is composed of atoms, which are joined together by covalent bonds. Many of the atoms also form other, weaker bonds with atoms from amino acids located much further along the chain, as shown in Figure 2.4. These weaker bonds are called intermolecular forces.

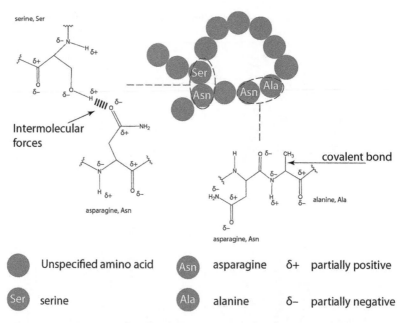

Figure 2.4   The atoms in amino acids are held together by strong covalent bonds. Other bonds can also form between amino acids that are not next door to each other in the chain. In this example, intermolecular forces attract the partially positive and negative regions of asparagine and serine together, causing the chain to form a loop.

Central to this, along with just about everything else in chemistry, is the attraction between protons and electrons. For example, two atoms right next to each other might share an electron each to form a covalent bond, such that the shared electrons attract each of their positive nuclei. But we also need to consider the attraction that exists inside a single atom. As shown in Figure 2.5, hydrogen has one proton and one electron, whilst oxygen has eight electrons and eight protons. As a result, oxygen holds its electrons much more tightly to itself than hydrogen does. At the same time, electrons repel each other, so larger elements, such as sulfur, have a weaker grip on their outermost electrons compared to oxygen. This differing attraction between an element's protons and electrons is referred to as electronegativity. As we saw in Chapter 1, when elements of different electronegativity form compounds, it leads to an uneven distribution of charge and these compounds develop little pockets of slightly positive or slightly negative charge.

Small compounds with this uneven distribution of charge are called polar. They have higher melting and boiling points than non-polar compounds of similar size because the partially positive regions of each molecule attract the partially negative regions of neighbouring molecules.

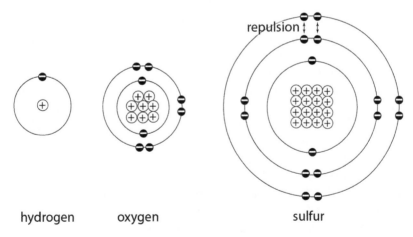

hydrogen   oxygen   sulfur

Figure 2.5   Oxygen holds its electrons to itself more tightly than hydrogen and sulfur. Attraction increases as protons and electrons become more numerous, but repulsion between electrons increases as new shells are added. (Neutrons are omitted from the diagram for clarity.)

In proteins, the situation is more complex. Small, polar molecules are drawn together by their positive and negative regions. But proteins are huge, consisting of long chains of amino acids, some of which are polar and others of which are non-polar. Proteins are big and flexible enough that their chain can fold itself so that the partially positive regions line up next to the partially negative regions.

This partly explains how a protein gains its unique three-dimensional shape. The genes dictate the sequence in which the amino acids should be linked and this linear arrangement of amino acids dictates the strict three-dimensional shape of the protein. Many factors contribute to this shape, as we will see in later chapters, but one of the key forces is the supplementary bonds mentioned above, known as intermolecular forces. The protein contorts itself so as to maximise the average strength of these intermolecular bonds between the differently charged regions of the amino acid chain. Proteins are rich with the elements nitrogen and oxygen, electron-hogging atoms that gain partial negative charges and, accordingly, when hydrogen atoms form bonds with nitrogen or oxygen, they become partially positive. Attraction between these oppositely charged regions draws myriad sections of the amino acid chain together until it is folded into its trademark shape. Friends and family members are held together by strong, covalent bonds but many form weaker bonds with the bank clerks and teachers located further down the amino acid chain.

This explains how proteins gain their characteristic shape but not how they change shape, which is the ability that enables them to do their job, just as myosin rocks its head back and forth like a beckoning finger. As we will see, there are many ways that a protein can change shape but one of the most common is when a stranger comes to town.

The American novel *Pollyanna* tells the story of an irrepressibly optimistic child who transforms an entire town. Not afraid to face criticism, she goes about encouraging the dispirited townsfolk to play the glad game, in which they search for the positives in situations that disappoint them. Hardest of all to thaw is her crotchety aunt and guardian, Polly. In time, she lifts the spirits of the entire town, drawing out the reclusive hermit Mr Pendleton, and her aunt even summons the courage to marry her former lover Dr Chilton. In essence, her optimism draws people together.

If the characters in *Pollyanna* were atoms in a protein, we can see how its entire shape would change. Perhaps the hermit plucks up the courage to ask out a widower and they form a relationship. This will have the effect not only of drawing together the two individuals but also the other individuals with whom they have pre-existing relationships, such as their family members. But the same rules apply in the atomic world. It is rare to pull an individual closer without drawing in their other friends and family. If one atom is drawn in a certain direction, it will drag its neighbours along too. All we need is a Pollyanna molecule to change the shape of the town.

The part of Pollyanna is played by a molecule called adenosine triphosphate (ATP). As we will see in Chapter 4, the compound is related to the nucleobase adenine, used in construction of DNA. Adenine, from the German for *gland*, was named by Albrecht Kossel in 1885 after he extracted it from the pancreas of an ox.[7] Figure 2.6 shows the structure of ATP. Three molecules of phosphoric acid, one of the pH-lowering ingredients in cola drinks, are strung together to make the triphosphate tail. The relevance of this is that acids are substances that release positive hydrogen ions. When the hydrogen ions desert the phosphate

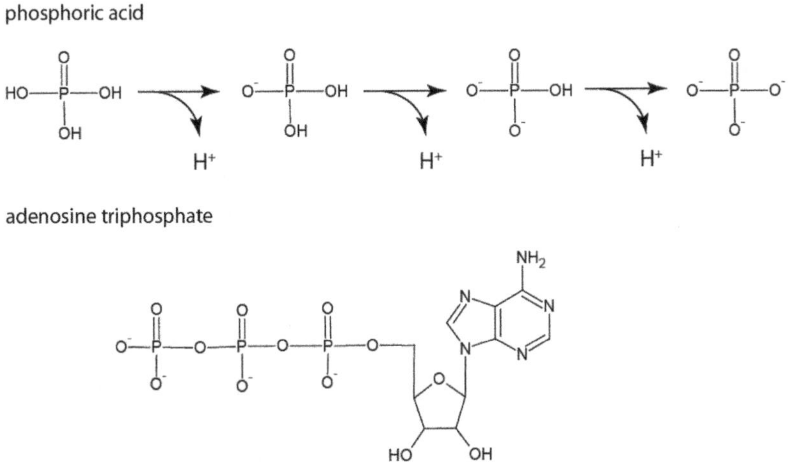

phosphoric acid

adenosine triphosphate

Figure 2.6 Hydrogen ions dissociate from phosphoric acid, which is structurally equivalent to the phosphate groups in ATP.

groups, they leave behind negatively charged oxygen atoms. Consequently, the tail of ATP has an unusually negative charge.

Two important features enable myosin to work with ATP. Firstly, myosin has two subsections, one of which swivels around a pivot. Secondly, it has a special docking site that perfectly complements ATP. Not only is it the ideal shape to accommodate ATP, but it also has a complementary electrostatic fingerprint. As we have seen, proteins are large enough to accommodate numerous regions of partially positive and partially negative charge. It would be no good if the binding site had a strongly negatively charged character in the place where the phosphate group sits. Since the phosphate group is also very negative, the two regions would repel each other and ATP would be pushed away from the binding site. So, both the shape and the charge distribution have to be complementary for a biomolecule to dock with a protein.

When myosin binds ATP, the consequences ripple through the entire protein. Researchers, including Stefan Fischer and Jeremy Smith, have suggested that once the ATP has bound to myosin, the ATP's negative phosphate group is held in close proximity to a partially positive hydrogen atom located between the 456th and the 457th amino acids in the myosin acid chain.[8] It is partially positive because it is bonded to an electronegative nitrogen atom, which is hogging their shared electrons. As shown in Figure 2.7, the hydrogen atom protruding from the acid chain is strongly drawn towards the oxygen atom of the ATP phosphate group and as they are pulled together the hydrogen atom drags its neighbours with it. Then, those neighbours drag their neighbours, which also drag their neighbours. Finally, the tension in one section of myosin pulls the other section right around its pivot, through an angle of about 60°. This is how the finger beckons.

This movement needs an energy supply. When we exercise for extended periods, exerted muscles start to burn as lactic acid is produced. This happens when we fail to provide the muscle cells with enough oxygen to oxidise the glucose fuel we get from our food. Instead, the muscles use a different method to extract the energy from the glucose, the only drawback being that it produces lactic acid, which causes the burning sensation. But how does the energy get from the glucose to the actin–myosin complex?

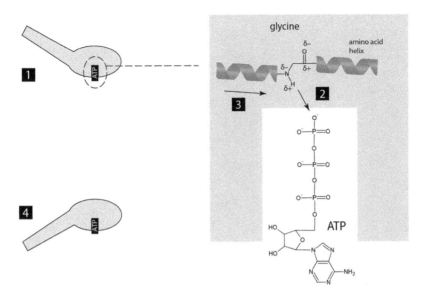

Figure 2.7 How ATP causes myosin to change shape. (1) ATP binds myosin. (2) The negatively charged tail of ATP attracts the partially positive hydrogen ion in the amino acid glycine towards itself. (3) As the glycine is drawn towards the ATP, it pulls along with it the neighbouring amino acids… (4) …causing the arm of myosin to turn around a pivot by an angle of $60°$.[8]

Binding myosin is a bit of a suicide mission for ATP. From our point of view, ATP triggers muscle-contracting activity in the myosin, which enables us to chew sandwiches, laugh, breathe and live. From its own point of view, ATP strides towards the executioner's block then raises the axe that will kill it. More scientifically speaking, myosin is transformed into the perfect agent to hydrolyse ATP, which effectively means lopping off its third phosphate group and spitting out the pieces: adenosine diphosphate (ADP) and a phosphate group.

This process acts as the energy supply. Hydrolysis refers to any process in which a compound is broken down by reaction with water. In this case, the bond between ATP's second and third phosphate groups is severed, and the liberated phosphate forms new bonds with atoms from the water molecule, as shown in Figure 2.8. More energy is released in the formation of the new bonds than is required to snip the old bond, so ATP hydrolysis is a significant source of energy for biomolecular events, including the contraction of muscles. Meanwhile, the resulting ADP cannot

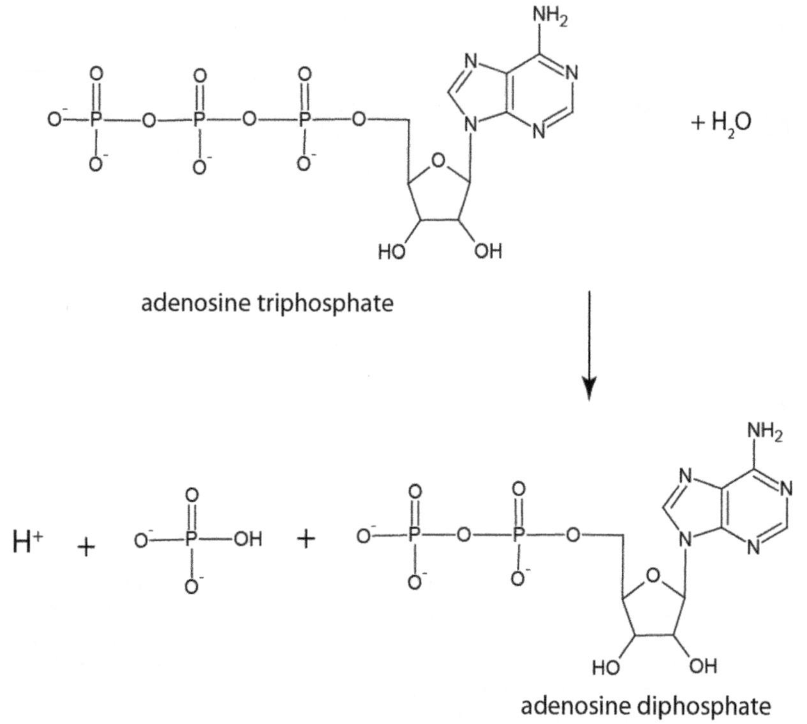

Figure 2.8   Hydrolysis of adenosine triphosphate (ATP) to adenosine diphosphate (ADP).

drive any more muscle contraction until a new phosphate group
is attached to it. This process of course requires energy, which is
supplied by the above-mentioned oxidation of glucose.

   With hydrolysis complete, the myosin returns to its original
shape. ATP is a good shape and a good electrostatic match for its
binding site in the myosin. But the resulting ADP and its severed
phosphate group have a different electrostatic fingerprint and
combined shape, which no longer matches the myosin binding
site. The ADP and phosphate group are ejected from the
protein. This means that the electronegative oxygen is no longer
attracting the partially positive hydrogen atom, so the latter
returns to its earlier position, as do all of its neighbours. The
myosin head rocks back to its original position.

   Now we have half of the story. Movement is not enough. In
order for the muscle cell to contract, the myosin head needs to
reach along the actin filament and haul it back several times,

so the other half of the story is what makes myosin intermittently sticky. At this stage we need to reintroduce myosin's partner actin. The word "sticky" is misleading because it suggests that myosin would attach itself to anything it touches like a strip of sticky tape. It would be more correct to say that myosin is intermittently attracted to actin. When the beckoning finger stretches forward, it is not attracted to the string, but when it pulls backwards, it is attracted to the string.

A protein does its jobs not only with its physical shape but also with its electrostatic fingerprint. We saw above that ATP has a docking station in myosin, a particular slot into which it can fit snuggly and more properly called a binding site. Not only is the shape of the binding site a good match, but its electrostatic character is also accommodating.

Myosin has another binding site, which is the perfect match for actin. This other binding site is located at the tip of its swivelling head, and both its shape and electrostatic fingerprint are complementary to actin. Just as ATP slots into its bespoke nook, so actin docks with this other binding site, so that when myosin pulls its head back down it hauls the actin rope. The final step of the puzzle is what makes myosin alternately grasp and release the actin.

Myosin has to choose between ATP and actin. When it binds ATP, it releases actin and *vice versa*. When there is no ATP bound, a group of atoms in the myosin head are arranged into the perfect shape, location and electrostatic fingerprint to attach to the actin filament. But when the electronegative oxygen atom in ATP tugs on the hydrogen atom, two changes occur in the myosin. Not only does it swivel its head, but it also rearranges the atoms in the actin-binding site so that they relinquish the actin. This is how the beckoning finger releases the string.

So, finally, we can see how the actin–myosin complex contracts muscles, as summarised in Figure 2.9. Myosin cyclically rocks its head back and forth. As it pushes its head forwards, it is not attracted to the actin, but when it pulls it back it is. Not only is the cycle repeated, but also the process is mirrored by multitudes of neighbouring myosin molecules in the same cell and others around it. Two important prerequisites exist: one is that there is enough ATP to fuel the process; the other is that a nerve signal has instructed the muscle to contract. This nerve signal releases a barrier that physically disables the actin–myosin complex.

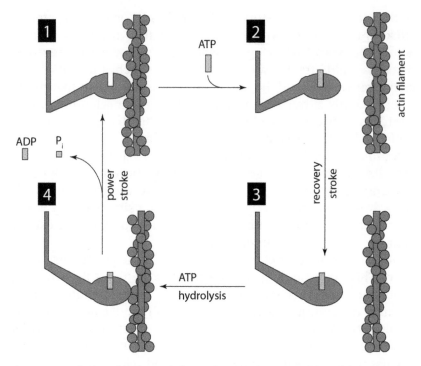

Figure 2.9 The mechanism of the actin–myosin contraction. (1) When the cycle starts, myosin is bound to the actin filament. (2) When the myosin binds ATP, it releases the actin filament and rocks its head forward. (3) Once the myosin has hydrolysed the ATP, it binds the actin filament again. (4) When the ADP and phosphate group (P_i) are ejected from the binding site, the myosin head rocks back to its original position.

So much for how we raise the chocolate to our mouth and chew it, but how does it signal its nutritional bounty to our taste buds?

2.2 HOW TASTE BUDS WORK

Taste buds are composed of taste cells, which collectively distinguish important nutrients. Taste cells each detect different examples of the same taste. For example, there are many sweet substances, including the well-known sugars, glucose, dextrose and sucrose, and all are detected by the same variety of taste cell. This is achieved with a variety of protein called a receptor.

There are two kinds of taste receptors. The sweet-, umami- and bitter-detecting variety are G-protein coupled receptors, which are shape-shifting proteins that transmit signals across cell membranes. However, researchers believe that sour- and salt-detecting receptors are simple ion channels.

2.2.1 Ion Channel Taste Receptors

Ion channel taste receptors detect ions that enter our mouths. This raises the question of what an ion is. An average adult human male carries about 1 kilogram of calcium in his body, not surprising perhaps as the metal's role in tooth and bone structure is well known. But for chemistry students this knowledge can clash with the fact that calcium metal reacts vigorously with water to produce flammable hydrogen gas. How can such a reactive metal form a healthy part of our diet? The answer is that our bodies contain ionic calcium rather than elemental calcium.

Ions are chemical species with different numbers of protons and electrons. ("Chemical species" is a catch-all term that can be applied to a variety of slightly different things that atoms can make, just as a sibling could be a brother or a sister. Unless stated the word "species" on its own will be the biological variety, referring to a specific type of organism, such as *Canis familiaris*, referring to household dogs.) Chapter 1 described how positively charged protons attract negatively charged electrons, even if the attraction plays out in a mind-bending way. In an atom of elemental calcium (that is, a piece of pure calcium) each atom has equal numbers of protons and electrons. The positive charge generated by the protons matches the negative charge generated by the electrons, making each atom charge neutral.

That all changes when calcium metal is mixed with water. During the reaction, each atom of calcium donates one electron to each of two hydrogen atoms, each of which is provided by separate water molecules. These electron-acquiring hydrogen atoms pair up into molecules of hydrogen gas, which bubble up out of the water, whilst each calcium atom now has two fewer electrons, meaning that it should no longer be called an atom but rather an ion. While the number of electrons has changed, the number of protons remains the same, meaning that each ion of calcium now has two more protons than electrons, giving it an overall +2 charge. Jettisoning these two electrons has many consequences. Firstly, the calcium is no longer so vigorously reactive, making it safe for human consumption. Secondly, the ions can dissolve in water, enabling them to exist in the watery medium of our bodies. Thirdly, these ions have a positive charge, which is critical to many of the cellular processes we will consider in this book.

Ion channels ferry ions across cell membranes. For reasons we will investigate more closely in Chapter 6, the characteristic charge of ions prevents them from moving through cell membranes. This results from the same chemistry that prevents water from mixing with cooking oil, which has a very similar molecular structure to the components of the membrane called lipids. Consequently, calcium ions can only get in and out of cells *via* ion channels.

The two ions detected by ion channel taste receptors are sodium and hydrogen. Acid-rich foods, like vinegar and citrus fruits, are sour because they contain acids, which are defined as chemical species that release positive hydrogen ions (H^{+}). On the tongue, these ions are thought to enter sour taste cells through ion channels. The key ingredient in table salt is sodium chloride, although salt shakers may contain a range of ions, including potassium, iodide, magnesium and so on. As with calcium above, the sodium in table salt is present in the form of ions, each of which has a charge of $+1$. Just as hydrogen ions activate sour cells, these sodium ions are thought to tumble through ion channels into the salty taste cells, as shown in Figure 2.10.

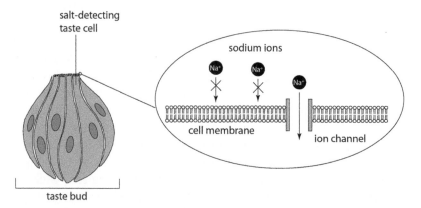

Figure 2.10   Ions are unable to cross cell membranes, except through ion channels.

In summary, our tongues register sour and salty tastes when molecules in the mouthful of food release ions that physically enter the taste cells through ion channels. How this invasion generates a signal to the brain is a matter we will consider in Chapter 3.

2.2.2 G-Protein Coupled Taste Receptors

The other three tastes are detected by G-protein coupled taste receptors. These are phenomenally important receptors that have many different roles around the body. It has been estimated that 50% of medicines work by activating these receptors. The elucidation of their structure earned the Nobel Prize in Chemistry for Robert Lefkowitz and Brian Kobilka in 2012.[9]

Receptors transmit signals across cell membranes. By straddling the membrane, these vital proteins maintain a presence both inside and outside of the cell. They are adapted to respond only to specific stimuli. As such, the receptor in a sweet-detecting cell would be activated by a sweet compound, such as glucose, but not by a bitter compound, such as cyanide. As with myosin, these receptors work by changing shape. Just as in muscle contraction, this conformation change alters the way they interact with other chemicals. The external portion of their body interacts with a stimulus outside the cell, but the resulting conformation change primes them to interact with other chemicals inside the cell. In this way, their activation triggers a cascade of biochemical events, launching a transmission to the brain signalling that sugar has been detected.

Taste and odour receptors are remarkable in the way they communicate with the outside world. Most of the signals transmitted by receptors are internal traffic, passing signals from one part of the body to another. But taste and odour are detected by a class of sensory receptors called chemoreceptors, which directly engage with chemical material from the outside world. Another example is the receptors embedded in our skin that detect venom injected *via* stings and bites from unpleasant creatures.

Taste is detected by chemoreceptors in the mouth, most of which are in taste cells in the tongue. When we take a mouthful of food, substances dissolve in our saliva and tumble into taste pores.

Any sugars in the food will be propelled this way and that by collisions with other molecules until they are thrown towards a sugar receptor. The molecule binds the receptor, triggering the conformation change, which launches a cascade of biochemical events inside the cell.

Taste receptors are much less discerning than other receptors. There has been an enduring regard for the idea of the "lock and key" in biochemistry. This has usually been applied to enzymes, another variety of protein. The idea is that there is one chemical and one chemical alone that an enzyme will bind with. But receptors vary in their pickiness. *Ligand* is a general term for any molecule that binds with a receptor. Some receptors demonstrate a great affinity for a single ligand, but sweet-tastant receptors have broader tastes, interacting with a range of sugars, such as maltose, dextrose, glucose and sucrose. This quirk meshes well with their visual representation. The receptor shown in Figure 2.11 resembles a television aerial, designed to detect a range of signals. But this broad palette can be tricked. Saccharine contains no calories, not being a sugar, but tastes

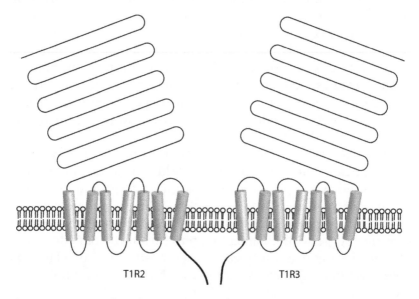

Figure 2.11 Taste-detecting receptors. The T1R2 and T1R3 couplet detect sweet substances. Note the weaving back and forth of the receptor across the membrane, characteristic of G-protein coupled receptors.

sweet because it happens to bind with the receptor.[10] Evolution has developed an imperfect net by which to catch a range of sugars but one that also snares the odd unwanted species.

Tastant receptors have unwieldy names and structures. The sweet-detecting receptors are called T1R2/T1R3,[3] while the umami-detecting receptors are called T1R1/T1R3. Each receptor comprises not one but two paired proteins. In common with all G-protein coupled receptors, their structures weave back and forth across the cell membrane, creating a profile rather like that of the Loch Ness Monster, his head and the odd coil of torso visible to viewers on land but the remainder submerged beyond view.

The umami receptor is similar to the sweet receptor because the paired T1R1/T1R3 proteins can bind with a range of amino acids. If this is so, then why is glutamate the flavour enhancer of choice? The umami receptor is just a better fit for glutamate than the rest. A whole slew of dresses may be size 8, but they will all fit slightly differently and one might just be a much better fit than the rest, as in this case. Although many amino acids bind with T1R1/T1R3, none elicits such an intense response as glutamate, which explains the enduring appeal of MSG.

2.2.3   What Happens Next?

When a sugar molecule binds a sweet-taste receptor it triggers a sequence of events that ultimately sends a signal to the brain. Signal cascades such as this are like the board game Mouse Trap, in which an elaborate contraption is constructed to catch a mouse. One player turns a crank, which rotates various gears, causing a spring-loaded lever to flick into a boot. The swinging footwear sends a ball down a slope and into a chute. Transmission continues through a seesaw, a hole in a bathtub and so on, until, eventually, a cage is knocked onto the mouse. The only difference is that the biological signal is transmitted much more efficiently than the board game mechanism, which frequently grounds to a halt.

Figure 2.12 summarises the process. The series of tediously named blobs represents different varieties of biomolecules, including proteins, enzymes and phospholipids. Shape is the currency between proteins. They must have complementary

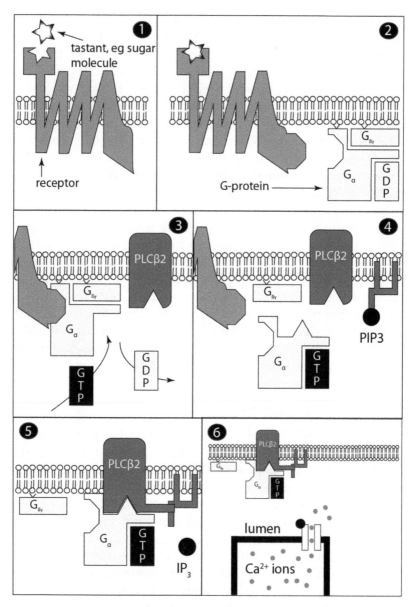

Figure 2.12 (1) Binding the tastant causes the receptor to change shape, which (2) enables it to interact with the G-protein. (3) This causes the G-protein to eject GDP, which is replaced by GTP. (4) Binding the GTP causes the G-protein to change shape, such that it is now able to bind with the phospholipase enzyme PLCβ2. (5) The bound PLCβ2 now severs the phospholipid PIP$_3$, detaching IP$_3$. (6) The detached IP$_3$ moves towards the lumen, where it opens the ion channel, releasing calcium ions into the cell.

shapes to interact, but interacting typically causes them to change shape. Whenever two things interact, one of them is changed into a new shape, enabling it to interact with the next object in the sequence. From the initial turning of the crank, this motley crew sends the signal down the line until the cage drops on the mouse.

When the sugar molecule binds the receptor, it triggers steps that lead to the release of calcium ions into the cell. Binding the sugar molecule causes the receptor to change shape, which enables it to interact with the G-protein moored next to it. This triggers a process reminiscent of the interaction between myosin and ATP. This time, guanosine triphosphate (GTP) is used, but it works just the same, releasing energy when its phosphate tail is clipped to produce guanosine diphosphate (GDP). The diagram shows a previously bound molecule of GDP being ejected so that GTP can take its place. This triggers another shape change, priming the G-protein to interact with the enzyme phospholipase Cβ2 (PLCβ2). This, in turn, lops the head off the neighbouring biomolecule, phosphatidylinositol (3,4,5)-trisphosphate (PIP$_3$). Ultimately, this severed head, named inositol trisphosphate (IP$_3$), floats over to the lumen, which the cell uses to store calcium ions. The IP$_3$ molecule opens the lumen's ion channel and thus releases calcium ions into the cell.[11]

2.3 WHAT THEY ALL HAVE IN COMMON

Changes to the balance of positive and negative charge underpin all the mechanisms we have explored. We have seen how muscles contract, how simple receptors detect salty and sour tastes, and how a more elaborate brand of receptor detects sweet, savoury and bitter tastes. All three cases require cells to be flooded with positive ions for reasons to be revealed in the next chapter.

This is most obvious with taste receptors. Salty and sour foods are characterised by the presence of positive ions, which directly enter the taste cell *via* ion channels. This, in turn, causes the overall positivity of the charge of the cell to increase. Bitter, sweet and savoury receptors use a more elaborate mechanism to achieve the same end. When the tastant binds the G-protein coupled receptor, it triggers a chain of events that ultimately releases calcium ions from a storage bay and thereby also increases the positivity of the charge inside the taste cell.

The situation with muscles is a little different. In this case, the influx of positive charge is what signals the muscles to contract. In order to eat the chocolate bar, it is first necessary to use muscular contraction to lift the bar to the mouth and then to chew it. The relevant muscles contract in response to signals emitted from our brain. When these signals reach the muscle cells, they trigger a chemical cascade like the one we have just seen. When positive calcium ions flood the cell, they unlock the barrier that prevents muscles from contracting when not in use.

Charge imbalance characterises many more biological processes than tasting and muscle contraction. In the next chapter we will see how the influx of positive ions launches and propagates signals all around the body *via* its control centre the brain, before considering the interplay between these signals and our actions.

The other question to be considered is whether the potent flavour sensation of chocolate is powerful enough to overcome rationality. We have seen in detail how the tongue detects the five different tastes. Chocolate activates all three of the nice tastes and, as with beer and coffee, the best selling brands are usually milk chocolate, in which the taste of the bitter alkaloid theobromine is almost completely masked by the other ingredients. Milk chocolate contains proteins a plenty to activate the umami receptors, as well as sugar and salt for their respective receptors.

Chocolate also presses other buttons, some of which are only just coming to be understood. A key part of chocolate production is the roasting of the beans, which releases a blend of aromatic chemicals similar to the fragrances of barbecued, fried or roasted meats, as well as baking bread. Chocolate also contains fat in the form of cocoa butter. For a long time, researchers believed fat's appeal was in its delectable consistency and new research suggests that fat may, in fact, form a sixth taste.

In 2005, Fabienne Laugerette and his team reported some interesting findings concerning another receptor aptly named FAT/CD36. FAT stands for fatty acid translocase, a reference to the breakdown products of fat's metabolism by the enzyme lipase, while CD36 stands for cluster of differentiation 36. Laugerette's team found that the transportation of the fatty acids into the cell triggers the release of calcium ions from the lumen,[12] just as with the G-protein coupled taste receptors. When the gene coding for FAT/CD36 was removed from the mouse genome, the rodents no longer demonstrated any preference for fat-containing foods.[13] In ancient China, Sun Tzu compared military strategy to cooking, saying that while there are but five tastes, their combinations are almost infinite, just as a handful of tactics can be combined to make limitless attack plans.[14] More than 2000 years later, it seems his assertion may have been overthrown. Fat just might be the sixth taste.

Thus do we understand the appeal of chocolate. Just how much control do we have over our arm as it grabs for the delicacy that has evolved over millennia to conquer just about every single one of our hot buttons? We will explore this question in the next chapter.

REFERENCES

1. S. D. Coe and M. D. Coe, *The True History of Chocolate*, Thames & Hudson Ltd, New York, 1st edn, 1996.
2. S. T. Beckett, *The Science of Chocolate*, Royal Society of Chemistry, Cambridge, UK, 1st edn, 2000.
3. E. R. Kandel, J. H. Schwartz, T. M. Jessell, S. A. Siegelbaum and A. J. Hudspeth, *Principles of Neural Science*, McGraw Hill, New York, 5th edn, 2013.
4. G. Morrot, F. Brochet and D. Dubourdieu, The Color of Odors, *Brain Lang.*, 2001, **79**(2), 309–320.
5. T. Hummel and A. Welge-Lüssen, Eds., *Taste and Smell: An Update*, Karger, Basel; New York, 2006.
6. J. Schwarcz, *Radar, Hula Hoops and Playful Pigs*, San Val, 2003.
7. "Adenine", Online Etymology Dictionary. Available at: http://www.etymonline.com/index.php?term=adenine. [Accessed: 25-Mar-2016.]
8. S. Fischer, B. Windshügel, D. Horak, K. C. Holmes and J. C. Smith, Structural Mechanism of the Recovery Stroke in the Myosin Molecular Motor, *Proc. Natl. Acad. Sci. U. S. A.*, 2005, **102**(19), 6873–6878.
9. "The Nobel Prize in Chemistry 2012", NobelPrize.org. Available at: http://www.nobelprize.org/nobel_prizes/chemistry/laureates/2012/. [Accessed: 08-May-2016.]
10. M. E. Frank and T. P. Hettinger, What the Tongue Tells the Brain about Taste, *Chem. Senses*, 2005, **30**(Suppl 1), i68–i69.

11. H. Lodish, A. Berk, C. A. Kaiser, M. Krieger, A. Bretscher, H. Ploegh, A. Amon and M. P. Scott, *Molecular Cell Biology: International Edition*, W. H. Freeman, 7th edn, 2012.

12. P. Degrace-Passilly and P. Besnard, CD36 and Taste of Fat, *Curr. Opin. Clin. Nutr. Metab. Care*, 2012, **15**(2), 107–111.

13. F. Laugerette, P. Passilly-Degrace, B. Patris, I. Niot, M. Febbraio, J.-P. Montmayeur and P. Besnard, CD36 Involvement in Orosensory Detection of Dietary Lipids, Spontaneous Fat Preference, and Digestive Secretions, *J. Clin. Invest.*, 2005, **115**(11), 3177–3184.

14. S. Tzu, *Art of War*, BN Publishing, 2010.

CHAPTER 3

The Chemistry of Pleasure

As so frequently happens with history, our modern problems with chocolate were pre-empted by its earlier advocates, the Aztecs. Father Durán was a Franciscan missionary who ventured to the Americas with the conquistadores. During his travels, he translated a collection of Aztec folk stories, one of which recounts an introspective fable. In this, the Aztec ruler Motecuhzoma Ilhuicamina dispatched an envoy back to the ancient homelands of his people. Having arrived in Aztlán, the travellers find their ancestors still alive and struggle to keep up as they are led up a hill to meet the goddess Coatlicue. She scorns their gifts and tells them their chocolate has made them unfit. Their guide leads them back down the hill and leaves them with this parting advice:

> *We become young when we wish. You have become old, you have become tired because of the chocolate you drink and because of the foods you eat. They have harmed and weakened you. You have been spoiled by those mantles, feathers and riches that you wear and that you have brought here. All of that has ruined you.*[1]

Over-indulgence is evidently not a new phenomenon. Could it be that the Aztecs shared our modern-day affliction of *chocoholism*? When this term was coined in the *Pasadena Independent* in 1961,[2]

The Chemistry of Human Nature
By Tom Husband
© Tom Husband 2017
Published by the Royal Society of Chemistry, www.rsc.org

the journalist probably did not mean it too seriously. But evidence is accumulating that people can become addicted, if not specifically to chocolate then to highly palatable foods in general.

One of the first scientists to argue that food could be addictive was the researcher Ann Kelley. Her discovery that the brain chemistry of junk-food-guzzling rats began to resemble that of heroin addicts emerged from her pioneering work on the body's internal reward system. It was during a school trip to Harvard University that her passion for science crystallised as she listened to a lecture about the rats that repeatedly pulled levers to stimulate electrodes implanted in the brain's pleasure centre. Her work transformed our understanding of this powerful part of the brain, the nucleus accumbens, and helped to identify the pivotal role it plays bridging our desires and deeds.

In the last chapter, we saw how our taste buds recognise different nutrients. We have receptors that react to different substances, but what makes us experience them as beneficial or harmful? This chapter will show how our brains reward us for eating those nutrients that serve the self-replicators, especially the fuels fat and sugar. Meanwhile, advances with the hypothesis of food addiction have shed light on the phenomenon that, while we are all rigged with the same taste buds, we differ in our appetites.

3.1 THE PLEASURE CIRCUIT

When our taste buds detect nutrients, they beam a signal along our nerves and into the brain. This signal is transmitted *via* neurons, the basic unit of neurology that do all of our sensing and processing *via* axons and dendrites—the connections between neurons. Estimates differ, but it is estimated that there are over 100 billion neurons in the human brain.

Figure 3.1 shows the main parts of a typical neuron. The cell body, known as the soma, contains the nucleus. Radiating out of the soma are the dendrites that receive information. If the neuron is signalled, this signal will be transmitted from another neuron's axon *via* connecting dendrites at its end, which connect to the soma. The majority of neurons have a myelin sheath insulating the axon, enabling increased signal transmission speed. This fatty coating gives the brain's white matter, the name given

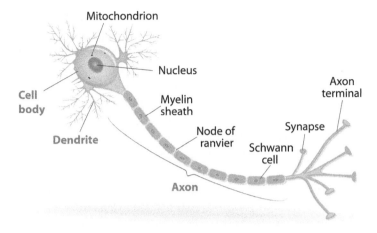

Figure 3.1 The typical structure of a neuron.
(Image from Shutterstock, artwork © Designua).

to the region comprised of axons like the wires in a computer, its characteristic appearance.

Neurons work in a similar way to the taste cells in Chapter 2. Both the G-protein taste receptors and the salt- and sour-detecting ion channels operate by flooding the cell with positively charged ions. This process snowballs and, if the signal is strong enough, chemical messengers are released, which carry the signal to the next cell in the chain. The electrical charge of these ions when moved is what propagates the signal.

3.1.1 Ions and Potential

The propagation of this signal relies on ions, which are supplied from minerals in our diet. As we saw in Chapter 1, ions are chemical species that contain different numbers of positive protons and negative electrons. Sodium ions have a +1 charge, each containing 11 protons and 10 electrons, while calcium ions have a +2 charge, with 20 protons and 18 electrons. These vital particles are distributed all throughout the body, both inside and outside of our various cells. In October, 2009, a young California mum died after drinking more than 7 litres of water in a competition called "Wee for Wii".[3] One of the major dangers of excessive water intake, in this case causing water intoxication, is that it dilutes the concentration of these important ions, which then stops the neurons from working.

The body's ability to herd these ions is a major part of cell function. Any ions contained in a glass of juice, such as hydrogen ions released from citric acid, will distribute themselves evenly throughout the liquid. The body, on the other hand, exploits an ability to segregate positive and negative ions so as to manipulate the potential difference between different regions.

Potential refers to the fact that, in the right circumstances, these segregated ions will move towards each other. Batteries operate on the same principle. Inside the negative terminal of a battery, electrons outnumber protons, whilst the converse is true at the positive terminal. As such, we say there is a potential difference between the two ends, which is also referred to as its voltage. If the two ends are joined by a wire, electrons will travel along it towards the positive terminal; in other words, they have the potential to move towards each other simply because of their mutual attraction. In a cell the potential difference refers to the difference in charge on the inside and the outside of the cell membrane.

Unlike in the glass of fruit juice, ions are rarely distributed evenly throughout our bodies. All cells have a resting potential of approximately 65 mV (ref. 4, p. 30), with the inside of the cell being the negative side and the outside being positive (it is perhaps with this in mind that the Wachowskis had the idea of using enslaved human beings as batteries in their film *The Matrix*). In other words, electrons outnumber protons inside the cell and *vice versa* outside it.

Cells achieve this resting potential by ejecting positive ions through ion channels. For reasons to be considered in Chapter 6, charged ions cannot pass freely through the membrane and instead they move through a wide array of ion channels, like the sodium ion channel that detected dietary salt in the previous chapter. When cells are not in use, ion channels automatically pump positive ions out of the cell in order to keep the inside of the cell relatively negative.

3.1.2 How Neurons in the Gustatory Nerve Transmit Sweet Signals

When we taste something sweet, a signal is beamed to the brain *via* the neurons composing the gustatory nerve. Processes in the taste cell trigger the release of positive ions into the neuron's dendrites. This ion influx snowballs until the cell is more positive on the inside than the outside in a process known as depolarisation. This triggers the release of chemical messengers called neurotransmitters from the far end of the neuron at the axon terminals. The neurotransmitters spread across the synaptic gap and activate the next neuron along the chain in a process that repeats until the signal reaches the brain.

The firing of a neuron is rather like the storming of a town hall by radical protestors. Inside the building, civic leaders and support staff are working as usual. As the protest comes to the boil, the protestors storm the front doors and flood into the lobby and, in the ensuing confusion, some of them are able to breach the reception and infiltrate the inside of the staff-only area. Their first move is to throw open the fire doors nearest to the reception, enabling more protestors to flood in from outside directly into the staff-only area. These new protestors fan out in all directions, with some carrying on down the corridor, lining the inside edge of the building. As soon as they reach the next fire doors, these are also flung open, allowing still more protestors to flock inside, continuing until all of the fire exits have been opened and the majority of protestors are in the building.

Aside from the human traffic, another change has occurred. Even if a few staff or councillors remain inside, we can assume that they are vastly outnumbered by the protestors, meaning that the town hall has gone from being more radical outside to being more radical inside. This reflects what happens to the distribution of electrical charge when a neuron fires.

3.1.2.1 Step 1: Activating the Neuron

When the sweet-taste cell detects the sugar molecule, the resulting signal cascade culminates in the release of molecules of adenosine triphosphate (ATP). Unlike its interactions with the muscle protein myosin, this time ATP acts as a neurotransmitter. The ATP binds with receptors on the cell membrane of the first neuron in the gustatory nerve, a process much simpler than the sugar-detection apparatus in the taste cell. The neuron's receptor is directly attached to an ion channel. When the receptor binds the ATP, the ion channel opens, ushering in sodium ions from outside the cell, as shown in Figure 3.2.

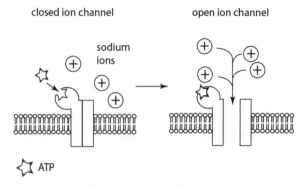

Figure 3.2   Acting as a neurotransmitter, ATP binds with the receptor, opening the ion channel to which it is directly attached.

3.1.2.2   Step 2: Propagating the Signal

The next stage is representative of processes that happen not only in neurons but also in the various varieties of taste cells. An initial inpouring of positive ions opens the floodgates for still more positive ions to rush the cell. This is the stage in the town-hall-storming scenario, when the first protestors rush in and throw open the fire escapes to let in their fellow protestors. The fire doors represent a particular variety of ion channels, which are voltage-gated.

Voltage-gated ion channels, as the name suggests, respond to changes in the potential (voltage) across the cell membrane (ref. 5, p. 383). The ion channel is a protein composed of various sub-units, two of which have a strong positive charge and, as such, are attracted to whichever side of the cell is more negative. With the cell in its resting state, the sub-units are attracted to the cell's negative interior, which is the closed position. But when positive ions flood the inside of the cell, these flanks will now be attracted to the cell's exterior. The protein's resulting shape change opens up a central channel, through which more positive ions can now enter the cell, as shown in Figure 3.3.

This is how the initial signal is amplified. Neurons have voltage-gated channels at regular intervals running right the way down the axon to the axon terminals. When positive ions enter the cell's dendrites, nearby voltage-gated ion channels open up. As more positive ions pour in through these newly opened channels, they open up more ion channels further along the

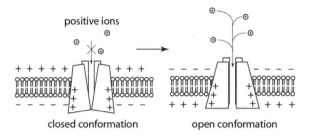

Figure 3.3 Voltage-gated ion channel. In the resting state the cell is more negative on the inside, which attracts the ion channel's positive flanks inwards. When positive ions enter the cell *via* other ion channels, the resulting reversal of the membrane polarity causes the flanks to be attracted towards the cell's exterior, so that even more positive ions are able to enter the cell.

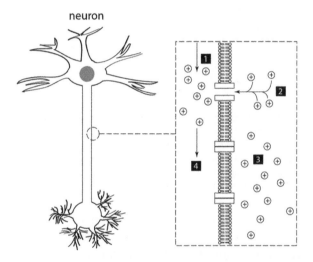

Figure 3.4 The influx of positive ions snowballs thanks to voltage-gated ion channels. (1) Positive ions admitted through other channels diffuse down the neuron's axon, generating sufficient charge to open a voltage-gated ion channel. (2) More positive ions can thus enter the cell. (3) Additional positive ions will also be able to enter once... (4) ... the latest batch of ions moves down the neuron and activates the next set of voltage-gated channels.

axon, as shown in Figure 3.4. This process repeats until the entire cell has been depolarised.

3.1.2.3 Step 3: Signalling the Next Neuron

Release of Neurotransmitters: Depolarising the neuron triggers the release of neurotransmitters from the axon terminals.

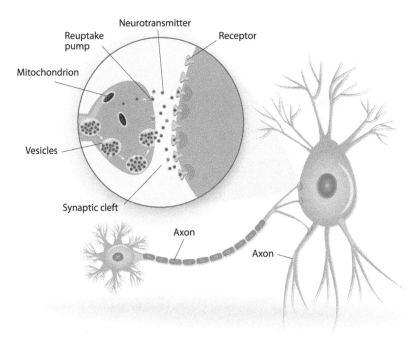

Figure 3.5   How signals are transferred between neurons. Bundles of neurotransmitters, called vesicles, are released into the synaptic cleft
between the neurons. After activating the post-synaptic neuron,
the neurotransmitters return to the pre-synaptic neuron *via* the
reuptake pump.
(Image from Shutterstock © Designua).

Figure 3.5 shows a synapse, the junction between the first, or
pre-synaptic, neuron and its signalled neighbour, the post-
synaptic neuron. Gathered just inside the axon terminal are a set
of bubbles called synaptic vesicles filled with neurotransmitters.
Depolarisation of the cell causes the vesicles to eject their
contents into the synaptic cleft. Once released, the neurotransmitters activate the post-synaptic neuron so that the signal
continues its journey to the brain.

Incoming ions cause the vesicles to jettison their cargo. Depolarisation of the neuron increases the positive charge inside
the cell. As ever, this change in potential triggers voltage-gated
ion channels, which, in the axon terminal, are the calcium-
admitting variety. The influx in calcium ions lights the fuse on a
poised vesicle, primed by the interaction of several proteins to
splurge at a moment's notice. The way the vesicles release the

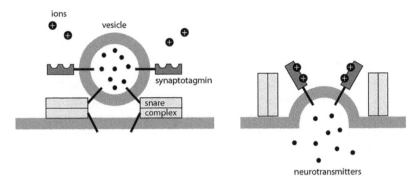

Figure 3.6 Release of neurotransmitters. The SNARE complex of proteins binds the vesicle to the membrane. When positive ions flood the neuron, they bind the synaptotagmin, which then interacts with the SNARE complex to trigger the release of the neurotransmitters.

neurotransmitters is quite intriguing as the vesicle membrane physically merges with the cell's membrane. Thomas Südhof has helped to identify a rich brigade of proteins involved in the process.[6] The trigger is called synaptotagmin. When depolarisation ushers in calcium ions, they bind to the synaptotagmin, causing it to change shape. Just as with myosin in the previous chapter, it is now primed to interact with a medley of proteins, known collectively as a SNARE complex. Prior to activation, the SNARE complex moors the vesicle to the membrane. Activation triggers additional shape-changing among the proteins, which ultimately enables the vesicle and the membrane to merge, as shown in Figure 3.6.

3.1.2.4   Step 4: Resetting the Machinery

An important feature of the neuron is the ability to reset itself. Once the neurotransmitters have been released, they bind the receptors of the post-synaptic neuron, propagating the signal. Then, the pre-synaptic neuron resets itself in preparation to fire again. This means that all of the neurotransmitters in the synapse have to be scooped up and bundled back into vesicles, which is achieved with a kind of transporter called a reuptake pump, as shown in Figure 3.5.

Transporters play an important role in our brains. Chapter 2 described the FAT/CD36 transporter, which draws fatty acids across the cell membrane into taste cells. Different varieties of transporters are also used to vacuum up neurotransmitters from the synaptic cleft. As we shall see, interfering with this activity can have profound effects on our mental state. The anti-depressant Prozac is an example of a selective serotonin reuptake inhibitor, which means that it prevents the reuptake of the neurotransmitter serotonin. This amplifies serotonin signalling, because the neurotransmitter molecules stay put for longer in the synapse, where they can continue tweaking the receptors of the post-synaptic neuron.

This is the basic mechanism of a neuron. Myriad variations distinguish the subtly different types, but the key points are the same: when a strong enough signal is detected in the dendrites, the cell is depolarised, triggering the release of neurotransmitters from the axon terminal. Neurons vary in the number of axon terminals they have and the number of dendrites. Also, different neurons will release and respond to different neurotransmitters. They may respond to one neurotransmitter but release another, and even single varieties of neurons often have more than one neurotransmitter that they are able to release.

3.1.3   Tasting Chocolate: The Next Step

Huge numbers of neurons in various brain regions respond when chocolate makes contact with our taste cells. The taste cells release the neurotransmitter ATP that signals the first neurons in the gustatory nerve. In fact, this is not a specific nerve but a loose term applied to any that innervate taste cells. Signals are passed along the chorda tympani and the glossopharyngeal nerves to the nucleus of a brain region called the solitary tract. Here, the signals are collated and transmitted to various regions of the brain, including the gustatory cortex, as well as a trio of neighbours called the thalamus, the hypothalamus and the amygdala.[5] Still other signals are diverted to the nucleus accumbens in the ventral striatum, which is where things get interesting because it is part of the brain's reward network.

The nucleus accumbens is directly adjacent to the brain's septum, the site of ground-breaking research in the 1950s. James Olds and Peter Milner implanted electrodes into the septal region of the brains of rats. Having had the brain region stimulated in an early experiment, the rats came back to the same place to do it again. This gave the researchers the idea of training the rodents to pull a lever so they could stimulate themselves. Not only did they learn to pull the lever, but they pulled it over and over again. Whatever the electrodes were doing, the rats wanted more of it.[7] Studies since have demonstrated that rats will choose the lever over food to the point of starvation. Had Olds and Milner located the part of the brain that causes pleasure?

Olds and Milner's seminal research led to the discovery of several interlinked brain regions, which together form a reward network. Other important areas in this circuit are the substantia nigra and the ventral tegmental area, both of which send output signals to the striatum, a structure in the centre of the brain composed of the caudate nucleus, the internal capsule and the putamen, as well as parts of the prefrontal cortex that sit at the very front of the brain.[8]

An important common factor of the neurons in these regions is that they are all dopaminergic, meaning that they primarily respond to and release the neurotransmitter dopamine. This is doubtless the reason why the release of dopamine was historically thought to cause the pleasurable sensation of reward, although now commonly regarded as an over-simplification. The current view is that dopamine guides us towards reward opportunities but that the actual pleasure is triggered by the release of opioids. This is a huge category of neurotransmitters, including the famous "natural pain killers", the endorphins (**endo**genous mor**phine**).

"Opioid" is an umbrella term that includes any plant-derived, synthesised or endogenous compound that interacts with opioid receptors. The opioids produced in our bodies are peptides. Like proteins, peptides are formed by linking specified amino acids together in a strict sequence, but they are much shorter than proteins. Along with endorphins, other common varieties are the dynorphins and enkephalins, an example of which is shown along with dopamine in Figure 3.7.

The systematic name of dopamine is 4-(2-aminoethyl)benzene-1,2-diol. In Figure 5.7 the hexagon represents a benzene ring formed of six carbon atoms joined on a single plane. The solid outer line and dashed inner line denote alternating single and

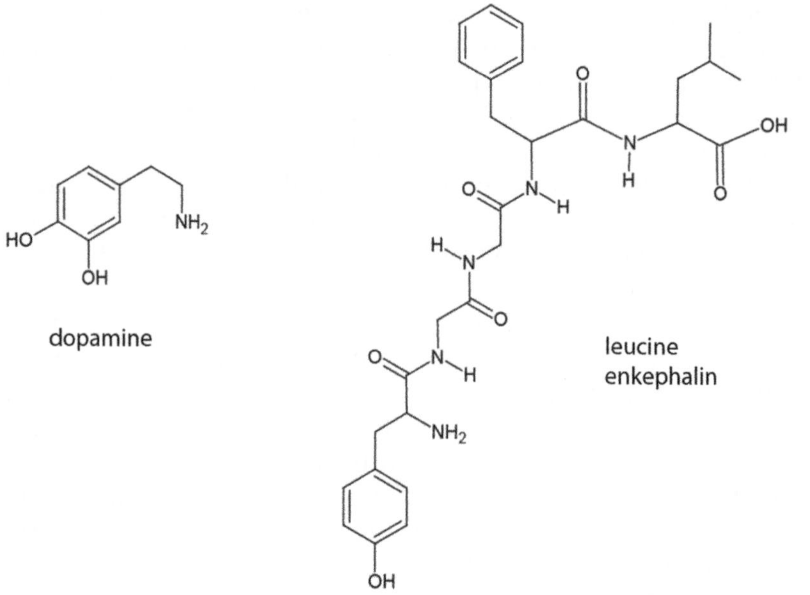

Figure 3.7   The chemical structures of dopamine and an example of an opioid peptide, leucine encephalin.

double bonds, but, actually, they collectively represent a set of roaming electrons that can move between the atoms and are hence described as delocalised. Finally, *amino* refers to the nitrogen-containing group at the end of the tail, and gives the common name (dopamine) its suffix.

While much is unclear about the roles of opioids and dopamine in neuroscience, what is certain is that they acts as neurotransmitters. In order for them to transmit signals from one neuron to the next, it must follow that there are receptors on the signalled synapses to detect them. In fact, dopamine has at least five receptors, all of which are G-protein coupled receptors (GPCRs), just like the receptor that detected the sugar molecule in the last chapter. This means that when dopamine binds to its receptor, that receptor changes shape, causing various neighbouring proteins also to change shape, releasing some things and picking up others, move around and so on, until the last toppling molecular domino opens an ion channel. Ions will storm the cell and, if there are enough of them, depolarise the neuron. This, in turn, will release its own neurotransmitters to transmit the signal on down the line. The five receptors are divided between two categories called D1-like receptors and D2-like receptors.[9] The even more diverse opioid receptors are chiefly divided into the subgroups mu, delta and kappa, and are also G-protein coupled. The cooperation of these and other neurons achieves phenomenal effects. Just the simple action of taking a mouthful of chocolate will cause millions of neurons to fire. Indeed, millions of neurons will have fired in anticipation before the first mouthful as our brains match incoming data with memories of chocolate and release dopamine to encourage us to repeat the experience. There are to the order of 100 billion neurons in our brains, which communicate *via* as many as 100 trillion synapses.[10] All of these neurons fire in the same basic way—detected neurotransmitters activate depolarisation—but somehow everything we sense and feel results from these interactions.

Of course, there are important differences. The parts of the brain, the combination of neurons (of which a colossal number are possible) and the neurotransmitters released are all thought to influence the resulting sensations.

It is amazing and inexplicable. Our brains concoct colour from electromagnetic radiation and convert alternating bands of high and low air pressure into beautiful symphonies. They reflect on this sensory data and imbue it with judgment that elicits complex emotions. They respond to different stimuli with warning pain or exquisite pleasure, and they distinguish what substances are usefully eaten and which best avoided. No one can truly explain how this palette of emotions blossoms from these crackling neurons as they Mexican-wave along their branching and merging pathways. Evolutionary psychologist Steven Pinker suggests we may never know, ranking it with quantum physics and what exists outside the universe in the realms of knowledge that is beyond our comprehension.[11] Ownership of this sophisticated organ is no guarantee of fathoming it.

The precise mechanism by which emotion emerges from neuronal interaction may flummox us always, but that is not to say that nothing useful can be understood about the brain. Much of our understanding of the reward network has arisen from research into addiction. As a result, our understanding of the condition has been revolutionised.

What qualifies a substance or behaviour as addictive? Traditionally, the condition was viewed in terms of dependence and withdrawal.[4] Addicts need their fix and the consequences of going without range from crankiness to intense physical discomfort, as vividly portrayed by Danny Boyle's film *Trainspotting* in which the cold-turkey-suffering Renton hallucinates dead babies walking across the ceiling whilst his guts are ravaged by cramps. But this old model of dependence and withdrawal cannot explain the tendency of reformed addicts to fall back into their old ways.[4]

This will make perfect sense to anyone who has been a smoker. The physiological withdrawal symptoms are supposed to finish within three months of smoking the last cigarette. But a week without is nothing to celebrate because the risk of falling back into smoking is colossal. In fact, for smokers and reformed smokers alike, there is one event bound to make the fingers twitch in unconscious reach for a cigarette: the sight of somebody else lighting up. Such visual cues are now thought to play a central role in addiction, along with smells and memories.[4] If a former drug addict sees an old partner in crime or hears a song that once accompanied drug taking, they will be at risk of spiralling back into addiction.

There is a very strong relationship between dopamine and addiction that is beautifully demonstrated by the effects of cocaine and amphetamine. Both of these drugs interfere with the dopamine transporter, the protein that draws neurotransmitters back into the neuron from which they were released. Consequently, when the brain releases a dopamine signal, the molecules remain in the synaptic cleft for longer, where they can continue to tweak the receptors of the post-synaptic neurons and hence amplify the signal.

This is a very interesting feature of proteins that crops up constantly in biochemistry. Proteins have very specific shapes, which, in theory, enable their exclusive relationships with other biomolecules. For example, while dopamine can activate a dopamine receptor it cannot take the place of ATP in fuelling the contraction of the actin–myosin complex in muscles. But now and then an imposter appears with a shape and electrostatic fingerprint that is complementary to a certain region of a protein. That region may be the protein's normal binding site, such as the part of the dopamine receptor into which dopamine inserts itself, or it may be some other part of the protein. Human inventions tend to have helpful casings that would prevent, say, rainwater spilling into the cylinder of an engine. But any atoms on the surface of a protein are fair game for mutual attraction with a suitable imposter that just happens to mesh. When that happens, the protein can change shape. This is shown when dopamine agonists make dopamine receptors react just as if they had bound dopamine itself, whilst at other times the imposter causes the protein to change into some other shape.

This is what happens with cocaine and amphetamine. When cocaine binds the dopamine transporter, it effectively blocks it and stops it working, whilst amphetamine uses its resemblance to dopamine to sneak through the transporter. Once inside, it serves extra helpings of dopamine into the synapse[4] *via* the breached dopamine transporter. The net result for each is similar: prolonged bouts of unusually high dopamine levels in the synaptic cleft, which amplify an intended signal or even generate a fake signal.

These two stimulants are not the only addictive drugs to interfere with dopamine signalling. Studies suggest that all addictive drugs cause spikes in the levels of dopamine in the spaces between cells of the nucleus accumbens.[4] This has been established using microdialysis, which is basically where you stick a needle into the subject's head and use it to draw substances directly out of the skull. Humans are seldom tested in this way, although it can happen in extreme cases, such as during brain surgery,[12] but controlled studies have been carried out on rodents.[13] Opiates hoodwink opioid receptors in a similar fashion. Drugs like morphine, produced from the extract of poppies, actually look markedly different from the opioid peptides our bodies produce. But the opiates share a common signature that enables them to activate our opioid receptors, triggering the same response as if they had been bound by the genuine article.

So, recreational drugs basically hijack the body's reward system. Like a child who cheats in a test to gain the prestige without the learning, a drug addict gets the neurotransmitter rush without doing anything that warrants a reward. When we consider the normal role of the reward system, it starts to explain the aspects of addiction that did not fit with the earlier models.

And dopamine signalling makes more sense when viewed not as the cause of pleasure but the motivation to seek pleasurable outcomes.

Kent Berridge at Michigan University has disentangled dopamine from the "liking" of a reward and suggests instead that it causes the "wanting" of a reward.[14] This has given rise to the idea of "dopamine for wanting, opioids for liking".[15] His theory suggests that the neurotransmitter does not make us enjoy things but rather motivates us to look for the things we enjoy. In support of this distinction a variety of studies have observed different creatures in the acts of *wanting* and *liking*. Mammals, including rats, chimpanzees and even human babies, demonstrate typical *liking* reactions, such as licking of the lips and rhythmically sticking their tongues out,[16] habits that correspond more with the release of opioids than dopamine. But the boundaries of each neurotransmitter's role in the two states are not completely clear cut; for example, opioids have recently been associated with the anticipation of a reward in rats.[17]

Berridge's ideas about dopamine and wanting are built on the findings of two previous pioneers. Pavlov was the trailblazing psychologist who conditioned his dogs to associate the ringing of a bell with the arrival of dinner. Afterwards, the dogs would salivate simply because they heard the bell, even if no dinner arrived. Wolfram Schultz took this work further. He trained monkeys to earn morsels of apple or juice by performing basic tasks, such as pulling a lever the moment that a light came on. Meanwhile, he measured the dopamine levels in the region of their brains comprising the substantia nigra and the ventral tegmental area,[18] and found something very interesting shown in Figure 3.8. Before they had been trained in the task, it was the appearance of the actual fruit that caused their dopamine levels to surge. But once they had mastered the task, the dopamine spiked as soon as the light came on, providing clues for what might have been afoot in the brains of Pavlov's duped dogs. But Schultz made another interesting discovery. If the fruit failed to materialise after the light was switched on, the trained monkeys showed a fall in dopamine levels at the exact time they had learned to expect the reward.[19]

This curious finding has since been rationalised as a *prediction error*, which is the brain's realisation that something has not

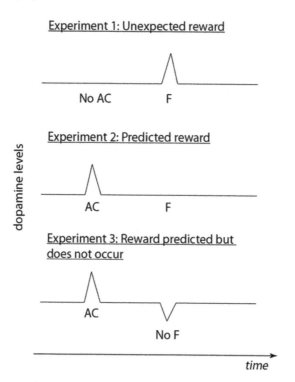

Figure 3.8 As primates learn to associate cues with the arrival of food (F), dopamine spikes shift from the arrival of the food to the arrival of the associated cue (AC). If the food is withheld, dopamine levels decrease at the moment the food was expected.
Adapted from W. Schultz, P. Dayan and P. R. Montague, A Neural Substrate of Prediction and Reward, *Science*, 1997, **275**(5306), 1593–1599.[20] Reprinted with permission from AAAS.

played out as expected. The largest dose of dopamine was paid out when the monkeys were surprised by the reward—they had not predicted it. Once they had associated the light with the fruit, a smaller dose of dopamine was released when the light came on, but the levels quickly returned to normal and showed not even a blip when the fruit actually arrived. In contrast, the other unpredictable outcome had the opposite effect. When the fruit never turned up, dopamine levels fell. One way to view these results is that the dopamine was helping the monkeys to learn how to get more fruit in the future. This interpretation has led to the idea that the neurotransmitter may act as a learning signal.[4]

One theory goes that drug addicts might start to link cues and fixes in a similar manner to the monkeys. Cocaine and amphetamine hijack the dopamine transporter, causing it to backfire so that the synapse is awash with the neurotransmitter. But dopamine is also implicated in the formation of long-term memories. One possible mechanism for this could involve the awkwardly named *cyclic AMP response element binding protein* (CREB), which is known to be activated by dopamine receptors. CREB acts as a transcription factor, meaning that it interacts with the body's DNA to influence which other proteins the body produces. It has also been associated with memory formation in various species, including fruit flies, marine snails and mice.[4] Just as the monkeys learned that the light signalled the arrival of food, so drug users may learn to repeat whatever it was they did to get their last fix, improving their chance of getting the next one.[21]

3.2   THE CASE FOR FOOD ADDICTION

A question yet to be solved is whether or not we can get addicted to food. The curious thing about this is that we are meant to eat. Our brains are adapted to reward us for eating and whether we see it as primarily beneficial to us or to our pesky self-replicators, we must eat in order to continue living. Whereas recreational drugs trick our brains, our reward for eating is rightfully paid. But while eating may be human nature, exploring the case for food addiction has been showing why we differ in our appetites.

The curiosity is how we can get addicted to substances we are supposed to ingest. In fact, some researchers argue that this definition is precisely the problem. The implications of a study led by Suzanne Dickson and Johannes Hebebrand arguably suggest the pathology has less in common with what drug addicts experience and more in common with the compulsion to update our Facebook status.

Headlines about scientific studies into Facebook addiction must rank as the most likely to be dismissed as media nonsense. But academics have already created a Facebook addiction scale in response to reports that young people are sleeping late, falling behind with their studies and even losing jobs as a result of their dependence.[22,23]

Far from being media hokum, it is not inconceivable that it could be classified as a genuine condition in the next few years. In 2013 it was officially agreed within the psychiatric community that behaviours, as well as substances, can be addictive. Gambling was included as the first recognised behavioural addiction in the fifth edition of the authoritative mental health handbook *The Diagnostic and Statistical Manual of Mental Disorders (DSM-V)*.[24] Other contenders, such as shopping, sex and Internet gaming, were not included, but that could change as evidence accumulates now that this paradigm shift has transformed addiction theory.

It is well established that gambling influences dopamine currents in the brain. The neurochemistry supports the role of dopamine as a motivator. One study showed how dopamine was released in stressful situations,[25] which does not fit with the previous belief that dopamine release directly triggers a pleasurable sensation. This was demonstrated in an ingenious study into gambling. Participants were invited to play a game, like a fruit machine with two spinning wheels instead of three. Researchers were particularly interested in near-misses, when one wheel might have a lemon on the win line, whereas the other wheel's lemon was either directly above or below it. Such outcomes are known to hoodwink gamblers into believing they are bearing down on a win, which encourages them to keep playing.

What the study found was that near-misses did indeed increase participants' desire to keep playing. Accordingly, brain scans showed elevated activity in the mesolimbic region, a dopamine-rich channel linking the reward network's ventral tegmental area with the nucleus accumbens. Moreover, players reported that the near-misses were more disappointing than the full misses. So, while the sensation was negative, the result was enhanced motivation. All of this fits well with the updated model that dopamine is not the source of the pleasure but rather the motivating factor that guides us towards opportunities for reward—in this case, the pleasure of a cash pay-out.

Dickson and Hebebrand have suggested that it is eating rather than food that is addictive.[26] More specifically, it is the behaviour of eating highly palatable food. The word "palatable" refers to the degree of pleasure that we experience when we eat a food. However, the size of the reward changes from person to person and depending on the context. As the philosopher Immanuel Kant put it, hunger is the best sauce, or in neuroscientific terms, the palatability of a food depends on the situation.

What is less subjective is the typical nature of a highly palatable food because they are invariably high in calories and, as such, it may be tempting to think that high-calorie foods could be designated as the substance of addiction. But calories and palatability do not always tally. Very few people would enjoy a block of butter as a snack despite its high calorific value. So, food cannot be an addictive substance because it is not one substance but many, all of which will have varying appeal to different people at different times. For these reasons, Dickson and Hebebrand argue that the food itself is not the addictive factor. They cite the brain chemistry that accompanies eating in support of their conclusion. As we saw in the last chapter, when we eat chocolate, the taste cells send a message to the brain signalling the detection of sugar, protein, salt and, if the hypothesis mentioned in Chapter 2 is confirmed, fat. This signal splits between brain regions, including the ventral striatum and nucleus accumbens. Inside these modules of the reward network our brains release soothing opioids. Ann Kelley was one of the first researchers to realise this. The young student who marvelled at Olds and Milner's experiments with the electrode-implanted rats grew up to blaze a trail in the neurochemistry of food addiction. She demonstrated that injections of enkephalin opioids into the nucleus accumbens of rats caused them to eat more[27] and also carried out investigations into the role of dopamine in eating. She even found evidence that a high-fat and high-sugar diet went as far as to change which genes were expressed in the cells of the rats.[28] She is fondly remembered as a bold pioneer and attentive mentor, who made great strides in this important area. Sadly, she died from cancer of the colon aged 53, having won a lifetime achievement award from the Society of Neuroscience in 2006,[29] as well as being invited to speak at a Nobel symposium in 2007. Even today, leading researchers in the field continue to acknowledge her impact as they pursue lines of research she originated.[27]

Rodents have also provided evidence that dopamine may motivate us to eat. Mice that have been bred to lack dopamine signalling die of starvation because it never occurs to them to venture out in search of food.[30] Although, interestingly, other studies have found that such mice can still experience pleasure when they do eat sugar,[31] supporting a superior role for opioids in the actual sensation of pleasure.

If we can get addicted to eating, surely we should demonstrate classic symptoms of addiction. When it comes to humans, scientists are not quite convinced that quitting sugar actually does lead to withdrawal symptoms, but studies have shown that animals display the corresponding symptoms when sugar is taken away.[32] But according to the latest revelations in the *DSM-V*, such symptoms are no longer necessary for a diagnosis of substance dependence anyway.

But while cravings may be imagined, eating addicts could demonstrate the same response to cues as drug addicts. Just as the whiff of smoke makes a smoker's fingers twitch for cigarettes, so the scent of toasting bread beckons one floating down the stairs like a character in a Disney film. And the biscuits are a lot less tempting stored safely in the cupboard than they are on a plate right in front of you. Even when we are stuffed full after a delicious meal, it is common to keep picking at leftovers if they are not tidied away.

Our powerlessness over such cues, posit Dickson and Hebebrand's team, could partly explain eating addiction. It is thought that these sights and smells trigger the release of dopamine—the neurotransmitter that makes us want. The problem is that people react differently to these motivational signals. Some find it easier than others to walk past those chocolate bars so thoughtfully laid out next to the supermarket checkout. Dickson and Hebebrand believe that biochemical differences could explain this variation in susceptibility.[26]

But individuals may *like* more as well as *want* more. Cues are thought to trigger cascades of dopamine that send us rushing for the confectionary, but people may also vary in their sensitivity to opioids. One study has already provided evidence that people with a common mutation of a particular gene may enjoy a bigger hit when they eat highly palatable foods,[15] which only increases their motivation to eat more of them.

The case for eating addiction is strengthened by parallels with substance addiction. Drug users, as well as many obese people, have fewer dopamine receptors than ordinary people, in both cases afflicting the striatal region of the brain.[32] Researchers argue that this may make them hyposensitive, meaning their senses are dulled to the usual sources of pleasure so that they must over-indulge to experience the same size hit as people with a normal number of receptors.[15] If there are fewer receptors for dopamine molecules to activate, then more dopamine molecules need to be released for the neurons to receive the normal signal strength. In other words, these unfortunates need to do more of whatever behaviour it is that triggers the release of dopamine, whether it be eating, drug taking, gambling or drinking.

The next question is what causes such people to have fewer dopamine receptors. One theory goes that individuals lose dopamine receptors as they gain weight, while another suggests that some people start off with fewer receptors and hence are more likely to overeat. Typically, evidence exists in support of both hypotheses. One study showed that rats lost dopamine receptors as they gained weight.[33] Another study uncovered a genetic predisposition to lower numbers of receptors; so-called binge eating participants were more likely to have a mutated version of a gene called *ankyrin repeat and kinase domain containing 1*, which caused them to express only three-quarters as many dopamine receptors as people with the non-mutated gene.[15] These data need not be viewed as conflicting; the complementary nature of these findings shows how debilitating food addiction might be. Certain unfortunate souls could be at greater risk of becoming obese and also have a harder time losing weight. And the double whammies do not stop there. Other genes were investigated in the above study led by Caroline Davis and James Kennedy. In addition to the mutation that predicted fewer dopamine receptors, many members of the binge-eating group had another mutation that predicted for unusually sensitive opioid receptors. These individuals, the researchers reasoned, experience more intense pleasure as their reward for eating, which motivates them to eat more.

The growing ability of neuroscience to dim and boost different neural circuits could lead to effective treatment. Research is ongoing into the use of drugs called receptor antagonists and their counterpart agonists (on which more will be discussed in Chapter 7). These mimics bind with receptors either to block traffic or substitute for it. Antagonists prevent neurotransmitters from binding with their usual receptor, whilst agonists activate the receptor in the neurotransmitters' absence. Opioid agonists have been shown to increase appetite for certain foods, while their antagonists have decreased it. Similar trials have been tried with dopamine circuits.[34] Such treatments could prove invaluable in the fight against obesity.

3.2.1 Where Next for Dopamine and Food Addiction?

Neither food nor eating addiction is assured a place in future volumes of the hallowed *DSM*. Questions surrounding the hypothetical pathology sit within the broader context of the questions about dopamine. As researchers queue up to say what the neurotransmitter isn't, there is less clamour to say what it is.

Berridge and fellow researcher Morten Kringelbach have questioned much thinking about the reward system and its association with dopamine. Having debunked the neuro-transmitter as a source of pleasure, they have fresh doubts about it as a learning signal. Concerns have arisen following studies that showed that mice could learn about rewards even if they had been genetically bred to lack dopamine signalling. Meanwhile, evidence has not been forthcoming that increasing dopamine levels leads to faster learning.[35] Even following its demotion from the starring role of pleasure, dopamine is struggling to find its niche.

But the problems do not stop there. Berridge and Kringelbach have even questioned the holy grail of reward. They find themselves unconvinced that the self-stimulating rats actually experienced any pleasure at all. In fact, the electrical stimulation of the brain region found by Olds and Milner encourages rats to engage in classically rewarded behaviours, like eating and sex, but the facial expression of rats so-motivated to eat gave no indication that they were taking more than the usual amount of enjoyment in the food.[35]

For a more compelling view of this heresy, we can look at the findings of human subjects. In some of the rare cases in which electrodes have been implanted in the pleasure centres of people, the results are not too different. These participants merrily clicked away at the button to self-stimulate, but the signs of pleasure were wreathed in ambiguity. One gentleman complained when the stimulating button was taken away, describing feelings of sexual arousal and even a "compulsion to masturbate". But while he *wanted* to masturbate he never orgasmed, which could be interpreted as the presence of desire but the absence of pleasure.[36]

Never mind the baby and the bathwater, a new voice wants to throw out the bathtub as well. Johann Hari argues in his book *Chasing the Scream* that biochemical mechanisms of addiction are virtually irrelevant, claiming that human contact is what pulls people out of the vortex of compulsive behaviour. He cites a study by Canadian researcher Bruce Alexander, who built a leisure world for rodents full of tube-linked nooks and crannies to explore. Residents of this complex, dubbed "Rat Park", were much less prone to drug addiction than rodents confined to small, bare cages.[37] Alexander reasoned that stimulation helped ward off addiction. Hari also notes the success of the drug policy in Portugal. Fifteen years ago, a chronic addiction problem prompted the government to decriminalise drugs and channel the reclaimed expense of jailing addicts into rehabilitating them. The move has been a resounding success. Instead of being shunned, addicts are welcomed back into society, provided with accommodation and helped to find paid work. Use of needle drugs has halved. Hari passionately argues that most need human connection.[38]

His work has been received by many with caution. A few years prior to the book's publication, Hari was found to have plagiarised other writers, subtly misrepresented interviewees and even spread rumours about rivals under an assumed alias,[39] crimes that led to him returning his 2008 Orwell Prize.[40] Reviewing for *The Guardian*, writer John Harris concludes that *Chasing the Scream* is "convincing" but "flawed".[39]

But even a most scornful critic of Hari's work has agreed with the central point of the need for human connection. Gary Christian accuses Hari of manipulating history to besmirch the leader of the US Federal Bureau of Narcotics, Harry Anslinger, as the originator of the war on drugs. While Hari argues that repressive public policy has stigmatised drug use, Christian offers compelling evidence to the contrary, citing data that while nearly half of Australians have experimented with illicit drugs at least

90% disapprove of the use of cannabis, heroin, cocaine and other drugs. But Christian concedes that the healing of addiction demands "interpersonal warmth and connection", which is broadly aligned with Hari's attitude.[41]

Hari's view is not incompatible with our biochemical understanding of human nature. Section 2 will show how human contact is underpinned by many of the same neural mechanisms explored in this chapter and the idea that drug use could substitute for meaningful social connection has a lot of merit.

So, is it possible to be a chocoholic? According to Dickson and Hebebrand, no. It is not the chocolate, or even the fats and sugars it contains, but rather the eating of any or all of an ever growing array of highly palatable foods to which people may be addicted. Whether or not the term is validated, the research will increase our understanding of why people eat different amounts. Meanwhile, it also adds to our understanding of human nature; the following chapters will consider why we evolved to experience pleasure when we eat.

REFERENCES

1. F. D. Durán, *The Aztecs. The History of the Indies of New Spain*, Cassell, London, printed in USA, 1964.
2. Definition of "chocoholic", *OED Online*. Oxford University Press.
3. Family of mother who died after drinking seven litres of water in radio contest for Nintendo Wii awarded £10m, *Mail Online*, 26-Mar-2016. Available at: http://www.dailymail.co.uk/news/article-1224051/Wee-For-Wii-water-drinking-contest-death-Jennifer-Stranges-family-awarded-10m.html. [Accessed: 26-Mar-2016.]
4. E. R. Kandel, J. H. Schwartz, T. M. Jessell, S. A. Siegelbaum, and A. J. Hudspeth, *Principles of Neural Science*, McGraw Hill, New York, 5th edn, 2013.
5. J. M. Berg, *Biochemistry / Jeremy M. Berg, John L. Tymoczko, Gregory J. Gatto, Jr., Lubert Stryer.*, W H Freeman & Company, a Macmillan Education Imprint, New York, 8th edn, 2015.
6. T. C. Südhof, A Molecular Machine for Neurotransmitter Release: Synaptotagmin and Beyond, *Nat. Med.*, 2013, **19**(10), 1227–1231.

7. J. Olds and P. Milner, Positive Reinforcement Produced by Electrical Stimulation of Septal Area and Other Regions of Rat Brain, *J. Comp. Physiol. Psychol.*, 1954, **47**(6), 419–427.

8. O. Arias-Carrión, M. Stamelou, E. Murillo-Rodríguez, M. Menéndez-González and E. Pöppel, Dopaminergic Reward System: A Short Integrative Review, *Int. Arch. Med.*, 2010, 3, 24.

9. Squire, *Fundamental Neuroscience*, Academic Press, Amsterdam; Boston, 4th edn, 2008.

10. E. J. Nestler, Transgenerational Epigenetic Contributions to Stress Responses: Fact or Fiction?, *PLOS Biol.*, 2016, **14**(3), e1002426.

11. S. Pinker, *The Blank Slate: The Modern Denial of Human Nature*, Penguin, London, New edn, 2003.

12. W. M. Abi-Saab, D. G. Maggs, T. Jones, R. Jacob, V. Srihari, J. Thompson, D. Kerr, P. Leone, J. H. Krystal, D. D. Spencer, M. J. During and R. S. Sherwin, Striking Differences in Glucose and Lactate Levels Between Brain Extracellular Fluid and Plasma in Conscious Human Subjects: Effects of Hyperglycemia and Hypoglycemia, *J. Cereb. Blood Flow Metab.*, 2002, **22**(3), 271–279.

13. R. Ito, J. W. Dalley, T. W. Robbins and B. J. Everitt, Dopamine Release in the Dorsal Striatum during Cocaine-Seeking Behavior under the Control of a Drug-Associated Cue, *J. Neurosci.*, 2002, **22**(14), 6247–6253.

14. K. C. Berridge, The Debate Over Dopamine's Role in Reward: The Case for Incentive Salience, *Psychopharmacology*, 2007, **191**(3), 391–431.

15. C. A. Davis, R. D. Levitan, C. Reid, J. C. Carter, A. S. Kaplan, K. A. Patte, N. King, C. Curtis and J. L. Kennedy, Dopamine for "Wanting" and Opioids for "Liking": A Comparison of Obese Adults with and without Binge Eating, *Obes. Silver Spring Md*, 2009, **17**(6), 1220–1225.

16. K. C. Berridge, T. E. Robinson and J. W. Aldridge, Dissecting Components of Reward: "Liking", "Wanting", and Learning, *Curr. Opin. Pharmacol.*, 2009, **9**(1), 65–73.

17. A. G. DiFeliceantonio, O. S. Mabrouk, R. T. Kennedy and K. C. Berridge, Enkephalin Surges in Dorsal Neostriatum as a Signal to Eat, *Curr. Biol.*, 2012, **22**(20), 1918–1924.

18. W. Schultz, Predictive Reward Signal of Dopamine Neurons, *J. Neurophysiol.*, 1998, **80**, 1–27.

19. W. Schultz, P. Dayan and P. R. Montague, A Neural Substrate of Prediction and Reward, *Science*, 1997, **275**(5306), 1593–1599.

20. W. Schultz, P. Dayan and P. R. Montague, A Neural Substrate of Prediction and Reward, *Science*, 1997, **275**(5306), 1593–1599.

21. M. E. Wolf, Addiction: Making the Connection Between Behavioral Changes and Neuronal Plasticity in Specific Pathways, *Mol. Interventions*, 2002, **2**(3), 146–157.

22. C. S. Andreassen, T. Torsheim, G. S. Brunborg and S. Pallesen, Development of a Facebook Addiction Scale, *Psychol. Rep.*, 2012, **110**(2), 501–517.

23. D. Karaiskos, E. Tzavellas, G. Balta and T. Paparrigopoulos, P02-232—Social Network Addiction: A New Clinical Disorder?, *Eur. Psychiatry*, 2010, **25**(Supplement 1), 855.

24. American Psychiatric Association, *Diagnostic and Statistical Manual of Mental Disorders*, American Psychiatric Association, 2013.

25. L. Clark, A. J. Lawrence, F. Astley-Jones and N. Gray, Gambling Near-Misses Enhance Motivation to Gamble and Recruit Win-Related Brain Circuitry, *Neuron*, 2009, **61**(3), 481–490.

26. J. Hebebrand, Ö. Albayrak, R. Adan, J. Antel, C. Dieguez, J. de Jong, G. Leng, J. Menzies, J. G. Mercer, M. Murphy, G. van der Plasse and S. L. Dickson, "Eating Addiction", Rather Than "Food Addiction", Better Captures Addictive-like Eating Behavior, *Neurosci. Biobehav. Rev.*, 2014, **47**, 295–306.

27. J. M. Richard, D. C. Castro, A. G. Difeliceantonio, M. J. F. Robinson and K. C. Berridge, Mapping Brain Circuits of Reward and Motivation: In the Footsteps of Ann Kelley, *Neurosci. Biobehav. Rev.*, 2013, **37**(9 Pt A), 1919–1931.

28. Fast Food "as Addictive as Heroin", *BBC*, 30-Jan-2003.

29. Ann Kelley, Studied Fast-Food's Effect on Brain, *The Boston Globe*. Available at: http://archive.boston.com/news/globe/obituaries/articles/2007/09/07/ann_kelley_studied_fast_foods_effect_on_brain/. [Accessed: 29-Mar-2016.]

30. M. S. Szczypka, K. Kwok, M. D. Brot, B. T. Marck, A. M. Matsumoto, B. A. Donahue and R. D. Palmiter, Dopamine Production in the Caudate Putamen Restores Feeding in Dopamine-deficient Mice, *Neuron*, 2001, **30**(3), 819–828.

31. C. M. Cannon and R. D. Palmiter, Reward without Dopamine, *J. Neurosci.*, 2003, **23**(34), 10827–10831.
32. K. D. Brownell and M. S. Gold, *Food and Addiction: A Comprehensive Handbook*, Oxford University Press, 2012.
33. P. K. Thanos, M. Michaelides, Y. K. Piyis, G.-J. Wang and N. D. Volkow, Food Restriction Markedly Increases Dopamine D2 Receptor (D2R) in a Rat Model of Obesity as Assessed with In-vivo muPET Imaging ([11C] Raclopride) and In-vitro ([3H] Spiperone) Autoradiography, *Synap. N. Y. N*, 2008, **62**(1), 50–61.
34. R. L. Corwin and F. H. Wojnicki, Baclofen, Raclopride, and Naltrexone Differentially Affect Intake of Fat and Sucrose Under Limited Access Conditions, *Behav. Pharmacol.*, 2009, **20**(5–6), 537–548.
35. K. C. Berridge and M. L. Kringelbach, Affective Neuroscience of pleasure: Reward in Humans and Animals, *Psychopharmacology*, 2008, **199**(3), 457–480.
36. R. Heath, Pleasure and Brain Activity in Man. Deep and Surface Electroencephalograms During Orgasm, *J. Nerv. Ment. Dis.*, 1972, **154**, 3–18.
37. B. K. Alexander, R. B. Coambs and P. F. Hadaway, The Effect of Housing and Gender on Morphine Self-administration in Rats, *Psychopharmacology*, 1978, **58**(2), 175–179.
38. J. Hari, The Likely Cause of Addiction Has Been Discovered, and It Is Not What You Think, *The Huffington Post*. Available at: http://www.huffingtonpost.com/johann-hari/the-real-cause-of-addicti_b_6506936.html. [Accessed: 23-Jan-2015.]
39. J. Harris, Chasing the Scream by Johann Hari Review—Taking on the War on Drugs, *The Guardian*, 09-Jan-2015. Available at: http://www.theguardian.com/books/2015/jan/09/chasing-the-scream-johann-hari-war-on-drugs. [Accessed: 05-May-2016.]
40. The Orwell Prize and Johann Hari, *The Orwell Prize*, 29-Sep-2011. Available at: http://theorwellprize.co.uk/news/the-orwell-prize-and-johann-hari/. [Accessed: 05-May-2016.]
41. G. Christian, Effacing the Scream: Confronting Drug Legalisation, *Quadrant*, 2016, **60**(4), 62.

CHAPTER 4

The Chemistry of Life's Origins

On the 28th September, 1969, at 10.58 in the morning,[1] the residents of Murchison in Victoria, Australia, were alerted to the presence of a huge, orange fireball hurtling out of the sky towards them. The flaming spectacle, trailed by a dull orange tail and blue smoke, was visible more than 400 km away. A sonic boom was emitted as the fireball exploded and scattered its scorching smithereens for miles around. The Murchison meteorite had landed.[2]

John Lovering was a professor of geology at Melbourne University at the time. He arrived on the scene to collect as many fragments of the meteorite as possible. His team found that the rock had ablated, meaning that the surface had melted, which had effectively sealed the contents inside[3] and, crucially, had preserved them from contamination. Lovering declared to the press: "This is almost as exciting as moon dust." For this was the year of the first moon landing and the professor had already been sent samples of the moon dust brought back by Neil Armstrong and his team.[2]

Decades of research was to show that the Murchison meteorite shed light on one of the most important scientific questions ever to be asked: how did life begin on Earth? When Charles Darwin rocked Victorian society with his theory of evolution, he was only able to suggest how complex life could emerge from simpler life.

The Chemistry of Human Nature
By Tom Husband
© Tom Husband 2017
Published by the Royal Society of Chemistry, www.rsc.org

But he offered far less information about how life appeared in the first place. The astronomical events of 28th September, 1969, now offered a clue. The Murchison meteorite was found to contain hundreds of amino acids—one of the very building blocks of life. And these building blocks of life had come to us from outer space. To use the scientific term, they were extraterrestrial.[3]

It was just a year before that the cult film *2001: A Space Odyssey* was released. Arthur C. Clarke's classic science fiction work suggested that intelligent life existed elsewhere in the universe. Moreover, these extraterrestrial beings wanted intelligent life to appear on Earth and accelerated the evolution of man's primate ancestors with a beguiling, black obelisk. Now the Murchison meteorite had presented scientific evidence that could be taken in support of the idea that we are indeed not alone in the universe and, soon, respected scientists were to make the case that alien intervention may have begun long before primates developed opposable digits.

In 1973 the scientists Leslie Orgel and Francis Crick published a paper suggesting that life was started on Earth by extraterrestrial beings. Both were well-established researchers. Crick was, of course, one member of the Nobel-Prize-winning dyad Watson and Crick, who, with Rosalind Franklin, had discovered the structure of DNA. Meanwhile, Orgel was the director of the Chemical Evolution Laboratory at the Salk Institute for Biological Studies. Together, they coined the term *directed panspermia*, the name of their article and the process by which intelligent life would deliberately send microorganisms into outer space in the hopes of seeding life on other suitable planets.[4]

The thrust of their argument was that the universe had been in existence for nine billion years before life began on Earth. This left ample time for intelligent civilisations to evolve on other planets and they argued that the technology required for the project was almost within the grasp of mankind. They envisaged that unmanned spacecrafts would deliver various microorganisms to whichever planets ended up in their path. The cargo would have to be kept at temperatures near absolute zero and a shield would be necessary to protect them from cosmic radiation.

The weakness of this theory concerned the motive of any civilisation that might undertake the task. They suggested only that it would be a display of the technological capability or "missionary zeal". One cannot help wondering that it was the possibility rather than the likelihood of the existence of extra-terrestrial life that compelled them to write the paper, as this concluding paragraph suggests:

> *These enquiries are not trivial, for if successful they could lead to others which would touch us more closely. Are the senders or their descendants still alive? Or have the hazards of 4 billion years been too much for them? Has their star inexorably warmed up and frizzled them, or were they able to colonise a different Solar System with a short-range spaceship? Have they perhaps destroyed themselves, either by too much aggression or too little? The difficulties of placing any form of life on another planetary system are so great that we are unlikely to be their sole descendants. Presumably they would have made many attempts to infect the galaxy. If the range of their rockets were small this might suggest that we have cousins on planets, which are not too distant. Perhaps the galaxy is lifeless except for a local village, of which we are one member.*[4]

The final sentence is particularly telling—the idea of a local village of alien life. Orgel went on to assist with the work of the Search for Extraterrestrial Intelligence (SETI), the organisation that beams messages out into space in the hopes of, some day, getting one back. It is tragic enough that with such a colossal universe in existence we are so powerless to explore it. The prospect of setting up a pen-pal-style arrangement with alien civilisations seems like the ideal compromise but, sadly, even that is not an especially likely prospect. With our current levels of technology, any message would take a year to transmit per light year of distance separating us from our correspondents. In short, it would only be in the event that aliens were found in this local village that would make communication realistic, let alone a visit in person.

Nowadays, researchers studying the origin of life are not too concerned with the theory of panspermia. Even if life did first appear on another planet, it doesn't solve the problem of how it came into existence in the first place. Orgel himself eventually conceded that he had been overly pessimistic about the potential for life to have evolved on Earth.

We may still be without evidence of extraterrestrial life, but the findings on the Murchison meteorite were still significant in that they strongly supported the idea that amino acids could be spontaneously produced in abiotic conditions, that is without the help of any biological organism. Living organisms, including humans, can and do make their own amino acids. Not only that, but we rely on them for just about every process that happens in our bodies. What was significant here was that the amino acids on the meteorite did not appear to have a biological source, unless they formed from the remains of some extraterrestrial lifeform, in which case you would expect to find other organic substances, as well as amino acids. Either way, the finding was significant.

The Murchison meteorite was not the first chapter in the search to explain the appearance of life on Earth. Charles Darwin's paradigm-obliterating *On the Origin of Species* offered explanation only as to how ancient organisms could evolve but not how they first appeared. In a letter to his friend J. D. Hooker he made his now famous comment that the precursor to all known life could have emerged from a stew of ammonia and phosphoric salts in some "warm little pond".[5] Since then, abiogenesis, the process by which the lifeless comes alive, has been a subject of inquiry that has attracted scientists of the highest ranks, including the Nobel laureates Francis Crick, Erwin Schrödinger and Richard Feynman. Darwin's is just one of many sound theories on the subject that are confounded by the need for supporting experimental data. In order to conclusively prove how life began on Earth, the scientific method requires that the process be recreated in the laboratory. As we shall see, such an event is a very long way off.

The puzzle of how life began bears much in common with the mystery of how the pyramids were constructed. For one thing, both conundrums have, at times, so defied explanation that certain commentators have suggested that such marvels can only have been achieved by visiting extraterrestrials. At a more practical level, the two processes can be subdivided as follows: where did the building blocks come from and how were they assembled? A notable difference, however, is that not all mysteries are as mysterious as others.

In fact, some would argue that the mystery of how the pyramids were constructed is already solved. Donald Redford, professor of classics and ancient Mediterranean studies at Pennsylvania State University in the US, has provided explanations for how the building blocks were produced, as well as how they were collectively arranged in a pyramidal form. An ingenious method was used to cut huge granite slabs with straight edges. Workers would cut small slots into which they would wedge wooden pegs. Next, they added water. The pegs absorbed the water, which caused them to expand. This expansion cleaved the rock with a straight edge. As to how these hefty slabs were hoisted into position on the pyramid under construction, Professor Redford has an explanation amply supported by the requisite experimental proof. "I usually show the sceptic a picture of 20 of my workers at an archaeological dig site pulling up a two-and-a-half-ton granite block. I know it's possible because I was on the ropes, too."[6] Case closed? A final point in common with our origins-of-life puzzle is that a mystery too ancient can never be completely solved, because we will never be able to say for sure what happened.

Perhaps the biggest difference between these mysteries is that the first lifeforms could not rely on the helping hands of long-suffering slaves but instead had to put themselves together. But what sounds like a stumbling block turns out to be more of a stepping stone. It is the manner of how nascent lifeforms self-assembled that sheds arguably the brightest light on the abiogenesis mystery.

Several years ago, I was discussing evolution with a friend and I noted how fortunate it was that early organisms had evolved the ability to reproduce. If we judge it very difficult for life to emerge—and our failure to recreate the phenomenon in the laboratory pays testament to the fact—we cannot expect it to happen too frequently. How lucky, then, that when it did happen, those first organisms learned to pass the torch. While King Midas turned everything he touched into gold, these early organisms shared the gift of life, transforming inorganic matter into additional lifeforms.

My point to my friend was that it was lucky that the first lifeforms had accidentally developed the ability to reproduce, such that life, having come into existence, stayed in existence. But I had it the wrong way round. It is the fact of reproduction, or more accurately, replication, that enabled life to exist at all. A key chapter in the emergence of life was the appearance of a new kind of molecule with a special characteristic. This kind of molecule had the ability to make copies of itself. It was the tendency of these molecules to self-replicate that many would argue represents the turning point between lifeless and living.

On which point, any attempt to explain how life began must necessarily consider what exactly life is. Abiogenesis researchers are very far from agreed on a definition of life. A thorough investigation of the competing definitions is beyond the scope of this text, but a recent and controversial offering suits our current purposes. Edward Trifonov suggests that: "life is self-reproduction with variations".[7] It may have been met with a storm of criticism, but the sparse definition is not without its merits. Kepa Ruiz-Mirazo, Juli Peretó and Alvaro Moreno offer a more detailed version: organisms are autonomous beings with chemical processing systems for metabolism, information coding and energy supply all encapsulated in a semi-permeable boundary.[8]

Before analysing the definition, it will be useful to consider exactly how a molecule might make a copy of itself. In fact, the proof is going on inside you as you read this sentence. In every one of your trillions of cells the DNA that makes you unique is making copies of itself. The difficulty for origins-of-life research is not to prove whether it happens but how it started happening. In fact, the matter is one of a whole farmyard of fowl and the eggs of their paradoxical origin. It is not just whether the chicken or the egg came first but which came first out of the duck, the goose, the turkey and all of their respective eggs. Our DNA is only able to make copies of itself in cooperation with an army of proteins, begging the question: was it proteins that first appeared and helped to assemble the first lengths of DNA? This sounds eminently logical until we note that the instructions for how to assemble the proteins from their own amino acid building blocks are stored in the DNA. DNA tells your cells which proteins to make and how, all the while relying on those very proteins to do its job.

Paradoxes seldom bear scrutiny. In the case of the actual chicken and the egg, our understanding of evolution is sufficient to state that what came first was an ancestor of the chicken, which evolved the ability to lay hard-shelled eggs in the interim period. In the case of self-replicating gene sequences, RNA is broadly held to be the ancestral forerunner to DNA. The so-called RNA world, in which all lifeforms stored their genetic information in RNA rather than DNA, was a far more widely accepted theory of the originators of the *directed panspermia* hypothesis, Crick and Orgel, who suggested the idea in separate papers at roughly the same time as a third researcher Carl Woese.[9–11]

RNA is an attractive candidate for the forerunner of DNA for various reasons, including that it is made of almost identical building blocks and, secondly, because, like DNA, it can store genetic information. Accordingly, much of the latest thinking on abiogenesis is based on how RNA self-replicates.

The basic concept can be considered in terms of a dance class. Imagine the teacher splits a large group into two halves. One half is asked to form a line along the wall of the dance hall. There are men and women in the line arranged in a random order. The remaining men and women are scattered around the rest of the dance hall. The teacher instructs them to find a partner and the men drift towards the nearest woman in the line, while the free women drift towards the nearest man in the line. As a result, the first line guides the production of a second line, which is a mirror image of the first, as shown in Figure 4.1.

Let's introduce another parameter. All of the dancers are either dog lovers or cat lovers. We assume (quite unreasonably) that dog lovers stick together, as do cat lovers. Again, the dance teacher instructs half of the students to form a line in random order. Next, as before, the first line guides the creation of a complementary copy. This time, dog-loving male dancers line up beside dog-loving female dancers, while cat-loving women line up by cat-loving men, as shown in Figure 4.2.

Finally, the process repeats. The dancers in the first line decide that, actually, they need a break, so the dance teacher ushers in another group of dancers, boasting the same polarised attitude toward canine *versus* feline pet ownership. Now, a third line of dancers forms next to the second line. This third line is a mirror image of the second line, which was itself a mirror image

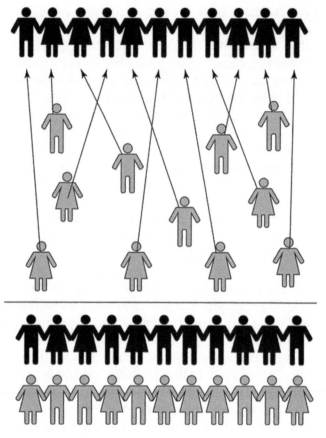

Figure 4.1   The first line guides the production of a second line, its own
mirror image.

of the first line. As such, as shown in Figure 4.3, the third line is a
replica of the first line. This is how RNA self-replicates.

Lengths of RNA are constructed from four different building
blocks called nucleotides. Figure 4.4 shows how each nucleotide
contains a ribose group, a phosphate group and a nucleobase.
The ribose and phosphate groups are common to all nucleotides,
but there are four unique nucleobases called cytosine, adenine,
guanine and uracil (ref. 12, p. 37). Like the men and women in
the dance class, they attract one another, but they are a little
choosier. Uracil always pairs up with adenine, while guanine has
to be with cytosine. (Note that DNA is composed of almost
identical building blocks, substituting deoxyribose for ribose

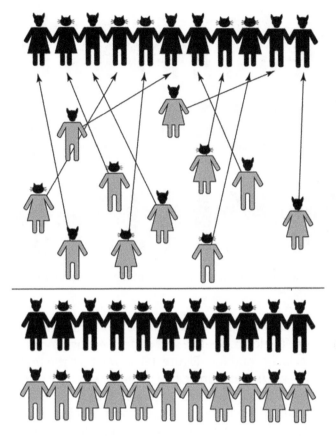

Figure 4.2 Dog-loving males and females pair up, as do cat-loving males and females.

and a different nucleobase called thymine for uracil; ref. 12, p. 119.)

Nucleobases selectively pair up because of their complementary electrostatic fingerprints. Chapter 1 showed how hydrogen atoms in one molecule of water attract the oxygen atoms in neighbouring water molecules. This is because the electrons of both elements preferentially huddle around the oxygen atoms, leaving the positive hydrogen nuclei comparatively exposed. Another element that hogs electrons is nitrogen. Nitrogen and oxygen are known as electronegative elements, while hydrogen is electropositive by comparison to both, and compounds rich in all three, such as the nucleobases, have the

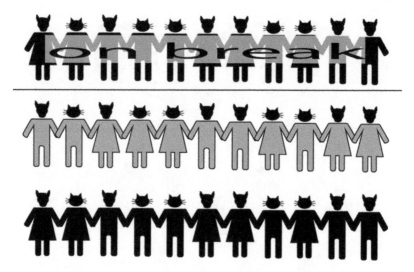

Figure 4.3 The third line—a mirror image of a mirror image—is a replica of the first line.

same uneven distribution of electrons that characterises proteins. This gives rise to electropositive and electronegative zones, denoted by the symbols $\delta+$ and $\delta-$. Figure 4.5 shows how the nucleobases line up with their electronegative regions next to the complementary base's electropositive regions. Uracil is less likely to pair with guanine because the latter's minus–plus arrangement clashes with the former's minus–plus–minus. Consequently, uracil pairs with the better-matched adenine and cytosine with guanine.

This selective pairing is what enables RNA chains to replicate themselves. An RNA chain consists of random sequences of nucleotides. As shown in Figure 4.6, the phosphate and ribose groups in the neighbouring nucleotides join up to make a backbone, which holds the nucleobases in a line. The complementary nature of the nucleobases explains two aspects of the phenomenon. First, it explains how an exact copy can be produced. Where, for example, a cytosine nucleobase is exposed on the chain, the nucleobase that will be most strongly attracted to it is guanine. Next, it is this attraction that provides an energy source for the process. It is precisely because of the attraction between the different pairs of the nucleobases that separate nucleotides will line themselves up alongside the chain and form a complementary chain. The two chains separate and now the

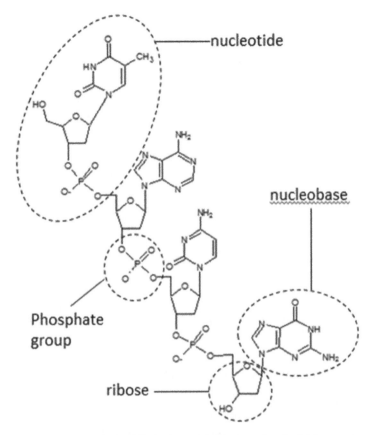

Figure 4.4 RNA consists of linked nucleotides, each of which can be subdivided into a ribose group, a phosphate group and a nucleobase.

new chain can do the same thing. The next complementary chain will be an exact copy of the original, just as the third dance line matched the first in Figure 4.3.

The tendency of RNA to produce copies of itself is highly significant, but it leaves many questions unanswered. For a start, where did RNA come from? This is a chicken and egg duo for which even the probable identity of a progenitor has yet to be confirmed. The RNA world hypothesis proposed RNA as a precursor to the DNA world to solve the chicken and the egg problem that DNA cannot self-replicate without proteins to help it. But RNA can only self-replicate in very basic ways unless it, too, has the support of proteins and other catalytic substances. The case is not solved; we have merely found a clue.

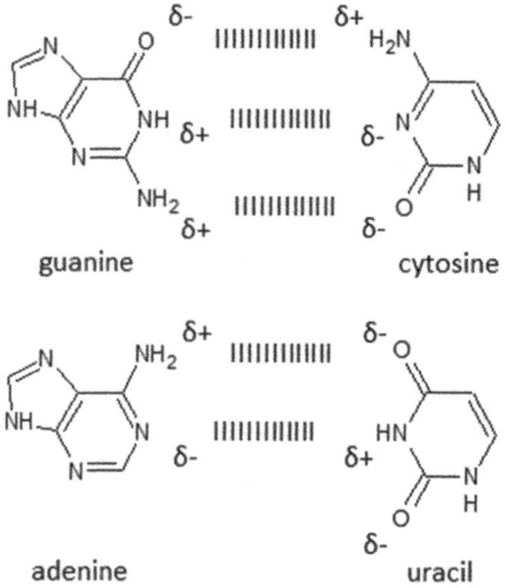

Figure 4.5   Pairings of the four RNA nucleobases. Intermolecular attraction, shown by dashed lines, exists between electronegative ($\delta-$) and electropositive ($\delta+$) regions.

4.1   WHERE DID RNA COME FROM?

The emergence of RNA from the primordial soup is, like seemingly every piece in the puzzle of life, simple in theory but harder in practice. In theory, nucleotides were produced by the random reactions between whatever simpler compounds would have flourished in those days. Then, all this assortment of nucleotides had to do was join up into long chains. The building blocks appeared and then they formed neat, orderly queues. These are elegant theories but a wealth of experimental evidence is needed to support them.

The appearance of nucleotide building blocks recalls the Murchison meteorite. The intercepted asteroid was teeming with amino acids, another important building block in the abiogenesis puzzle. This, in turn, recalls the landmark experiment of Stanley Miller, which it is practically illegal to omit from any discussion of the origins of life. Miller discharged electrical sparks in a flask containing gases meant to recreate the primordial atmosphere: ammonia (NH_3), water (H_2O), hydrogen (H_2) and methane (CH_4). Afterwards, he detected amino acids in the

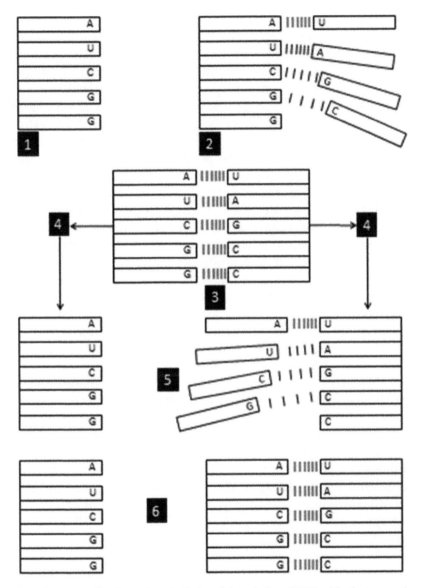

Figure 4.6 Replication of RNA chains. (1) A chain of RNA with the sequence adenosine, uracil, cytosine, guanine, guanine, *etc.* (2) Free-floating nucleotides are attracted to their base pairs in the RNA and line up alongside them. (3) A new complementary chain has formed. (4) The two chains separate. (5) More free-floating nucleotides line up in exclusive pairings with the complementary RNA chain. (6) The next RNA chain created is identical to the original chain, which has thus been copied.

reaction vessel.[13] It is quite reasonable to suppose that these simple compounds would have been in existence because atoms react and make compounds, it's just what they do. The problem for Miller was that his model atmosphere was ultimately judged to be unrealistic. For one thing, hydrogen is so light that it just drifts up and out into space, so it would not have been present in the necessary amounts. Nevertheless, the infamous Miller–Urey experiment and the findings on the Murchison meteorite both show that amino acids can be produced in abiotic conditions, that is without a causative biological process.

The meteorite and Miller's experiment prove that amino acids can be produced abiotically, but what about nucleobases? The spontaneous appearance of RNA requires that the requisite building blocks would be available. In fact, the Murchison meteorite not only contained amino acids but also nucleobases, including uracil.[14] The fact that such biologically relevant compounds originated in space and were delivered to Earth has some exciting implications, which we will revisit later on. Meanwhile, experimental work has confirmed that the nucleobases can be produced in the laboratory using conditions and reagents that could have existed in the primordial soup.[15]

Nucleobases may have been produced in the laboratory, but that is not enough to form an RNA chain. In order for that to occur, each of the four nucleobases needs to be attached to a ribose group and a phosphate group. These, too, have been produced in laboratory conditions designed to recreate primordial Earth but therein is precipitated the next problem: scientists have found it extraordinarily difficult to get the three to join up. You can put the three parts of a nucleotide in a flask, but they have proven only slightly more inclined to self-assembly than flat pack furniture.[15]

As harsh conditions will accelerate evolution in the wild, so tough problems provoke novel solutions. The thinking of how to produce nucleotides has recently been turned on its head. Researchers have realised that rather than making the separate parts and then struggling to fit them together, it might make more sense to just make the nucleotide directly.

Imagine two different paths to a cake, both shown in Figure 4.7. The first is that you separately make the two layers of sponge, the cream filling and the icing sugar case. Slip the filling

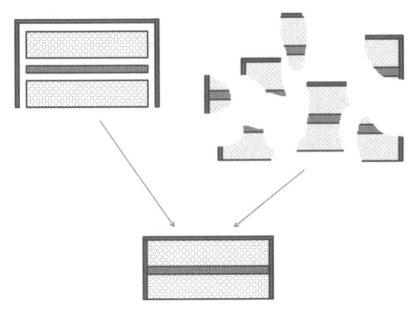

Figure 4.7 Two ways to make a cake: (1) make the sponge, cream filling and icing layers separately, and then join them together; (2) jury-rig icing- and cream-daubed chunks of sponge into a cake.

between the sponge layers, pop the case on top and the cake is complete. But how about this: from an assortment of misshapen fragments, each formed of a chunk of sponge daubed indiscriminately with a bit of icing here and a bit of cream filling there, you jury-rig the cake by cobbling the fragments together so that all the icing ends up on the outside and all the cream filling in the middle. It's not how you would expect to do it but with the right chunks combined in the right way, it could be done.

This is a way to conceptualise the shift in thinking on the spontaneous production of nucleotides. In 2009 scientists John Sutherland, Matthew Powner and Béatrice Gerland produced a compound almost identical to the cytosine nucleotide from a mixture of simple compounds. Instead of trying to coax separate ribose, phosphate and nucleobase groups into joining together, they took five compounds with a similar overall blend of elements and heated it. Figure 4.8 shows how the mixture of glycolaldehyde, glyceraldehyde, cyanamide, cyanoacetylene and phosphate produced β-ribocytidine-2′,3′-cyclic phosphate, which differs by only a few atoms from the cytosine nucleotide, called

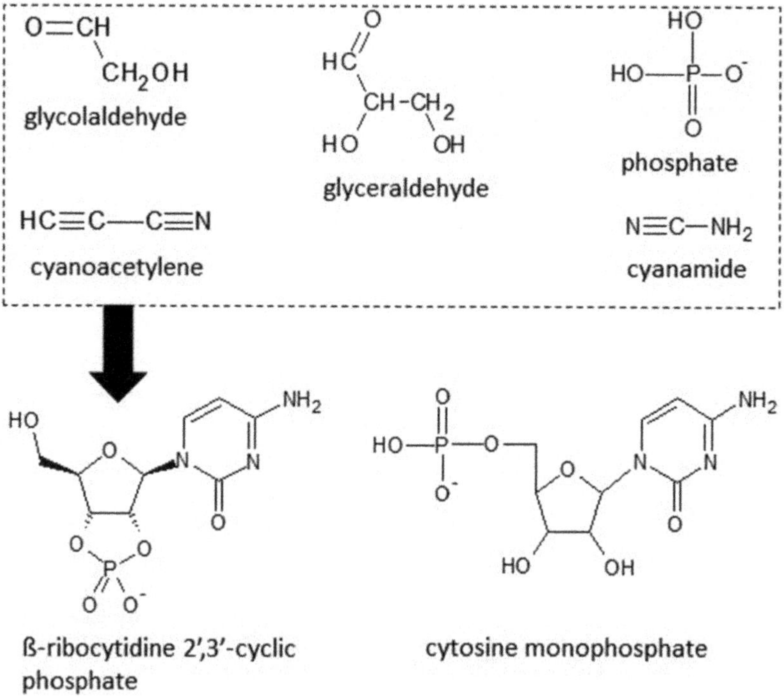

O=CH
\
CH₂OH
glycolaldehyde

glyceraldehyde

HC≡C—C≡N
cyanoacetylene

phosphate

N≡C—NH₂
cyanamide

ß-ribocytidine 2′,3′-cyclic
phosphate

cytosine monophosphate

Figure 4.8 Five simple compounds react together to form β-ribocytidine-
2′,3′-cyclic phosphate, which closely resembles the RNA nucleotide
cytosine monophosphate.

cytosine monophosphate.[16] Their findings have been hailed as a
landmark breakthrough to a problem that has dogged researchers for 20 years. The team won $150 000 to fund future
research from the Origins of Life Project for their important
work.[17]

Sutherland and his team have achieved a major breakthrough
but so many questions remain unanswered. The other three
nucleotides (uracil, adenine and guanine) are still absent from
the latest prebiotic cocktail and, even if they were present, the
means by which to link them into an RNA chain remains to be
uncovered. It is hard enough to synthesise building blocks in
conditions representative of the primordial soup, let alone to get
them to join up to each other. There is also the matter of
handedness (formally termed *chirality*), the fact that laboratory
processes invariably produce a mixture of "left-handed" and
"right-handed" nucleotides and amino acids. In our bodies the

situation is quite different; amino acids are exclusively left-handed, while nucleobases are all right-handed.[18] Not only do we need to know how nucleotides form, we also need to know why their descendants are exclusively right-handed.

These difficulties have led to suggestions that there may be even another precursor to the RNA world. One of the most imaginative suggestions was the idea that silicate crystals, available from clay, could have stored genetic information.[19] One strength of the theory is that montmorillonite clay catalyses the polymerisation of nucleotides.[15] Another possibility is the precursor PNA, which stands for peptide nucleic acid. In this version, peptides—chains of linked amino acids—form the backbone to which nucleobases attach themselves. This theory has many supporting pieces of evidence. First of all, the Murchison meteorite contained PNA-related molecules. Also, separate fragments of PNA can be induced to line up alongside complementary strands of DNA.[15] Much contemporary research is engaged in a search for simpler versions of the nucleobases or the phosphate backbone.

Another important issue is the matter of how a self-replicating molecule, impressive though it may be, suddenly becomes a single-celled organism. In answering this question, we will consider another revelation: that evolution started long before there were organisms to evolve.

4.2   CHEMICAL EVOLUTION

The respected origins-of-life researcher Addy Pross has recently formalised a remarkable hypothesis: that biological evolution is merely the secondary stage of a process that starts at the chemical level. This builds on decades of research in a young field of science called *systems chemistry*. It has long been established that replicating molecules of RNA demonstrate characteristics of evolution. Pross has now suggested a theory to explain these observations, along with his own take on a definition of life: *a self-sustaining, kinetically stable, dynamic reaction network derived from the replication reaction.* We will now examine each component of this definition in turn.[20]

The crux of Pross' theory is dynamic kinetic stability, which goes some way to solve a conundrum of life first observed by Erwin Schrödinger. The famous physicist noted a conflict between the chemistry of life and entropy, a universal process described by the immutable law of nature known as the second law of thermodynamics.[21] (See the Appendix for a detailed explanation.) In simple terms, this law states that the universe is always getting more chaotic. Consider how the elements were produced. Billions of years ago, there was only one element, the simplest element, hydrogen. It clumped together into colossal stars, in the core of which the force of gravity exerted so much pressure on the atoms that they started fusing into new, larger elements: nitrogen, carbon and so on. Chaos was already in full swing, but then these stars would pass a tipping point, triggering a supernova—an astronomical explosion flinging a cosmic cocktail of elements deep into space, which would ultimately form new clumps called planets, asteroids and so on. You could wait your whole life to witness a supernova, but here's something you will never see—a supernova going backwards, in which the elements fly back to some central point and split into smaller and smaller atoms until all that remains is a star composed of pure hydrogen. The reason is that this would not only require huge amounts of energy but also because the system would have become less chaotic. Even if the energy were available, the statistical likelihood that the components would all gather up into a more orderly system makes winning the lottery seem as easy as breathing.

But then consider photosynthesis. The process is energy ravenous and yet produces highly ordered starch granules from carbon dioxide and water. This is an oversimplification of the process, but it is a useful demonstration of the apparent conflict between the chemistry in organisms and the second law of thermodynamics. Organisms are *fantastically* ordered. In each of the trillion cells in your body exists a length of DNA that has been even more neatly packed than the suitcase of a holiday maker who has succeeded in cramming not only the kitchen sink but also the entire contents of three wardrobes, along with the complete toy and DVD collections of his six children. How is this possible?

Here, Schrödinger invoked an important caveat to the second law: the rules only apply in a *closed system*.[21] If we sealed a person into an airtight box, it would soon decay into thermodynamic equilibrium—Schrödinger's euphemism for death.

And this was his point: *"The device by which an organism maintains itself stationary at a fairly high level of orderliness ... really consists in continually sucking orderliness from its environment."* It is by consuming the starchy products of photosynthesis and excreting them in a less ordered form that we are able to sustain life. Schrödinger argued that we owed this ability to DNA in some way and it seems he was right.

Organisms do not obey the law of thermodynamics in the conventional sense, but laws must be obeyed and, according to Pross, they do so by displaying dynamic kinetic stability.[20] Rivers have served as a metaphor for life since ancient times, thanks to the dynamic character they share with humans. One can find the same river day after day, but the water flowing through it is constantly changing. The same is true of a waterfall. The stunning Niagara Falls have been in existence for some 12 000 years and, in that sense, they are a permanent feature. However, there are two things on which they rely absolutely: gravity and a water supply. If gravity did not provide the energy to make them fall, they would cease to exist, and the same would be true if no water was available to fall. As Pross notes, this is the perfect metaphor for life—it can only stay in existence with constant supplies of both energy and ordered material. Moreover, the waterfall can be said to demonstrate dynamic kinetic stability.

Now we turn to the next feature of Pross' definition of life, the replication reaction. This recalls Trifonov's definition of life as self-reproduction with variations. We have seen the self-reproduction, the tendency of RNA chains to produce copies of themselves, but said nothing of variation. It has been known for some time that, when RNA chains self-replicate, they do not always produce faithful replicas.[22] Sometimes, mistakes are made during the replication of a chain and the result is a slightly different chain. When this happens, we say a mutation has occurred. Nevertheless, this new chemical species will also start self-replicating, at which point, two possible outcomes exist: the new chain will produce copies of itself either *faster* or *slower* than the original chain.

Remember that both the original chain and the new chain rely on the presence of free nucleotides in order to construct copies of themselves. If one chain is replicating faster than the other, it will use up all of the free nucleotides before the other one has a chance. To use Schrödinger's phrase again, whichever is best at sucking orderliness from its environment will drive the other into extinction. Sound familiar? In other words, both chemical species compete for the same resources and the fitter one survives. Self-replicating RNA chains demonstrate survival of the fittest.

Pross' contribution was to state the well-known biological phenomenon in chemical terms.[20] How do you quantify which species is "the fittest"? The argument could be made that the cheetah is absolutely the fittest organism around but not if you plunge it into the sea and chain it to the edge of a scalding, hydrothermal vent. In that situation, extremophiles are un-questionably the fittest organisms in existence, indeed because they are the *only* organisms in existence. Pross argues that a population of self-replicating species, whether it be the bio-logical variety, such as cheetahs and extremophiles, or the chemical variety, such as RNA chains, will maximise fitness by maximising dynamic kinetic stability. Individual cheetahs will die, but if the population persists then the *species* must be fit enough to survive. So, it is the species that is stable rather than the individual and the fittest species, according to Pross, is the one with the largest- and longest-surviving population. It is remarkable that this behaviour is demonstrated both by self-replicating RNA chains, as well as organisms.

4.3   COSMIC ORIGINS

If these breakthroughs feel significant then we only need glance at the plethora of unsolved riddles to fall crashing back down to Earth. It is easier to show what RNA evolved *into* than what it evolved *from*. Also, where did it happen? Success with mont-morillonite clay points to the importance of location. Was it in Darwin's warm little pond, the scorching little ponds of hydro-thermal vents, or in the vicinity of the renaissance material montmorillonite? But the Murchison meteorite may tell us more than just what it is possible to produce in abiotic conditions.

It is estimated that one million tonnes of carbon were de-livered to Earth *every year* in the form of meteorites during the heavy bombardment phase of Earth's early years, when the pla-net was being liberally pebble-dashed with stray asteroids. Since amino acids and nucleobases have been detected on various meteorites, including the Murchison, there is every chance that these ancient asteroids delivered a similar payload. This raises the question of whether life could have emerged *without* these cosmic deposits.

It can seem fanciful to suggest that life somehow depends for existence on events that occurred in outer space. Indeed, the idea that life *started* in space and came to Earth—the so-called panspermia hypothesis—leaves unanswered the mystery of its origins. But *pseudopanspermia* may prove not only realistic but essential to abiogenesis. This is the formal term for an idea already implied, that the emergence of life on Earth depended on the delivery of biological precursors, such as amino acids and nucleobases, from elsewhere in the universe. If this does seem fanciful, we only need remind ourselves that we are all made of stars, each of us composed of those atoms forged in hydrogen stars more than 4.6 billion years ago. Earth's interaction with the cosmos is just as dynamic as the rushing Niagara Falls.

Scientists are becoming increasingly convinced that Mars may have played a key role.[23] It has long been held that life could only come about in the so-called *habitable zone* of any solar system. This is the section of a solar system in which planets are at a suitable temperature for the planet's water supply to be in the liquid state. Too hot, and all the water turns to gas, too cold, and it would be ice. All known organisms depend on water to act as the medium in which the many reactions that support life take place. But now it has been suggested that water is too "corrosive" and that the reactions discussed above, in which RNA chains were formed and began self-replicating, could not have taken place in the damp environment of early Earth. Researcher Joseph Kirschvink is increasingly convinced that these first steps towards life were taken on the drier red planet and then transferred to Earth *via* some kind of asteroid activity.

There is something fitting about the idea that man descended from a galactic collaboration. It could be the latest link to be uncovered in a chain of events linking man's presence on Earth to the days of Darwin's warm little pond. Even the slightest variation to many of these links could have steered our planet to a human-free destiny. If any of the ice ages had been differently timed or of different severity, if the meteor hadn't wiped out the dinosaurs, if the universal forces of nature were of minutely different strengths, then we might not have made it. Could it have been a singular—and singularly unrepeatable—celestial transfer that enabled life to begin on Earth? Perhaps the pioneering astronauts in the Mars One mission will uncover crucial evidence in support of the theory. Even in that event, many more questions will remain before the greatest mystery of all is solved.

REFERENCES

1. The Meteoritical Society, Meteoritical Bulletin: Entry for Murchison. Available at: http://www.lpi.usra.edu/meteor/metbull.php?code=16875. [Accessed: 13-Apr-2015.]
2. D. A. Henry, "Star Dust Memories"—A Brief History of the Murchison Carbonaceous Chondrite, *Publ. Astron. Soc. Aust.*, 2003, **20**(4), vii–ix.
3. John Lovering, The Murchison Meteorite Story: Melbourne Museum. Available at: http://museumvictoria.com.au/melbournemuseum/discoverycentre/dynamic-earth/videos/the-murchison-meteorite-story/. [Accessed: 13-Apr-2015.]
4. F. Crick and L. Orgel, Directed Panspermia, *Icarus*, 1973, **19**, 341–346.
5. F. Darwin, *The Life and Letters of Charles Darwin*, John Murray, London, 1887.
6. Penn State University, How Were the Egyptian Pyramids Built?, *ScienceDaily*. Available at: http://www.sciencedaily.com/releases/2008/03/080328104302.htm. [Accessed: 13-Apr-2015.]
7. E. N. Trifonov, The Origin of the Genetic Code and of the Earliest Oligopeptides, *Res. Microbiol.*, 2009, **160**(7), 481–486.
8. K. Ruiz-Mirazo, J. Peretó and A. Moreno, A Universal Definition of Life: Autonomy and Open-ended Evolution, *Origins Life Evol. Biospheres*, 2004, **34**(3), 323–346.
9. F. H. C. Crick, The Origin of the Genetic Code, *J. Mol. Biol.*, 1968, **38**(3), 367–379.
10. L. E. Orgel, Evolution of the Genetic Apparatus, *J. Mol. Biol.*, 1968, **38**(3), 381–393.
11. C. Woese, *The Genetic Code: The Molecular Basis for Genetic Expression*, Harper & Row, New York, 1967.
12. H. Lodish, A. Berk, C. A. Kaiser, M. Krieger, A. Bretscher, H. Ploegh, A. Amon and M. P. Scott, *Molecular Cell Biology: International Edition*, W. H. Freeman, 7th edn, 2012.

13. S. L. Miller, A Production of Amino Acids Under Possible Primitive Earth Conditions, *Science*, 1953, **117**(3046), 528–529.
14. Z. Martins, O. Botta, M. L. Fogel, M. A. Sephton, D. P. Glavin, J. S. Watson, J. P. Dworkin, A. W. Schwartz and P. Ehrenfreund, Extraterrestrial Nucleobases in the Murchison Meteorite, *Earth Planet. Sci. Lett.*, 2008, **270**(1–2), 130–136.
15. K. Ruiz-Mirazo, C. Briones and A. de la Escosura, Prebiotic Systems Chemistry: New Perspectives for the Origins of Life, *Chem. Rev.*, 2014, **114**(1), 285–366.
16. M. W. Powner, B. Gerland and J. D. Sutherland, Synthesis of Activated Pyrimidine Ribonucleotides in Prebiotically Plausible Conditions, *Nature*, 2009, **459**(7244), 239–242.
17. Winners of Research Awards Announced, *Origin of Life Challenge: How Did Life Begin?* Available at: http://originlife. org/?q=first. [Accessed: 13-Apr-2015.]
18. L. Orgel, Darwinism at the Very Beginning of Life, *New Sci.*, 1982, **94**(1301), 149–151.
19. A. G. Cairns-Smith, *Genetic Takeover and the Mineral Origins of Life / A.G. Cairns-Smith*, Cambridge University Press, Cambridge, 1982.
20. A. Pross, *What is Life?: How Chemistry Becomes Biology*, Oxford University Press, 2014.
21. E. Schrödinger, *What is Life?* Cambridge University Press, Cambridge, UK, 1944.
22. *The Chemistry of Life's Origins*, Eds. J. M. Greenberg, C. X. Mendoza-Gómez and V. Pirronello, Springer Netherlands, Dordrecht, 1993.
23. C. Barras, No More Primal Soup: Creating Life Without Water, *New Sci.*, 2014, (2965).

CHAPTER 5

The Chemistry of Evolution

Accounts differ as to the snappiness of Thomas Huxley's withering put down in his legendary debate with Archbishop Samuel Wilberforce. The zoologist's public defence of the theory of evolution earned him the nickname *Darwin's Bulldog*. When the Archbishop smugly asked his opponent if the ape from which he descended was on his grandmother's or his grandfather's side, posterity failed to record *verbatim* his response, but it was along the lines of: *"I'd rather be descended from an ape than ridicule ideas out of cowardice to face them"*.[1]

If Wilberforce found descendancy from primates so objectionable, he would have really struggled with later discoveries. This was in June, 1860, when Louis Pasteur was on the brink of proving a 300-year-old hypothesis that germs spread disease. How would the Archbishop have coped with the knowledge that apes are the most intelligent of our ancestors, in a trajectory that reaches past bacteria to Darwin's warm little pond? In fact, it is more correct to say that we are descended from bacteria-like organisms, not least because no one can go back two billion years in time to check. Nevertheless, this chapter will take us all the way from self-replicating molecules to reproducing humans, exploring how the evolution of pleasure enabled the development of choice.

The Chemistry of Human Nature
By Tom Husband
© Tom Husband 2017
Published by the Royal Society of Chemistry, www.rsc.org

Many of the details are hazy and some are inexplicable. The overview is as follows: self-replicating molecules encased themselves in primitive cell membranes, known as micelles, to produce single-celled organisms. These self-replicating molecules evolved into a cookbook for a group of life-sustaining molecules called proteins. One variety of proteins eventually bound single-celled organisms into the clusters that became multi-cellular lifeforms. As these complexified, sex evolved, producing distinct genders that teamed up in a trade-off between accelerated evolution and reduced proliferation.

5.1 FROM SELF-REPLICATORS TO ORGANISMS

5.1.1 Cell Membranes

Research still continues into the evolution of cell membranes and, in arguably one of the more promising areas of research, experiments have demonstrated how they can form spontaneously in conditions representative of the primordial soup. Moreover, they are generating new directions for the research into the trickier pieces of the puzzle.

The membrane itself is a permeable casing that encapsulates the contents of a cell. As we will see in Chapter 6, they are structurally very similar to cooking oil, which makes them a nifty way to limit how easily different substances can enter the cell.

Recent research has demonstrated that micelles can form spontaneously. (A micelle is shown in Figure 5.1.) The membranes of our cells are formed from complex lipids—organic molecules that resemble cooking oil in molecular structure. Their spontaneous creation in abiotic conditions has been judged unlikely (ref. 2, p. 289). However, among the bounty detected in the Murchison meteorite, introduced last chapter, there was a series of compounds that could have served as basic lipids, including long-chain carboxylic acids and polycyclic aromatic hydrocarbons (ref. 2, p. 290). Experiments have shown that such lipids spontaneously aggregate into membrane-like spheres on montmorillonite (ref. 2, p. 293), yet another use for the versatile clay.

At some point, such a process presumably encapsulated self-replicators in these early membranes, an important step towards the emergence of life. Some may argue that a self-replicator in a

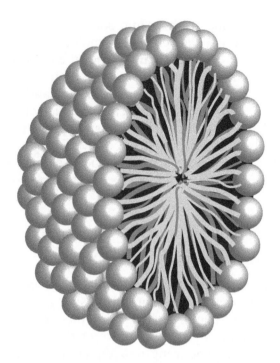

Figure 5.1 A micelle. Lipids spontaneously aggregate into spheres.
(Image from Shutterstock, artwork © Magnetix).

membrane already constitutes a single-celled organism, but
would it meet the necessary criteria? Two characteristics of life
are nutrition and excretion, which present problems for this
simplistic model.

Membranes are critical for excluding unwanted substances
but allowing passage of biologically useful compounds. As we
have seen, self-replicators require a constant input of fuel and
building supplies, so it is essential that the membrane admits
them. This aspect of nutrition has been greatly facilitated by the
evolution of transporters—the membrane-straddling devices
introduced in Chapter 2 that ferry useful substances across
the membrane, just one of many ways in which proteins have
advanced the interests of the self-replicators.

5.1.2 Proteins

Proteins have been a boon for the self-replicators. We have
already considered the paradox of whether genes preceded

proteins or *vice versa*. Certainly, it is true that self-replicators cannot self-replicate without assistance. Exactly how that service was rendered in the past remains unclear. But the modern day self-replicators—our genes—rely on a support team of protein helpers. Proteins or protein-like substances may predate self-replicators, but one thing is certain: self-replicators evolved into instructional manuals for making proteins.

Many examples of modern-day proteins have already been introduced. Chapter 2 described the protein myosin, which, with its partner actin, enables muscles to contract. Before we consider our agency over this process, it is useful to appreciate how our bodies make it in the first place. Myosin marries different conceptions of the word protein. In everyday language the word "protein" refers to the desirable nutrient, abundant in foods such as fish and meat. The reason for its desirability is that our bodies use proteins from food in order to construct our own proteins. Meat actually contains myosin and when we eat it our bodies diligently dismantle the protein only to reassemble it in our muscle cells.

Like all proteins, myosin is composed of building blocks called amino acids. These compounds are defined by so-called functional groups, one of which we have met already: the NH_2 group that gives dopamine its suffix. The other characterising group accounts for the sourness of vinegar: the COOH group. Referred to as a carboxylic acid group, it releases hydrogen ions that activate the sour-detecting taste cells described in Chapter 2. Moreover, this tendency to release hydrogen ions is mirrored by the amine group's ability to accept them, giving rise to the existence of zwitterions, a single chemical species containing both a positive and negative charge (ref. 3, p. 29), as shown in Figure 5.2. This is another way that attractive forces arise between different amino acids, helping to form the protein's hallmark shape. Many of them were named after what they were discovered in, such as asparagine being isolated from asparagus juice in France in 1806.

Our genes are recipes that dictate the sequence for amino acids to be connected to build a protein. The process is broken up into two phases, transcription and translation. Transcription is the production of a protein blueprint and translation is how the body transforms the blueprint into an actual protein.

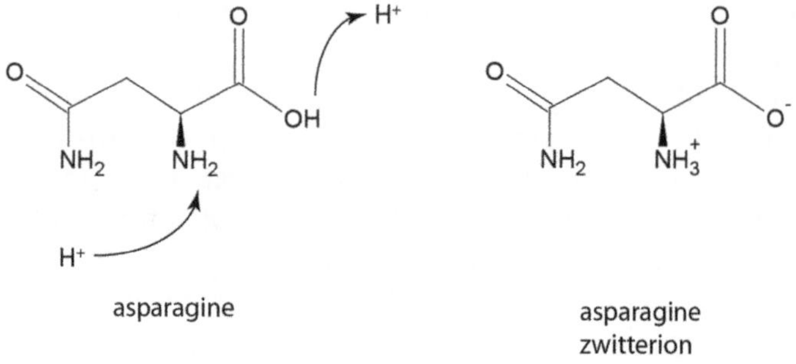

asparagine

asparagine
zwitterion

Figure 5.2   Amino acids are defined by the presence of a carboxylic acid group
(COOH) and an amine group (NH_2) attached to a central carbon
atom. Note that this example contains two amine groups. The
COOH group can release hydrogen ions, while the amine group
can accept them. When both happen, a zwitterion forms.

Transcription is almost identical to the first phase of self-
replication. Chapter 4 explained how a strand of nucleobases
induces a complementary strand of free-floating nucleobases to
line up next to it in mirror image. Each of its nucleobases attracts
the complementary nucleobase, all of which are then joined together to form a complementary strand. The process is repeated
so that the second complementary strand is a replica of the
original strand. There are two key differences between self-
replication and transcription. First, transcription stops with the
production of the complementary strand. Secondly, the strand is
composed not of DNA but of RNA.

Transcription involves several proteins that interact over different stretches of DNA. These proteins collaborate to recognise
the desired stretch of DNA, unfold the famous double helix and
then run off the complementary strand. For now, we are mainly
concerned with the action of a protein called RNA polymerase. As
shown in Figure 5.3, when the desired gene has been identified
and the two coils of DNA have been unfolded, RNA polymerase
clamps itself to the gene and then skates along it like a monorail.
As it does so, it coordinates the assembly of the complementary
strand. If it skates over a guanine nucleobase in the gene, it adds
the complementary cytosine nucleobase to the new strand and
vice versa. Likewise, if the RNA polymerase encounters a thymine
nucleobase, an adenosine is added to the new strand, but the
same is not true in reverse. As we saw in Chapter 4, RNA is

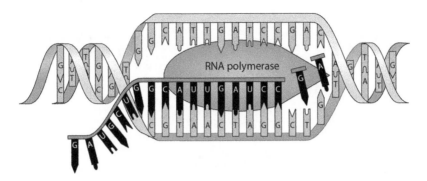

Figure 5.3 RNA polymerase scans one of the DNA strands, stringing complementary nucleobases into a complementary strand.

composed of adenosine, cytosine, guanine and uracil, whereas DNA uses thymine in place of uracil. As such, if the RNA polymerase encounters an adenosine nucleobase, it will add uracil in place of thymine in the chain. The complementary strand grows in this fashion until the RNA polymerase runs into a special sequence of nucleobases that acts as a stop signal: UAG, UAA or UGA in RNA, or TAG, TAA or TGA in DNA. This complementary strand is referred to as messenger RNA (mRNA) for the obvious reason that it acts as a messenger from the primary genetic information source to the protein-building machinery (ref. 3, p. 864).

The mRNA is then used in the process of translation. Once transcription is completed, the mRNA is separated from the RNA polymerase and delivered to a ribosome, a protein factory in the cytoplasm outside the cell's nucleus. Having attached to the ribosome, the sequence of nucleobases is translated into a sequence of amino acids on the basis of their electrostatic fingerprints as follows: cytosine is a perfect electrostatic match for guanine, and adenosine for both thymine and uracil. During the translation, this same principle is scaled up to select individual amino acids, facilitated by another kind of RNA called transfer RNA, or tRNA.[3]

As shown in Figure 5.4, to one end of a molecule of tRNA is attached a particular amino acid, while the other end is equipped with a "fingerprint reader". But rather than reading the electrostatic fingerprint of a single nucleobase, tRNA is attracted to triplets of nucleobases called codons. If, for example, the

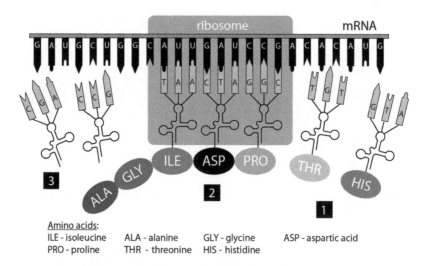

Figure 5.4 Translation of mRNA (1) Amino-acid-laden molecules of tRNA approach ribosome; (2) amino acids are transferred from the tRNA to the growing chain of amino acids; (3) unladen tRNA molecules move away from the ribosome.

fingerprint reader detects cytosine, adenine and uracil (as shown by the triplet furthest right in Figure 5.4), the combined fingerprint of the three nucleobases is a perfect match for the variety of tRNA that carries the amino acid histidine. As such, only that tRNA can bind to the strand of mRNA. This process continues with each codon attracting the corresponding tRNA, each of them lining up on the mRNA, before their amino acids are detached from the tRNA and linked into a long chain of amino acids. In this way, the mRNA dictates the precise sequence in which amino acids should be linked in order to create the corresponding protein.[3]

The completed protein rolls off the production line only after several more stages. The sequence in which the amino acids are connected is referred to as its primary structure. Next, it undergoes a process called folding, which is when it assumes its specific trademark shape and, subsequently, certain modifications may be added or it may even be teamed up with another protein.

A huge transition has occurred between those first fledgling lifeforms and the species populating the Earth today. Nowadays, all known organisms code proteins using DNA, although clearly

RNA has been retained in various auxiliary roles. But the likelihood is that early lifeforms began coding protein production with a precursor to DNA, which was probably RNA. Not only is it true that modern-day DNA is unable to replicate itself without help from proteins, but it is also unable to code those proteins without help from other proteins. This suggests that proteins predate DNA, but even then their appearance is still wreathed in mystery.

Now we can consider the relationship between shortcuts and evolution. In Chapter 2 we saw how the self-replicators evolved by taking shortcuts. More specifically, this kind of shortcut is an example of a mutation, a copying error between one strand and its replica. If 100 self-replicators are all bidding for the finite supply of building materials and energy, the ones that survive are the ones that replicate more rapidly, and if one self-replicator stumbles over a shortcut, such as a mutation occurring that produces a faster self-replicator, the new version will now copy itself more rapidly, exploiting the finite resources more efficiently and forcing the other, slower replicators out of existence.

But things get more complicated once we factor in protein-coding. It is becoming clearer how a bacteria-like organism may have emerged from the primordial soup. Part of the solution is that a self-replicator encases itself in a lipid membrane, possibly helped along by a catalytic blob of montmorillonite clay. Then, self-replicators begin to code the production of proteins. This is a huge step forward as some of those proteins were able to serve the interests of the self-replicators. Anything they did that provided readier access to the critical components of fuel and building supplies would give the parent self-replicator an advantage over its competitors.

This is where the term "shortcut" really becomes inadequate. If we consider the self-replicators in isolation from their protein-coding activities, we expect the faster ones to breed out the slower ones, meaning that all shortcuts serve their interests. But they were not isolated. We assume that at some point it was true that they were both replicating and coding proteins, which has consequences because, now, shortcuts not only affect the rate of replication but also the recipe for the proteins.

Changing the code means changing the protein, which could help or hinder the replicator's fitness. While a shortcut may enable faster replication, it could also ruin the recipe for a previously beneficial protein. As such, the term "mutation" is much more suitable, which refers simply to a copying error between two strands of the genetic material, whether it be DNA or RNA, or during transcription between the two.

Once again, mutations can be good or bad (or neutral). If we consider our emerging bacterium, there are some things it would find useful. The membrane is good at keeping out toxic substances that may damage its self-replicating interior. It would also be useful if it could stockpile fuel and keep it from its rivals. Sugars make good fuels, but their typical blend of oxygen and carbon makes for an uneven distribution of electrons, which bars entrance *via* the membrane. But suppose a mutation produces a transporter that can straddle the membrane and ferry sugar molecules inside the cell. This bacterium now has a significant advantage over its rivals. Its fitness, that is, its ability to produce offspring, increases and soon it has driven rival self-replicators out of existence. But now all bacteria have that protein, so it is no longer an advantage. Meanwhile, another mutation might ruin the recipe for this useful protein. The bacterium in question will fall behind in the race and the mutation will be lost from the gene pool.

The great arms race of evolution starts here. Self-replicators race to exploit the finite materials on which they subsist—fuel and building materials. Sometimes, shortcuts help them replicate faster, which gives them an advantage over rivals. But they also code proteins, which can be very helpful. Some mutations invent new proteins, some of which may be very helpful. Other mutations invent harmful proteins or ruin the recipes for helpful ones. Only the recipes for the helpful proteins are preserved because they help the corresponding self-replicators to breed inferior models out of existence. Evolution has begun.

5.2   FROM REPLICATION TO REPRODUCTION

While many gaps are left to fill, the current hypothesis presents a plausible explanation for how inorganic matter came alive. Between then and now, an interesting event is the schism between self-replication and reproduction. The theory goes that early self-replicators needed a constant supply of fuel and building materials to keep replicating. We have seen the powerful forces that keep us literally feeding those demands, but what of our other urges?

5.2.1 Single-celled Organisms

Even in the simplest organisms, self-replication and reproduction are quite distinct. It is almost certain that the earliest organisms reproduced in a similar way to modern-day microorganisms *via* cell division. In bacteria this process starts with the replication of all genetic material, followed by the cellular machinery (ref. 4, p. 242) before the cell can divide. Even just this last stage requires over 20 different proteins (ref. 4, p. 246), some of which make the cell longer, while others collaborate to make a divisome—a self-constricting ring structure that cleaves the cell in the middle.

Is a bacterium a self-replicator? Up to now, the expression has described the molecules that made copies of themselves, whereas the bacterium is a whole organism. But since its genetic code is replicated, along with its arsenal of proteins and cellular apparatus, it seems reasonable to say that reproduction and self-replication are broadly synonymous in this case. However, Richard Dawkins has argued that the bacterium itself is not a replicator. The genome of an asexually reproducing organism may be considered a replicator but not the organism itself.[5]

Already, we see an interesting phenomenon. It takes about 20 min for a bacterium to undergo division (ref. 4, p. 246). The self-replicator (its genetic code) still gets to self-replicate, but now, every time it does, it needs to wait until various other things have been replicated before it can go again. In that sense, donning the armour of the membrane and exploiting its protein toolbox has slowed down its operation. On the other hand, any remaining naked self-replicators, which have not developed membranes and proteins, would now be prey for their more evolved rivals. It is like an emerging artist who accumulates agents and managers; his share of the takings dwindles, but he ends up richer because the support staff help him sell more.

5.2.2 Multicellular Organisms

The schism widens with the emergence of multicellular organisms. Once cells group together the process of reproduction looks even less like self-replication.

A non-negotiable feature of multicellular lifeforms is that their cells attach to one another. This could and probably did happen in a variety of ways. Nowadays, there are many ways cells adhere to each other,[6] including a whole class of proteins called adherens. Chapter 9 will introduce a particular example called cadherins, which help to bind cells together in the developing human embryo. Glycoproteins, hybrids of carbohydrates and proteins, are known to bind fungi cells together, which,[6] given that syrup is sticky and that proteins are routinely used to make glue, is not too surprising.

But a new degree of separation has inserted itself between replication and reproduction. In single-celled organisms, self-replication of the genome and reproduction are coincidental *via* the process of cell division. In the first multicellular organisms, self-replication is coincidental with cell division but no longer with reproduction. Consider a specimen of seaweed. If its cells divide more rapidly than they self-terminate, the seaweed grows. Each instance of cell division entails replication of the genome, but the seaweed has not produced any offspring.

Arguably, the most basic form of reproduction in seaweed is fragmentation.[7] The species *Rhodochorton purpureum*, better known as red algae, proliferates this way. Pieces of the plant are literally broken off by pressures, such as waves. These fragments then sprout shoots and adhesive filaments called rhizoids, which bind them to rocks. Self-replication and cell division continue, but since two separate organisms now exist, reproduction can also be said to have occurred.

Researchers Richard K. Grosberg and Richard R. Strathmann argue that multicellular life only develops when it serves the interests of the self-replicators.[8] The actual process of self-replication is much the same as in single-celled organisms. Whenever a cell divides, self-replication of the parent cell's genetic material has occurred and, as with single-celled organisms, self-replication can occur as frequently as cell division. Moreover, the collaboration between the cells allows them to divide labour, which can be invaluable. For example, cells cannot simultaneously engage in photosynthesis and nitrogen fixation (the process by which nitrogen is plucked from the air and reacted with organic compounds to make vital substances, like amino acids and nucleobases). As such, single-celled cyanobacteria have to choose one method or the other, but in multicellular cyanobacteria different cells can specialise and share the spoils.

5.2.3 Sexual Reproduction

This is where schism turns to a chasm. We can be thoroughly sure that sexual reproduction exists—in a vast range of species—but it is much harder to explain how it evolved. For decades, scientists have pondered how to resolve the apparent contradiction that the process basically halves reproductive efficiency. How on Earth can sexually reproducing species dominate when it takes two organisms to make a child instead of just one? Meanwhile, at the level of the self-replicators, sexual reproduction represents an enormous fail; a parent who reproduces does not replicate their genetic material in their progeny but rather splices half of it with that of another member of their species.

Darwin's theory is often distilled into the idea of the survival of the fittest, which thus demands a definition of fitness. As we saw in the last chapter, origins-of-life researcher Addy Pross equates the fitness of self-replicators with the dynamic kinetic stability of the chemical species; the larger and longer lasting their population, the greater the fitness. Meanwhile, at the level of organisms, fitness can be defined as the number of offspring left by individual members of a species.[9] Absolute fitness is the average number of offspring per member of a group, while relative fitness ranks individuals according to their specific number of children.[9]

A bloodthirsty example concerns the work of controversial anthropologist Napoleon Chagnon. In a contentious attack on the noble savage doctrine, he gathered data showing that violence promoted fitness of members of the warring Yanomamö hunter gatherer society of Venezuela.[10] According to his data, the tribesmen who killed the most members of rival clans also went on to have the most children. But another research group, led by Stephen Beckerman and Kathryn Long, found the opposite in an even more violent society—the Waorani people of Ecuador.[10] Whereas 30% of Yanomamö males were killed during inter-clan warfare, the figure for the Waorani people was 42%, but it was the less aggressive Waorani males who sired the most children.

The definitions can be applied as follows: we would expect the absolute fitness of the Yanomamö to exceed that of the Waorani simply by virtue of the fact that a greater proportion of them stay alive to have children. Meanwhile, relative fitness peaks with aggressive character in the Yanomamö but with non-aggressive character in the Waorani. Beckerman and Long's suggested explanation is that the Yanomamö people broker truces long enough for war heroes to raise children, while warfare between the Waorani tribes is unrelenting.[10]

Chagnon was criticised for implying a genetic basis to his findings. He was perceived to suggest that natural and/or sexual selection aggregated violence-promoting alleles in the more prolific Yanomamö warriors.[10] This idea will be considered in Chapter 12, but the concepts of alleles and sexual selection are relevant now.

Allele is a broad term referring to variants of genes,[11] about which the famous example of eye colour is instructive. The gene *OCA2* codes a transporter for the amino acid tyrosine, which the body transforms into the brown pigment melanin. Another gene, *HERC2*, dictates how much OCA2 the body expresses, but some historic mutation has generated a slightly different version of this gene, which is now commonly expressed among people of European descent. The mutated form is less efficient at driving production of OCA2, meaning fewer transporters get made, meaning that less tyrosine is imported such that reduced melanin availability results in blue eyes.[12] These two variants of *HERC2* are examples of alleles.

Alleles appear at corresponding positions of paired chromosomes. Chromosomes are the separate volumes of the encyclopaedia of protein recipes constituting an individual's genome.[11] Following sexual reproduction, a human child inherits 23 chromosomes from its father and another 23 from the mother, but the function of each duplicates that of its partner. Our body has a choice of two alleles for every gene it expresses, which leads to the phenomenon of dominant and recessive traits. A person with brown eyes may or may not also carry the blue polymorphism of HERC2 because the body defaults to the brown-yielding dominant version. In order for blue eyes to develop, an individual must have the same copy of blue HERC2 on each of the chromosomes in the 15th pair.

Alleles and fitness are central to the mystery of sexual reproduction, with advantages from the former hypothesised to compensate for disadvantages to the latter.

Sexual reproduction cuts the fitness rate in two. When each female relies on a male to breed, they produce half as many offspring per individual. The great evolutionary biologist John Maynard Smith considered how a sexually reproducing species would be affected if a genetic mutation produced a female that could only reproduce asexually.[13] Based on the assumption that both varieties of female would produce the same number of eggs, the asexual mutant would produce the same number of offspring but without the need for male assistance. Per individual, her output would be twice that of her sexually reproducing peers, meaning that her descendants would soon drive sexuals out of existence. He described this as the two-fold cost of sexual reproduction.

Viewed in terms of the self-replicators, things do not appear any better. These molecules have been called replicators because they replicate themselves. Putting aside for one moment the loss of fidelity over time, sexual reproduction represents an immediate sacrifice. George Williams, a contemporary of Maynard Smith, noted that in order for a parent to replicate its entire genome in its offspring, it needs to produce at least two children (ref. 14, p. 10). But an asexually reproducing organism can achieve the feat with one child.

At first glance, it seems as if a gulf now separates the processes of self-replication and reproduction, but the thesis of this book is that our modern-day behaviour echoes the requirements of the self-replicators that brought life into existence. And as we shall see in Section 2, our brains are hard-wired to seek mating opportunities in which for us to reproduce, but how does this serve the interests of the self-replicators? When they have access to sufficient fuel and building materials, they can continue forging their replicas. Those that remain in existence today—such as the genes busily copying themselves in your cells as you read—are those whose ancestors chanced to evolve features that enabled them to out-replicate their peers.

We cannot equate sexual reproduction to self-replication. The difference between a human child and one of its parents is colossal compared to the difference between a bacterium and its daughter cell. For the bacterium, the terms self-replication and reproduction mean almost the same thing, whereas for humans, a child is never a replica of its parent for the simple reason that it cannot be a replica of both of its parents. However, if we look closely, we can see that sexual reproduction still facilitates a tremendous amount of self-replication of genes.

The point is that we die. At the moment, we reach Schrödinger's *chemical equilibrium,* the self-replicative tendencies of 30 trillion sets of DNA abruptly cease. They have had a good innings. An adult body contains about 30 trillion cells and, in a lifetime, they will undergo cell division a total of 1000 trillion times,[15] meaning we can expect our DNA to replicate itself 1000 trillion times. But when we die that comes to a halt. The only way for self-replication to continue is if we happened to have one or more children. However, as we have seen, the DNA in those children will not be a replica of the DNA in the body of the expiring parent.

Does it matter if the genome in a child is an imperfect replica of the genome of each parent? The term self-replicator is somewhat of a misnomer. I would not be sitting here typing this if my ancestral self-replicators had done exactly what they said on the tin. It was thanks to the copying errors—the mutations—that they were able to evolve proteins that increased their share of the finite resources of fuel and building materials. A self-replicator does not have obsessive-compulsive-type tendencies, which cause it to have a breakdown upon producing an imperfect copy. We might better call them self-impersonators or self-mimickers. And, even then, there is a problem with the ambiguity of the prefix *self.* They might produce a copy of them*selves,* but they do not do it *by* themselves. Perhaps the best term for what they do would be *assisted self-mimicry.* Herein lies the appeal of Trifonov's definition of life: self-reproduction *with variation.*

To what extent, then, does the genome of a human child mimic that of each parent? Better than it looks at first glance. It is true that each parent contributes just half of its genome, but there is a lot of duplication between the stuff that gets left out. All humans share about 99% of the same DNA in common. To put it more scientifically, if you straightened out the genomes of any two humans and laid them together side by side, only about 1 in 200 base pairs would actually be different.[3] It is as if a lecture-dodging student creates a complete set of notes by photocopying the first half of one classmate's notes and the second half of another's. Both classmates attended the same course so, while their notes will differ, they summarise the same concepts. So, some of what is left out is replicated *de facto* simply by virtue of the fact that the other parent carries almost identical genetic material.

The real question is the extent to which the sexually reproductive trade-off advances the interests of the self-replicators. Again, this has been very difficult to pin down. It is clear that sexual reproduction exists, so either there is a fault with our evolutionary theory or we have not been able to explain how it applies to this readily observable phenomenon. Species that can reproduce both sexually and asexually, such as the strawberry plant, tend to opt for sexual reproduction when signs indicate that the environment is about to change.[14] Meanwhile, Maynard Smith proposed a variety of potential advantages, including that sexual reproduction might enable more rapid evolution, that it may dodge the accumulation of harmful mutations, or that it promotes the repair of damaged DNA.[16] These can be understood within the context of meiosis, to be introduced below.

Sexual reproduction required the evolution of a new kind of cell division. The variety described for bacteria is mitotic division, where one cell divides into two, each of which contains a complete copy of the genome contained in the original cell. On the other hand, sexual reproduction proceeds by meiosis. In this version the genome is first replicated and then the resulting two sets are distributed between four daughter cells called gametes. From the chromosomal viewpoint, 23 pairs are replicated, producing 92 single chromosomes, which are divided equally between the four gametes.[17]

One advantage of meiosis was demonstrated by the scientist Barbara McClintock. Most scientists would be content with just one of her discoveries, but she made enough to win a Nobel Prize. She bruised the ego of one grad-school supervisor when she identified discrete parts of corn chromosomes in three days, something he had been attempting for much longer.

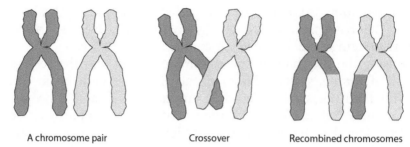

A chromosome pair   Crossover   Recombined chromosomes

Figure 5.5   Crossover during meiosis. Before placement in separate sex cells, the legs of a chromosome pair cross over. The overlapping genetic material is detached and spliced into the opposite chromosome.

Having completed her PhD, it was with her own graduate student, Harriet Creighton, that she made another discovery.[18] They bred together two strains of corn, one of which was waxy with purple kernels and another that shared neither trait. When they investigated the offspring, they found some that had one trait but not the other, which was curious because the genes for both were located in the same chromosome of the parent plant.[19] What they had observed was a process called crossing over, in which parts of paired chromosomes switch places, as shown in Figure 5.5.

The discovery of crossover led to the realisation that genes are shuffled during meiosis. The 92 chromosomes are not simply distributed between the four gametes. Directly prior to the first cell division, paired chromosomes swap a length of genetic material between themselves.[20] What had happened in McClintock's experiment was that the chromosome containing the genes for waxiness and purple kernels had swapped one or the other gene with the corresponding allele of its partner.

The shuffling of genes becomes beneficial when peacocks start strutting around with longer feathers, which fortuitously douses another puzzle. Scientists have also struggled to explain why organisms continue to show variation in their sexually desirable traits when they are selected precisely for those traits. If females select long-feathered peacocks, evolutionary theory suggests that all males should end up with similarly long feathers. Yet peacocks continue to display considerable variation in their plumage.[20]

Enter the genic capture hypothesis. What this argues is that the length of the peacock's feathers is not indicative solely of some subset of feather-growing genes but rather provides a proxy for the overall quality of its genes. Essentially, the capacity to grow the best plumage requires that the male is in optimum condition, a state that places demands on all of its genes, not only those directly associated with producing feathers.[20]

Support for this hypothesis has been provided *via* a long-running study into flour beetles. The 10-year study, led by Alyson J. Lumley and Matthew J. G. Gage, compared two populations of the insect.[21] In one population, males competed for female mates in the normal way, with nine males to each female. In the other population there was no selection; each male was paired with a female and left to reproduce. After breeding in these different ways for 50 generations, the experimenters changed tack and switched to inbreeding.[22] Each generation, they selected siblings from each population and mated them. What they found was striking. The beetles from the population that had undergone sexual selection endured 20 generations of inbreeding, but the other population went extinct after just 10 generations. They concluded that the genetic health of the sexually selective population was far higher than that of the other population, which fits with the genic capture hypothesis.

These findings are reminiscent of the view forwarded by Addy Pross in the last chapter. He suggests that it is societies of self-replicators rather than individuals that demonstrate stability. In the same way, sexually reproducing organisms collaborate to optimise genetic health of the population. Self-replicators have evolved an insurance policy. If a copying error produces deleterious genes in one individual, they can be flushed out of the gene pool when meiosis aggregates them in one gamete. Meanwhile, zygotes that combine a high proportion of the best genes going will mature into specimens that produce many offspring. The indiscriminate replication of DNA continues apace in our individual cells, but at the critical moment of reproduction, it is the competitive selection of genes that secures their place in future generations.

Species thrive when mutations produce advantageous proteins. Unfortunately, most mutations are more harmful than helpful.[23] This is how the considerable cost of sexual repro-duction justifies itself: although it halves the number of baby-producing members of a species, it enables the population to shed the harmful mutations while retaining the helpful ones. Our self-replicators collaborate to produce the best specimens in the cells that they can continue self-replicating within, albeit on a genome that is a perfect replica of neither parent.

5.3 DEVELOPING CHOICE

One notable set of mutations bequeathed bacteria their system of propulsion, which is useful for our other line of inquiry. Certain species of bacteria have a set of swishing tails called flagella, which they use to move away from danger and towards food.[4] There is more than a passing resemblance to the bio-chemical engines that drive the sperm cells of many species, including humans, towards the female ova. In fact, the tails of sperm are also called flagella, but we must be cautious in our comparison. Imagine two cars, one with a combustion engine and a steel frame, the other rendered from carbon fibre housing an electrical motor. Clearly, the two vehicles achieve the same function, even while constructed from different materials and to different designs. Sperm and bacteria demonstrate a similar distinction in the mechanism of their flagella (ref. 24, p. 10).

The analogy is not adequately cautious to capture the differ-ences in their motion. Both sperm and bacteria demonstrate chemotaxis, in which they follow a trail of molecular bread-crumbs towards their goals. In fact, no one is completely sure how sperm cells guide themselves as they seem to respond not only to chemicals but also to additional cues, such as changes in temperature.[25] A recent study even concluded that they "slither" along surfaces, such as the lining of the fallopian tubes.[26] Bac-teria do something altogether different, which we will consider momentarily. A key reason for their difference in motion is that, while sperm flick their flagella from side to side,[27] bacterial flagella rotate (ref. 3, p. 1026). In spite of these many differences, it is still notable that both sperm and bacteria use flagella to propel themselves in beneficial directions.

There is an elegant eccentricity to the motion of bacteria. Imagine the captain of a trawler looking for a school of fish.

He has no idea where they are, so he takes the following course of action. He spins the ship's wheel and lets it come to rest at random and, accordingly, the ship changes course. If he sees a fish in the ship's new found path, he sticks to that course. If not, he spins the wheel again. This is how bacteria seek food, by picking a random direction and sticking to each new path only on the condition that it bears fruit.[28] But notice how the word "picking" has crept in there, slyly suggesting that the bacteria *decide* on a course of action. Of course, these brainless blobs cannot decide! They are utterly at the mercy of their environment.

Bacteria proceed by rotating their flagella in opposite directions. When they rotate counter-clockwise, the flagella combine into a single bundle, which propels them directly forwards. When they rotate clockwise, they tumble around at random. Flagella are spiralled like the helices of DNA and their threading clashes with clockwise rotation, which consequently splays the bundle and sets all the flagella flailing at cross purposes (ref. 3, p. 1028), as shown in Figure 5.6. The resulting tumbling randomly assigns a new path.

As shown in Figure 5.7, if a bacterium encounters a beneficial nutrient, such as the sugar ribose, it heads straight; if it does not encounter a nutrient, it resumes its tumbling. But bacteria do not have brains or eyes. They do not see something they know to be beneficial and make a decision to pursue it. Rather, their interaction with sugar triggers inevitable biochemical events that dictate the direction in which its flagella rotate.

This happens *via* a signal cascade similar to those we have met previously. A receptor straddles the bacterial membrane, its activation triggers a conformation change and this shapeshifting echoes down the domino rally of companion proteins to act as a toggle switch. The signal dictates the direction of motion of the molecular motor that rotates the flagella. This unit is composed of several molecules, each comprising a band of proteins called FliN, FliM and FliG (ref. 4, p. 532).

Ribose manipulates the molecular motors, as shown in Figure 5.8. Bacteria express various receptors, which detect both nutrients and poisons. The ribose-detecting receptor is called Trg and it forms a trio with two more proteins inside the cell called CheA and CheW. CheA is in the business of stripping phosphate groups from passing molecules of ATP. It emerges

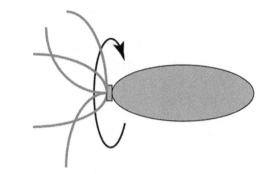

clockwise rotation splays flagella

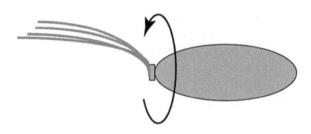

anticlockwise rotation binds flagella

Figure 5.6   When bacterial flagella rotate anticlockwise they form a bundle, but clockwise rotation splays the flagella.

from the encounter with the phosphate group stuck to itself but then quickly passes it on to another protein called CheY. In its phosphorylated state CheY binds to the motor proteins and causes them to rotate clockwise, which is tumble mode.[29]

The direction of rotation is toggled by disturbing the balance between the different forms of CheY. The phosphate group is a bit of a hot potato, with each hand that touches it quickly tossing it to the next person. No sooner has the CheA attached the phosphate group to itself than it transfers it to CheY. But as soon as CheY is phosphorylated, it almost immediately chops the phosphate group off and passes it on. For the brief moment that the CheY is holding the phosphate hot potato, it binds to the motor proteins and causes them to rotate clockwise. As soon as a molecule of CheY has dispatched its phosphate group, it stops promoting clockwise rotation. As such, the Trg–CheA–CheW

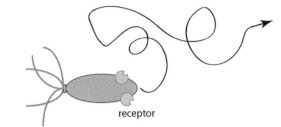

receptor

with no sugar in vicinity, bacteria enter "tumble mode"

sugar

when sugar molecules bind receptors, bacteria head straight

Figure 5.7 Bacteria move in a straight line when their receptors bind sugar molecules, but tumble haphazardly in the absence of sugar.

triad has to maintain a steady supply of phosphorylated molecules of CheY in order for the flagella to continue rotating clockwise.

Ribose effectively kills the circuit. When a molecule of ribose binds the receptor, it triggers a change in shape, which ripples through the Trg–CheA–CheW triad, with the effect that the CheA can no longer phosphorylate CheY. Any phosphorylated CheY molecules in the vicinity will soon have removed their phosphate groups, and since the triad is not providing any new molecules of phosphorylated CheY, their stocks dwindle. Soon, they are in the minority, meaning that the molecular motors resume their anti-clockwise rotation and the bacterium heads in whichever direction it was left pointing by its stint in tumble mode.[29]

This is absolutely a case of the tail wagging the dog. When there is no ribose around, the tumble circuit kicks in and the bacterium is tossed about for a while before heading in a new direction. If the bacterium encounters sugar on this new path, the sugar molecules will keep the tumble circuit switched off so that the cell stays on the same course. Once the ribose runs out, the tumble circuit is reactivated so as to randomly select a new path. Meanwhile, when the bacterium binds a toxin, the opposite

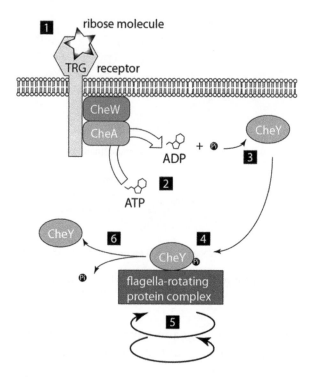

Figure 5.8 How ribose reverses the direction of flagella rotation. (1) Ribose binds the Trg receptor. (2) The resulting conformation change primes protein CheA to strip a phosphate group from ATP. (3) Protein CheY binds a phosphate group. (4) The phosphorylated CheY binds to the flagella-rotating protein complex, which... (5)... reverses the direction of rotation. (6) CheY undergoes a reaction to remove its own phosphate group, at which point the flagella motor will resume its previous direction of rotation.

happens, such that it tumbles around in search of a new path away from the poison.

The benefits of this circuit to the self-replicators are obvious. With its tumbling explorations, the bacterium will ingest more sugar than if it stayed still. Meanwhile, the sugar fuels continued self-replication.

Moreover, the randomised advance of the bacterium echoes the iterative process of evolution itself. Each protein is encoded by a particular sequence of nucleobases, but each sequence was stumbled over by chance, just as the bacterium seeks out sugar by randomising the direction of its advance. The earliest organisms would not have had this tumbling propulsion system, but new proteins kept being invented by the mutation-exploiting self-replicators. Many of them may have been useless, but, over time, the retained portion of these randomly generated proteins eventually meshed into this propulsion ensemble. Since it enables the bacterium to ingest more sugar, the organism will outbid its rivals for the finite building materials, not only producing descendants more rapidly but bequeathing them with the genetic recipe to make their own propulsion systems.

5.4   INTRODUCING PLEASURE

Humans are obviously different from bacteria. One important difference is that we are not wagged by our tails or, at the very least, we enjoy that illusion. Going back to the chocolate bar and the muscles, we are aware—on many levels—that the chocolate bar contains beneficial nutrients, yet we can choose not to pick it up and eat it. So, if the chocolate bar sits beside an apple, we might choose to take the healthy option instead. It may not complement something else we are ingesting, such as beer, or we might be watching our weight. We might (an infrequent occurrence for me) feel too full to eat the chocolate. Species lower down the food chain should also factor in the risk of predators. A mouse should ditch a juicy looking seed if it sits within the shadow of a large magpie. As we shall see later on, fruit flies will steep their larvae in hooch if they see a particular breed of wasp nearby.

At some point, species evolved freedom of movement. The movement of bacteria derives from the presence or absence of recognised nutrients and toxins. Their path is, therefore, utterly determined by what surrounds it. But animals have the ability to move around. They have eyes that recognise potential mates, sources of nutrition and hazards. Smells also help them to evade danger and find food and mates.

But how do the self-replicators keep a leash on species once they evolve the freedom to move around? This freedom must be tethered to the underlying directive that if the mouse foregoes the one seed, it must make sure to find another meal elsewhere. In other words, animals evolved greater autonomy over their movements relative to potential sources of nutrition and mates, but those newfound freedoms will still make use of the self-replicators. Animals were given freedom in how to skin the cat but not the freedom to leave it unskinned.

How to achieve this? It is basically like that childhood seeking game, where you're getting warmer, warmer still, colder. In the nervous system a module developed that rewards "good" behaviour and punishes "bad" behaviour. "Good" behaviour triggers a pleasurable sensation, while "bad" behaviour causes pain. Good behaviour serves the self-replicators: eating provides fuel and building materials, and mating extends opportunities for self-replication *via* reproduction. Bad behaviour imperils the organism and, by extension, the self-replicators.

The revered psychoanalyst Sigmund Freud began his deliberations in *Beyond the Pleasure Principle* with the contemplation of a mere vesicle.[30] His use of the word was broadly synonymous with the term "single-celled organism", distinct from the parcels used to dispatch neurotransmitters that we met in Chapter 3. Nevertheless, he was far-sighted enough to appreciate that organisms far lower than humans might experience pleasure. So, how low can we go?

Back in 2012, a group of researchers erupted into laughter as a longshot paid off. The team, led by Galit Shohat-Ophir and Ulrike Heberlein, had spent several days observing animal behaviour and another several days interpreting data. With the numbers finally crunched, they realised the data supported their hypothesis: rejected fruit flies hit the bottle.[31]

In sensible terms, the research team found an inverse correlation between the consumption of ethanol, the intoxicating ingredient in alcoholic drinks, and the frequency of mating. The more male fruit flies were rejected by females, the more ethanol they consumed.[32]

This lead to a media storm of anthropomorphic frenzy. Fruit flies were just like humans! If they couldn't get their ends away, they drowned their sorrows with hooch. Cartoon likenesses appeared of literal bar flies, weeping on each other's shoulders, coping with rejection human-style. At least one group of researchers was quick to criticise the misleading frivolity of such interpretations, furnishing readers with some background information about fruit flies and alcohol.[33]

Fruit flies are no strangers to alcohol. They demonstrate a preference for mating in rotting fruit, 3 to 5% of which is composed of ethanol. Like us, they can use the substance as a fuel.[33] Also like us, the substance causes behavioural changes, which we associate with drunkenness. Initial consumption leads to hyperactive behaviour, which manifests as erratic and rapid flight. Continued consumption dulls their senses, causing them to bump into surfaces, as well as each other.[31] Eventually, they pass out. Many sociobiologists would bridle at the description of these collective behaviours as drunkenness, because the term applies to a human phenomenon with many additional characteristics. But there is certainly a resemblance in this case.

Dr Shohat-Ophir and her colleagues decided to exploit the fruit fly's familiarity with ethanol. Although the effects on the nervous system are much more poorly understood than amphetamines and cocaine, there is a consensus that it hijacks the same neural reward circuitry. In humans, ethanol elevates levels of that reward kingpin dopamine.[34] They hypothesised that the compound would play a similar role in fruit flies.

Collectively, their evidence suggests that the fruit flies kept their reward levels topped up with ethanol when sex was not available. Sex is a highly pleasurable activity and, as we will see in Section 2, this arises from its activation of the reward system. When the male fruit flies were prevented from mating with females, they looked for another way to satisfy the cravings of their reward network.

Delving deeper, the team found further evidence by investigating the mechanism of this reward pay-out.[32] They investigated a protein with a human homologue, which is a close variant of one of our proteins that has a similar function. We express neuropeptide Y, while fruit flies express neuropeptide F (NPF). The term *neuropeptide* might equally be applied to the opioid peptides we met in Chapter 3. Studies have shown that high levels of neuropeptide Y are accompanied by a sense of contentment in humans, with a corresponding lack of reward-seeking behaviour. On the other hand, low levels of the biochemical predict stressed or depressive moods,[31] which encourage the person to activate their reward system. Shohat-Ophir's team hypothesise that the neuropeptide is like a thermostat, which keeps the level of reward consistent and, if the levels drop, the individual is more likely to engage in reward-seeking behaviour and *vice versa*.[31]

As such, flies with high levels of NPF ought to lay off the alcohol, which is exactly what the team found. First of all, they selected a control group of wild flies. Next, they genetically modified an experimental group so that the flies would express a particular gene in neurons activated by NPF.[31] This extra gene, called *dTRPA1*, is another homolog of a gene expressed in humans, called *transient receptor potential cation channel subfamily V member 1* (*TRPV1*), which we will consider in detail in Chapter 8. The protein is normally used to signal changes in temperature so that the organism can avoid dangerous levels of heat. By splicing the genes together, Shohat-Ophir's team modified the NPF neurons so that they were activated by heat. In this way, they were able to directly activate the neurons simply by increasing the temperature. Sure enough, when the flies were exposed to increased temperature, the mutants lost their appetite for ethanol, while the control group's intake was unaffected.[32] This provides further proof that ethanol activates the NPF neurons, thereby activating the reward system.

This is consistent with our ecological understanding of fruit flies. As the media painted a picture of lovesick *Drosophila* hitting the sauce, another research group was quick to redress the balance. Palestina Guevara Fiore and John Endler point out that *Drosophila* can use ethanol as fuel, which could improve their mating potential by, for example, giving them more energy to complete their courting displays.[33] As such, it is no surprise that the species has evolved to reward ethanol intake.[33] And that's not the only reason. Earlier, we considered how the reward network can also use pain to ward an organism away from dangers, such as predators. Since Shohat-Ophir's experiment, evidence has emerged that ethanol's reward incentive for fruit flies helps them to avoid the parasitic interests of one of the insect world's nastiest species.

There seem to be no depths to which wasps will not sink. Working as a waiter in my youth, customers often saw me fishing drowning wasps from a jug of syrup, which we used to serve alongside fresh-pressed lemon juice. Since I had a shaved head at the time, many would ask me if I was a Buddhist. Let me go on record and say that I regret saving those wasps. Their only saving grace is when they happen to use their powers of evil to consume other pests. Many is the time I have seen a wasp staggering about as the summer heat dwindles, thinking I should just leave them to a quiet death. But then they invariably stumble over and sting me as if with nothing but malice in their heart. But I have it lucky compared to some poor souls, including cockroaches, caterpillars and our friends, *Drosophila*. The Braconidae and Ampulicidae families of wasp boast species that kidnap caterpillars[35] and cockroaches,[36] respectively, disabling them with their sting before laying eggs on their paralysed bodies. As the larvae emerge, they devour their prey alive. Meanwhile, the *Leptopilina heterotoma* species of the Figitidae family, a smaller brand of wasp, hijack *Drosophila* pupae in a similar vein.[37]

Research in a different laboratory found an interesting link between *Drosophila* reward, ethanol and these hideous wasps. When it comes to booze, wasps are light-weights compared to fruit flies. As such, a key reason why fruit flies are inclined to lay their eggs in decomposing fruit is because the ethanol content is considerably more toxic to the parasitic wasps. In short, *Drosophila* are more likely to survive gestation steeped in hooch because it poisons the larval wasps. And this fact is not lost on the mother flies. A study led by Balint Z. Kacsoh and Todd A. Schlenke found that pregnant fruit flies were twice as likely to lay their eggs in ethanol if they saw a *Leptopilina* wasp lurking nearby. Not only that, but the team also recorded lower levels of NPF when they sighted the wasps.[37]

There is a growing body of evidence supporting the hypothesis that fruit flies are guided by a reward system analogous to ours. Not only do our reward systems affect us in similar ways, but there are similar mechanisms at work. Neuropeptides Y and F are structurally very similar and, also, a recent study has confirmed that dopamine is active in the *Drosophila* network, as well as ours.[38]

For me, this is both surprising and expected. Even simple organisms like fruit flies are far more complex than bacteria. Evolution has uncoupled nutrient detection and movement with a brain capable of cost–benefit analysis. The bacterium is fated to alternately flail and cruise, thrashing about until a randomly incurred sugar molecule temporarily rights its course. But fruit flies will perish without the ability to dodge malevolent wasps. Even these tiny invertebrates have the ability to suspend gratification where danger counsels patience. Unlike bacteria, they can decide to move away from food sources in the short term. But, ultimately, their decision-making process is tempered by two underlying directives: feed and mate. In this way, they sustain the thermodynamic demands of keeping the replicators replicating. This is expected. What surprises me is that fruit flies can experience pleasure.

Scientists have now applied Berridge's language from Chapter 3 to insect reward systems. But, as researchers Clint J. Perry and Andrew B. Barron note,[39] it is much harder to ascribe the notion of "liking" to an insect. While mammals commonly lick their lips when experiencing hedonic reward, the closest indicator in the insect world is the bee's waggle dance, which Perry and Barron acknowledge is not much use for gauging the pleasure of other insects, including fruit flies. All this poses fascinating questions for how a pleasurable sensation manifests from the shuttling of electrical charges around neurons, which are alike in structure and function in all species. But since the mystery seems no clearer in humans, why should it be any easier to explain in *Drosophila*? Certainly, the pleasure we take is likely to have a far more complex flavour, but it seems we are no different from the very lowliest of animals when it comes to the strings pulled by the self-replicators. Shohat-Ophir believes that all animals experience some kind of pleasure. Moreover, we experience pleasure when we serve the interests of the self-replicators.

REFERENCES

1. Thomas Henry Huxley (1825–1895). Available at: http://www. ucmp.berkeley.edu/history/thuxley.html. [Accessed: 09-Oct-2015.]
2. K. Ruiz-Mirazo, C. Briones and A. de la Escosura, Prebiotic Systems Chemistry: New Perspectives for the Origins of Life, *Chem. Rev.*, 2014, **114**(1), 285–366.
3. J. M. Berg, in *Biochemistry*, ed. Jeremy M. Berg, John L. Tymoczko, Gregory J. Gatto, Jr. and Lubert Stryer, W. H. Freeman & Company, a Macmillan Education Imprint, New York, 8th edn., 2015.
4. M. Schaechter, in *Desk Encyclopedia of Microbiology*, ed. Moselio Schaechter, Academic, London, 2nd edn, 2009.

5. R. Dawkins, in *The Extended Phenotype: The Gene as the Unit of Selection*, ed. Richard Dawkins, Freeman, Oxford, 1981.
6. K. J. Niklas and S. A. Newman, The Origins of Multicellular Organisms: Multicellular Origins, *Evol. Dev.*, 2013, **15**(1), 41–52.
7. A. M. Breeman and B. W. Hoeksema, Vegetative Propagation of the Red Alga Rhodochorton purpureum by Means of Fragments that Escape Digestion by Herbivores, *Mar. Ecol.: Prog. Ser.*, 1987, **35**, 197–201.
8. R. K. Grosberg and R. R. Strathmann, The Evolution of Multicellularity: A Minor Major Transition?, *Annu. Rev. Ecol. Evol. Syst.*, 2007, **38**(1), 621–654.
9. A. J. F. Griffiths, in *Introduction to Genetic Analysis*, ed. Anthony J. F. Griffiths, Susan R. Wessler, Sean B. Carroll and John Doebley, WHFreeman & Company, New York, 11th edn., 2015.
10. S. Beckerman, P. I. Erickson, J. Yost, J. Regalado, L. Jaramillo, C. Sparks, M. Iromenga and K. Long, Life Histories, Blood Revenge, and Reproductive Success Among the Waorani of Ecuador, *Proc. Natl. Acad. Sci.*, 2009, **106**(20), 8134–8139.
11. H. Lodish, A. Berk, C. A. Kaiser, M. Krieger, A. Bretscher, H. Ploegh, A. Amon and M. P. Scott, *Molecular Cell Biology: International Edition*, W. H. Freeman, 7th edn, 2012.
12. R. A. Sturm, D. L. Duffy, Z. Z. Zhao, F. P. N. Leite, M. S. Stark, N. K. Hayward, N. G. Martin and G. W. Montgomery, A Single SNP in an Evolutionary Conserved Region within Intron 86 of the HERC2 Gene Determines Human Blue-Brown Eye Color, *Am. J. Hum. Genet.*, 2008, **82**(2), 424–431.
13. J. M. Smith, *The Evolution of Sex*, Cambridge University Press, Cambridge, 1978.
14. G. C. Williams, *Sex and Evolution*, Princeton University Press, Princeton, NJ, 1975.
15. D. Quammen, Contagious Cancer: The Evolution of a Killer, *Harper's*, Apr-2008.
16. J. M. Smith, in *The Origins of Life: From the Birth of Life to the Origin of Language*, ed. John Maynard Smith and Eörs Szathmáry, Oxford University Press, Oxford, 1999.
17. *Encyclopedia of Biological Chemistry*, ed. William J. Lennarz, Stony Brook University, Stony Brook, NY, USA, M. Daniel

Lane, Johns Hopkins University School of Medicine, Baltimore, MD, USA ed. W. J. Lennarz and M. Lane, Academic Press, M-Q. Amsterdam, 2nd edn, 2013, vol. 3.

18. S. Ravindran, Barbara McClintock and the Discovery of Jumping Genes, *Proc. Natl. Acad. Sci.*, 2012, **109**(50), 20198–20199.

19. R. Swaby, *Headstrong: 52 Women Who Changed Science–and the World*, Broadway Books, New York, 2015.

20. J. L. Tomkins, J. Radwan, J. S. Kotiaho and T. Tregenza, Genic Capture and Resolving the lek Paradox, *Trends Ecol. Evol.*, 2004, **19**(6), 323–328.

21. A. J. Lumley, Ł. Michalczyk, J. J. N. Kitson, L. G. Spurgin, C. A. Morrison, J. L. Godwin, M. E. Dickinson, O. Y. Martin, B. C. Emerson, T. Chapman and M. J. G. Gage, Sexual Selection Protects Against Extinction, *Nature*, 2015, **522**(7557), 470–473.

22. Population Benefits of Sexual Selection Explain the Existence of Males, *Phys.org*. Available at: http://phys.org/news/2015-05-population-benefits-sexual-males.html. [Accessed: 30-Mar-2016.].

23. P. D. Keightley and M. Lynch, Toward a Realistic Model of Mutations Affecting Fitness, *Evolution*, 2003, **57**(3), 683–685.

24. D. Voet and J. G. Voet, *Biochemistry*, Wiley, 4th edn, 2011.

25. A. Bahat and M. Eisenbach, Sperm Thermotaxis, *Mol. Cell. Endocrinol.*, 2006, **252**(1–2), 115–119.

26. R. Nosrati, A. Driouchi, C. M. Yip and D. Sinton, Two-dimensional Slither Swimming of Sperm Within a Micrometre of a Surface, *Nat. Commun.*, 2015, **6**, 8703.

27. K. Inaba, Sperm Flagella: Comparative and Phylogenetic Perspectives of Protein Components, *Mol. Hum. Reprod.*, 2011, **17**(8), 524–538.

28. C. K. Mathews, K. E. van Holde, D. R. Appling and S. J. Anthony-Cahill, *Biochemistry*, Prentice Hall, Toronto, 4th U.S. edn, 2013.

29. T. W. Grebe and J. Stock, Bacterial Chemotaxis: The Five Sensors of a Bacterium, *Curr. Biol.*, 1998, **8**(5), R154–R157.

30. S. Freud, *Beyond the Pleasure Principle and Other Writings*, Penguin, London, 2003.

31. Galit Shohat-Ophir, *Interview*, 21-Jan-2016.

32. G. Shohat-Ophir, K. R. Kaun, R. Azanchi, H. Mohammed and U. Heberlein, Sexual Deprivation Increases Ethanol Intake in Drosophila, *Science*, 2012, **335**(6074), 1351–1355.

33. P. Guevara-Fiore and J. A. Endler, Male Sexual Behaviour and Ethanol Consumption from an Evolutionary Perspective: A Comment on Sexual Deprivation Increases Ethanol Intake in *Drosophila*, *Fly (Austin)*, 2014, **8**(4), 234–236.

34. I. Boileau, J.-M. Assaad, R. O. Pihl, C. Benkelfat, M. Leyton, M. Diksic, R. E. Tremblay and A. Dagher, Alcohol Promotes Dopamine Release in the Human Nucleus Accumbens, *Synap. N. Y. N*, 2003, **49**(4), 226–231.

35. J. S. C. Wiskerke and L. E. Vet, Foraging for Solitarily and Gregariously Feeding Caterpillars: A Comparison of Two Related Parasitoid Species (Hymenoptera: Braconidae), *J. Insect Behav.*, 1994, **7**(5), 585–603.

36. L. M. Lebeck, A Review of the Hymenopterous Natural Enemies of Cockroaches With Emphasis on Biological Control, *Entomophaga*, 1991, **36**(3), 335–352.

37. B. Z. Kacsoh, Z. R. Lynch, N. T. Mortimer and T. A. Schlenke, Fruit Flies Medicate Offspring After Seeing Parasites, *Science*, 2013, **339**(6122), 947–950.

38. C. J. Burke, W. Huetteroth, D. Owald, E. Perisse, M. J. Krashes, G. Das, D. Gohl, M. Silies, S. Certel and S. Waddell, Layered Reward Signalling Through Octopamine and Dopamine in Drosophila, *Nature*, 2012, **492**(7429), 433–437.

39. C. J. Perry and A. B. Barron, Neural Mechanisms of Reward in Insects, *Annu. Rev. Entomol.*, 2013, **58**(1), 543–562.

Section 1: Concluding Remarks

So, just how do all of these circuits fit together? We have considered how our taste cells distinguish the nutritional content of food, the data from which the brain processes to guide us towards the beneficial nutrients with a rewarding, pleasurable sensation. Meanwhile, our grey matter learns to associate cues with memories so that the sight of the chocolate bar is enough to set the reward network grinding into action. Thus, it is that our brains have encouraged us to eat the food even before the wrapper is off. Just how much control do we exercise over the actin–myosin complex, whose contraction enables us to pick up, open and chew the chocolate bar? We imagine that we are not like the bacterium, having its tail wagged by the molecules composing its immediate environment, but if we choose to avoid one chocolate bar, we will surely eat something else in the near future.

The existence of pleasure is intriguing. On the one hand we might reason that the very presence of pleasure frees us from the bondage of the bacterium's wholly determined existence. It cannot be that our brain automatically reaches for the chocolate bar based on its interpretation of the relevant input data; otherwise what role would pleasure play? Surely this is where the freedom of our will sits—in the choice of how to get that pleasure? But we're not quite out of the woods yet.

In *Elbow Room: The Varieties of Free Will Worth Wanting*, philosopher Daniel Dennett makes a fascinating point about yet another vicious brand of wasp.[1] *Sphex ichneumoneus* is in the Apoidea superfamily, along with the cockroach-devouring Ampulicidae wasps. This bug kidnaps a cricket and leads it off to its lair, where it will be consumed alive by the wasp's hatching larvae.

But the biologist Jean-Henri Fabre found out something very interesting about the behaviour of *Sphex*.[2] *Sphex* first stings the cricket and leads it to its lair. Having arrived, it momentarily leaves its prey outside the entrance before going in to check that everything is shipshape indoors. Next, it comes back to lead the cricket inside to its chilling doom. But Fabre wondered how the wasp would react if the script was changed. When it went inside, he picked up the cricket and moved it away. When the wasp came out, it went in search of the cricket. Having found its missing prey, the wasp led the cricket back to the mouth of its lair before repeating its earlier inspection. While the wasp was once again checking everything was just so inside its grim dungeon, Fabre again moved the cricket away from the entrance. The whole pantomime repeated again. Each time Fabre moved the cricket, the wasp fetched it, left it just outside and returned indoors for its last minute checks. This process was repeated 40 times.

To what extent, then, does pleasure free us from the bondage of determined choice? If Dr Shohat-Ophir is correct, the wasp, like all other animals, experiences pleasure. Yet, it is plainly obvious from Fabre's findings that *Sphex* lacks the ability to deviate from its script. While pleasure may guide its actions, the sequence seems more to resemble the motion of a train along a track, rather than a car driver choosing between the different routes to his destination. The mere existence of pleasure does not appear quite sufficient to emancipate us from servitude to the self-replicators.

Daniel Dennett argues that we are more developed than the *Sphex* wasp, but there is an irony about this story. Fred Keijzer, an associate professor of philosophy at the University of Groningen, has hinted that the behaviour of a long line of cognitive scientists resembles the bug more than they might like to admit.[2] It seems that Fabre's experiment failed to meet the gold standard of scientific inquiry; when others carried out the experiment, the wasp's behaviour did not always get stuck in a permanent loop. In fact, Fabre admitted as much in his own account of the inaugural experiment. He declared the information in the very paragraph of a passage raided by a long succession of science communicators, yet it was ignored. Meanwhile, the faulty anecdote has been endlessly repeated as if its perpetrators are stuck in the very sphexish loop from which they seek to emancipate themselves. The next section will consider which others of our puppet strings are pulled by the brain's pleasure centre.

REFERENCES

1. D. C. Dennett, *Elbow Room: The Varieties of Free Will Worth Wanting*, MIT Press, Cambridge, Massachusetts, London, England, New edn, 2015.
2. F. Keijzer, The Sphex Story: How the Cognitive Sciences Kept Repeating an Old and Questionable Anecdote, *Philos. Psychol.*, **26**(4), 502–519, 2013.

Section 2:
Love and Relationships

Introduction

If the pleasure we take in eating motivates us to provide the self-replicators with fuel and building supplies, what do our other pleasures achieve? Love is one of the greatest pleasures that we can enjoy as humans and, sure enough, it serves an invaluable role for the self-replicators.

Section 2 explores the role of love in the process of reproduction. Two key players in elucidating the chemistry of love are Michael Liebowitz and Helen Fisher. Although their theoretical models differ, they both distinguish between lust and romantic love, while also considering the role of attachment. Chemical candidates have been suggested for each of these states of the heart and mind.

Chapter 6 considers the hormones and neurotransmitters that stoke the fires of lust, with particular emphasis on testosterone and dopamine. Chapter 7 investigates the neurotransmitters that drive the heart-stopping delirium of romantic love. Chapter 8 explores what happens when the turbulent waters of romance wash us up on the calmer shores of companionship. Now a new vanguard of chemicals takes charge of our senses, switching obsession for compassion. In this more altruistic mode we are ready to lavish our attention on the wailing progeny born of love's flightier beginnings. Finally, Chapter 9 introduces some of the fundamental processes that enable a single fertilised ovum to grow into a baby.

The Chemistry of Human Nature
By Tom Husband
© Tom Husband 2017
Published by the Royal Society of Chemistry, www.rsc.org

This idealised version of events shows how the self-replicators ensure their future. When we die, we can no longer maintain the twin streams of fuel and building materials on the self-replicators in our cells. But if this chemical stew has had its time-honoured effect, it is in the fruits of our loins that the ancient replication process continues.

CHAPTER 6

The Chemistry of Lust

Back at the start of the 21st century, older men on both sides of the Atlantic began queueing up for testosterone supplements.[1,2] There were various symptoms they hoped to eliminate, such as depression, fatigue and irritability. But, of course, what they really wanted was to boost their libido and put a bit more zing in the pickle. Curiously, this sudden enthusiasm resembles events that occurred just before the turn of the 20th century. Back then, a similar set of ailments had men thronging to receive a remedy dubbed the "elixir of life" by the press. Droves of gullible men paid for treatments that were mere placebos, many, sadly, with their lives.

Man's desperation to restore his waning libido is nothing new. What is new is that science arrived to challenge superstition, although this is not immediately beneficial to those in need of treatment. A little knowledge is a dangerous thing to a charlatan, or rather to their patients, who are duped by the words of great men, twisted to serve immoral ends. Nevertheless, the elixir of life was an important landmark on the way to a deeper understanding of human nature and was the first step towards the chemistry of lust.

Questions still remain, fanning the embers of residual superstition as ever they do, but testosterone plays a critical role

The Chemistry of Human Nature
By Tom Husband
© Tom Husband 2017
Published by the Royal Society of Chemistry, www.rsc.org

in lust, just as lust plays a critical role in romantic love. Anthropologist Dr Helen Fisher argues that lust is the indiscriminate urge for sexual gratification. This restless itch bids you scratch it before any sweetheart has been identified. But lust sparks the flames of attraction, the longing for a more discerning emotional union. This secondary phase, argues Fisher, allows individuals to streamline their mating operation by focusing all their attention on that one special someone. Love begins with lust, and lust begins with sex hormones, including testosterone. This is true for men and women.

In a way, the elixir of life was the first testosterone replacement therapy (TRT), but it was very different from its modern-day descendent. Both contained testosterone, but the dosage of the elixir was too low to have a biological effect. Also, that olden-day remedy wasn't as convenient as the attractively packaged gels that today's run-down man can rub into his shoulders. No, the luckier of these men had to ingest a filtered solution of pulverised animal testicles. Their unfortunate counterparts had to receive testicular implants. Though amounting to little more than a skin graft,[3] these were lethally dangerous. A particularly unscrupulous charlatan called Brinkley ran up a death toll of up to several hundred patients with his transplant of goat testicles.[4]

Then, as now, legitimate science was invoked to justify the treatments. And then, as now, the ambiguity of the data lent itself to profitable interpretations. It all started with a visionary but controversial scientist by the name of Charles-Édouard Brown-Séquard. In his eighth decade the decorated researcher shocked both the scientific community and the public with the findings of an experiment he had performed on himself. This was completely in character for the maverick scientist. During his life, he swallowed a piece of sponge on a string, which he drew back up to investigate his gastric juices. He swallowed small samples of the vomit of cholera patients to induce mild symptoms that he treated with laudanum (a solution of opium in alcohol). On one occasion, he was found unconscious having covered himself head to toe in varnish in a bid to investigate the function of the skin. But it was his very last experiment that was to provoke furore.[4] On July 20th, 1889, Brown-Séquard published an article in the *Lancet* describing the effects of injecting medicine made from animal testicles. His rationale was as follows:

"It is particularly well known that eunuchs are characterised by their general debility and their lack of intellectual and physical activity. There is no medical man who does not know also how much the mind and body of men ... are affected by sexual abuse or by masturbation. Besides, it is well know that seminal losses ... produce a mental and physical debility which is in proportion to their frequency. These facts and many others have led to the generally admitted view that in the seminal fluid, as secreted by the testicles, a substance or several substances exist which, entering the blood by resorption, have a most essential use in giving strength to the nervous system and to other parts ... It is known that well-organised men, especially from twenty to thirty-five years of age, who remain absolutely free from sexual intercourse or any other causes of expenditure of seminal fluid, are in a state of excitement, giving them a great, although abnormal, physical and mental activity."[5]

Brown-Séquard described the ingredients and the effects of the potion as follows. He mixed testicular blood, semen and juice extracted from the crushed testicles of dogs or guinea pigs. He was lyrical about the transformation that he attributed to this concoction. He claimed that his intellectual capacity and energy had been restored, that the jet of his urine was more powerful and that even when constipated he had a great "improvement with regard to the expulsion of fecal matters".[5] He was not so insensible as to ignore the possibility of the placebo effect and consequently urged his scientific brethren to try injecting the mixture themselves,[3] a suggestion that many indeed took up, including the great Louis Pasteur.[4]

Unfortunately, his methods provoked widespread derision and contempt. Although the press dubbed his remedy the "elixir of life", scientists distanced themselves. Antivivisectionists were in uproar about the cruelty he was inflicting on animals. Naturally, the church was disgusted by his candour on sexual matters, especially when Brown-Séquard suggested that men may wish to masturbate but stop short of ejaculating, in order to maintain their vigour.[4] This was an era when masturbation was deeply frowned upon, never mind that so-called "hysterical" women were being relieved of their symptoms by genital massage.[6]

Brown-Séquard was an unconventional man. His biographer Michael Aminoff, a physician and scientific researcher like Brown-Séquard himself, has told his story beautifully. He argues that the great maverick almost certainly suffered from bipolar disorder, as evidenced by alternating spells of depression and periods of manic activity, during which he would work for up to 19 hours a day, subsisting on endless cups of coffee so as to avoid the inconvenience and expense of eating. In fact, Brown-Séquard was a visionary, as well as a medalled hero to the people of his hometown in Mauritius. It was here that he sampled the vomit from cholera patients as he laboured—free of charge—to alleviate suffering during an epidemic of the disease.[4]

It turned out that Brown-Séquard's hunch about testicular function was absolutely correct. Various sex hormones, amongst which testosterone predominates, are produced in the male gonads and their biological role is strongly implicated in virility, as well as a slew of other functions. Several decades later, in 1939, Adolf Butenandt and Leopold Ružička received the Nobel Prize for their synthesis of testosterone,[3] a landmark on the road to TRT. Even in his lifetime, Brown-Séquard witnessed the use of glandular extracts to treat other conditions, such as thyroid problems. Aminoff believes that these alternative rudimentary hormone therapies all stemmed directly from Brown-Séquard's work and, as such, titles him the Father of Endocrinology.[4]

In spite of his insight, the rejuvenating effects of Brown-Séquard's remedy were almost certainly down to the placebo effect. While nearly 2000 patients worldwide reported positive effects of the treatment,[4] no one ever thought to conduct a blind trial. In 2002 a follow-up Australian study recreated the potion according to Brown-Séquard's recipe, revealing the testosterone content would have been less than 1 milligram, far below the necessary dose for a biological effect.[7]

But what of the modern day equivalents distilled from over 100 years of research? There is no doubt that hormone replacement therapy is a boon to people with inadequate testosterone, but prescribing it to reverse the effects of ageing is as futile now as it ever was. Sadly, positioning oneself on the continuum adjoining these disparate complaints is not a straight forward matter, especially when doctors and pharmaceutical companies throw their biases into the arena.

Researchers Lisa Schwartz and Steven Woloshin of the Dartmouth Institute fear that testosterone supplements are being prescribed too liberally.[8] They argue that the treatment is being recommended for a wide range of symptoms, many of which are natural by-products of ageing. This sudden enthusiasm for the medicine is no coincidence, suggest Schwartz and Woloshin, but rather the result of a far-reaching awareness campaign run by the producers of the leading brand of supplements, *Androgel*. Abbott Laboratories, the previous incarnation of a pharmaceutical company since renamed AbbVie,[9] ran the *Low "T"* campaign using websites, advertisements and ghost-written articles, all of which recommend the supplements not only for decreased libido and erectile dysfunction but also for ailments like flagging energy levels, mood changes and weight gain.

Diagnosis is the issue. Hormone replacement therapy is absolutely the correct prescription, provided that hypogonadism is the correct diagnosis. This condition is where sufferers produce little, if any, testosterone and can be readily identified with a simple blood test. But up to a quarter of men who are prescribed TRT in the US have not had this baseline test.[10] It is not simply that the TRT may be ineffective at treating such symptoms[8] but rather the supplements have been linked with a five-fold risk in cardiovascular problems, such as heart attacks and strokes.[11] Schwartz and Woloshin have voiced strong concerns that the *Low "T"* campaign is no more than an uncontrolled experiment, which exposes subjects to huge risks in the form of a treatment unlikely to help with problems that may be nothing to do with testosterone levels.[8]

It's not hard to see why men would allow themselves to be guinea pigs in this huge uncontrolled experiment. In fact, the experiment has been ongoing for much longer than a century. For millennia, man has tried to medicate for his waning libido with a host of remedies, the rationale for many of which was pure superstition. Even in recent times, people were still taking crushed rhino horn purely because the poached appendage resembles an erect phallus.[12]

Since Brown-Séquard came along, things have been a lot more scientific. He paved the way to a discovery that many people had already intuited—lust has an origin in the genitals. But, in spite of this discovery, testosterone supplements are still unreliable for boosting libido. What scientists now believe is that a minimum testosterone level is ample for a functioning system. Victims of hypogonadism fall below this threshold and, hence, typically demonstrate low libido and erectile dysfunction. Happily, these problems can be effectively treated with testosterone supplements (ref. 13, p. 298). But giving supplements to men with normal levels of testosterone does not appreciably increase their sex drive or their ability to get erections.[14]

This seems to clash with received wisdom on the subject. Surging hormones are blamed for the incessant sexual cravings that afflict adolescents. Steroid-taking athletes are said to be unusually frisky. (Arnold Schwarzenegger took steroids and made no secret of his appetite for women. Indeed, several accusations that he acted inappropriately on such urges nearly cost him his campaign to be the governor of California.) Also, men are said to experience a decline in testosterone from the age of roughly 30 onwards, which correlates with a decline in libido.

This just goes to underline the complexity of biological systems. While such observations may well correlate with levels of testosterone, there is no guarantee that the one change *causes* the other. A mantra among scientists is that correlation is not the same as causation. Tyler Vigen has picked some comically unrelated correlates from public statistics, including that the divorce rates in Maine rose and fell in almost perfect tandem with the consumption of margarine over the same period.[15] An irresponsible analyst might infer that butter substitutes wreck marriages, but the point is that correlation is often, but not always, coincidence. The discovery of a correlation can guide researchers towards a causal relationship but not without doing more science first.

Age is characterised by a general deterioration of the body, so there are many other possible explanations for the accompanying loss of vigour. Teenagers undergo an overhaul of the wiring in their brains, featuring proliferation followed by the pruning of neurons[16] that plays a critical role in sexual behaviour (ref. 13, p. 37). Meanwhile, exercise has been shown to boost the sex drive,[17] which may shed some light on the situation with athletes taking steroids.

Not for the first time, the story seems to be that no one knows exactly what the story is. This is normal, but there is a broad consensus that at least some testosterone is needed for men to enjoy a healthy libido even though exceeding the recommended dose won't crank up cravings to ever dizzier heights.

6.1 ANDROGENS AND WOMEN

Women are not much more straightforward than men in their response to testosterone. There is evidence that testosterone levels have a relationship with sex drive but only in some women. This may explain why some—but not all—women suddenly get turned off the delights of bedroom escapades when they go on the pill. Oral contraceptives have been shown to reduce testosterone levels,[18] but some studies have also found that, for some women, this reduction coincided with a substantial increase in sexual desire.[19]

Another factor often discussed is the menstrual cycle. Some women have increased arousal just before or even during their periods, while most peak in their desire around ovulation.[14] A curious distinction in studies of both men and women is the frequency of masturbation compared to what is sometimes drily referred to as "dyadic sex", in other words sex with an actual person. One study found that fluctuating menstrual testosterone levels made no difference to how often the women were intimate with their partners, although it did find that surging testosterone made them masturbate more often.[20]

This points to an obvious problem with measuring sex drive. Wanting to have sex and then actually getting it are two profoundly different things, in stark contradiction to masturbation, where everyone enjoys a far higher degree of autonomy in realising their desires. Researchers measure myriad indicators, such as how often participants wanted sex, how often they had sexual thoughts, how often men got involuntary erections at night, the quality of orgasm, the frequency of orgasm, how much the sex was desired and how much it was enjoyed. Another confounder is that women are more complex than men in how they show arousal. As any man remembers from his awkward teenage years, being turned on is an embarrassing beacon to anyone within eyeshot. A swollen clitoris is much less of a giveaway, as is the flow of blood and the strength of its pulse in the vaginal wall but both can be used to measure desire.

As a result, the treacherous waters of research into female testosterone supplements are tricky to navigate. To date, studies have shown that androgen supplements caused "an increase in satisfying sexual events" for premenopausal women with sex problems,[21] a stronger vaginal pulse in healthy women,[22] and greater sexual desire, more frequent orgasms and more pleasurable sex.[23] This is in contrast to various studies that found no significant impact on sex drive.[14]

Things get clearer for older ladies. Scientists agree that postmenopausal women who supplement their oestrogen therapy with testosterone experience improved sexual function, the only drawback being that some of the women also develop spots or sprout facial hair.[24] There is also a broad consensus that testosterone supplements increase sexual desire in women with the self-explanatorily titled hypoactive sexual desire disorder.[25] This could be considered equivalent to the proviso that men ought only to be prescribed testosterone if they have hypogonadism, although the conditions are not equivalent.

As with men, we can be fairly certain that androgens, and especially testosterone, play a role in female desire. It may be true that a threshold exists for women, below which desire dwindles, but above which additional testosterone does nothing to fan its flames. Little evidence exists in support of this theory, but it is clear that sex hormones play a role in female sexuality.

6.2 HOW TESTOSTERONE WORKS

If testosterone fuels desire, how is it that boosting the body's natural supplies does not ratchet up lust? This is rather like asking why more symphonies do not get played if a concert hall is supplemented with extra conductors. It is true that diversifying the conductors may have the same effect on the repertoire, but each musician can only play one piece at a time. Testosterone is, certainly for men, a prerequisite for desire but so are many other chemicals. The role of testosterone is not unlike a conductor and the chemistry it directs is as beautiful as any symphony.

Testosterone enhances libido by manipulating the flow of neurotransmitters. To do so it works across the borders of multiple cells, pairing up with allies to run off copies of genes that are used to stockpile enzymes. These enzymes pluck nitrogen atoms from passing amino acids and use them to forge short-lived neurotransmitters that can roam between cells, altering their settings to adjust their sensitivity to incoming signals, as well as the strength of their outgoing transmissions.

Testosterone does not work alone but alongside several similar steroids. In everyday language the word "steroid" usually describes drugs taken by athletes to help develop muscles. They do this by directing the body to make more proteins, in particular the actin and myosin muscle components we met in Chapter 2. This demonstrates a typical feature of hormones: the ability to interact with DNA and thereby influence the proteins that get made. Four examples of hormones are shown in Figure 6.1. Most hydrogen atoms have been left out, according to the

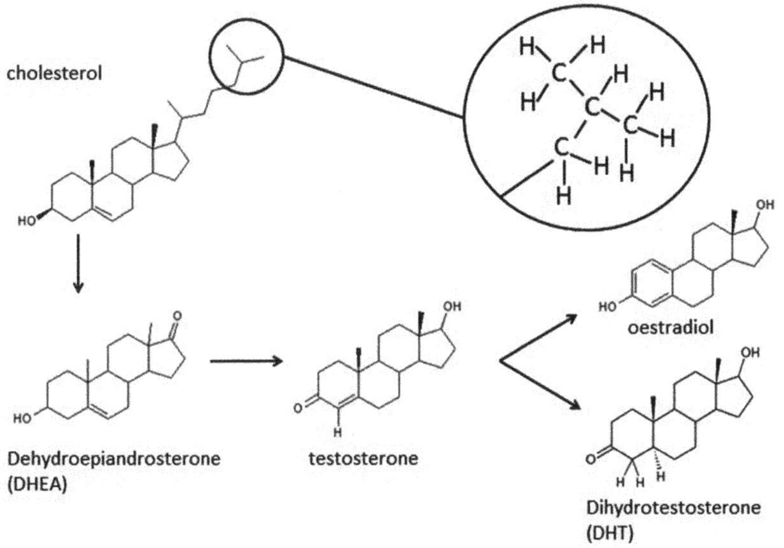

Figure 6.1 Cholesterol is the starting point to make various androgens, including dehydroepiandrosterone (DHEA), testosterone, dihydrotestosterone (DHT) and oestradiol.[26]

diagrammatic conventions detailed in Chapter 2, but some have been left in to emphasise the relevance of the prefixes *dehydro-*, indicating *no hydrogen where usually expected*, and *dihydro-*, meaning *two hydrogen atoms where testosterone has only one*. Note the distinctive hexagon and pentagon motifs, representing cyclohexane and cyclopentane, respectively, a common characteristic of steroids.[26]

A chief source of sex hormones is cholesterol, a word synonymous with bad health, despite its essential role in our biochemistry. It is converted into testosterone in five steps, each requiring a different enzyme. Figure 6.1 shows how testosterone is produced from cholesterol *via* dehydroepiandrosterone (DHEA). Still another enzyme can transform testosterone into oestradiol. Testosterone, dihydrotestosterone (DHT) and DHEA are androgens, sex hormones chiefly involved with male development, while oestradiol belongs to the female-developing oestrogens. In spite of this gender segregation, all four play roles in people of both sexes.[26] That said, male users of anabolic steroids have been known to start growing breasts, where their bodies have converted excess testosterone into oestrogens.[27]

Testosterone is unusual among the compounds we have met so far because it does not need a transporter to enter cells. It roams freely through cell membranes (ref. 26, p. 6), not unlike the way that butchers can move through hanging chain screens, while leaving unwelcome insects on the other side. These dangling chains are a good analogy for the cell membrane, which, as shown in Figure 6.2, is composed of hydrophobic tails moored to hydrophilic heads.

Hydrophobic substances are described as non-polar, while hydrophilic substances are polar. The polarity of a substance describes how evenly the electrons are distributed between the positive nuclei. In Chapter 1 we saw that the electrons in water molecules preferentially huddle around the more positive nucleus of the oxygen atom, leaving the hydrogen nuclei comparatively exposed. Such molecules are said to be polar and flock together because their uneven distribution of electrons gives rise to partially positive and partially negative regions.

The hydrophobic effect is the name given to the phenomenon that prevents polar and non-polar substances from mixing, the classic example being cooking oil and water. (See the Appendix

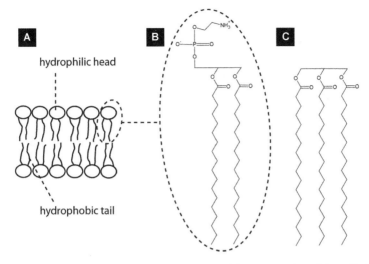

Figure 6.2 The resemblance between the lipid bilayer and cooking oil. (A) A typical schematic diagram of the lipid bilayer. (B) Exploded view of the structure of phosphatidylethanolamine, a common example of a membrane lipid. (C) A generic cooking oil.[27]

for a detailed explanation.) The reason these liquids do not mix is complicated. It has often been said that they are immiscible because the water molecules are more attracted to each other than to the oil molecules. It is true that the attraction is greater between water molecules because they can form hydrogen bonds, the weak force attracting the partially positive hydrogen atoms to the partially negative oxygen atoms in neighbouring molecules. It is also true that the molecules of the cooking oil have only a very limited ability to form hydrogen bonds. However, oil and water molecules can still attract each other *via* weaker intermolecular forces known as dispersion forces. So, the comparative attraction between the different molecules does not explain the immiscibility of oil and water, but water's ability to form hydrogen bonds is still relevant to the real explanation.

These hydrogen bonds have a huge impact on the behaviour of water molecules. In ice each water molecule is bonded to four neighbouring water molecules in a pattern that repeats throughout the crystal, as shown in Figure 6.3. All of the molecules are held at an equal distance from one another in an arrangement called a lattice. In cold water, molecules are still

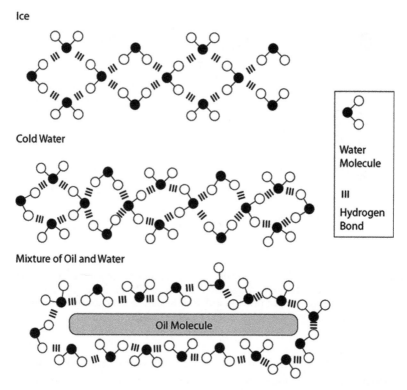

Figure 6.3   In ice, the water molecules adhere to a strict pattern that runs
throughout the crystal. Water molecules in liquid water are still
held in a lattice but the pattern is distorted with hydrogen bonds
that vary in length. When an oil molecule mixes with water, the
H_2O molecules are forced to form a clathrate around the oil
molecule, which entails an unfavourable increase in order.

thought to be bonded to four neighbours, but the arrangement is
literally more fluid.[28] Some hydrogen bonds will be longer than
others and there will not be a strict pattern that dictates their
arrangement. But when a molecule of a hydrophobic substance,
such as cooking oil, is thrust into water, the lattice-like structure
is disturbed. The water molecules maintain their hydrogen
bonds, but to do so they are forced to form a cage around the
hydrophobic molecule known as a clathrate (from the Greek
klēthra, meaning bars,[29] such as those that might form a cage).

The formation of these cages is thermodynamically
unfavourable. In Chapter 4 we considered the second law of
thermodynamics, which dictates that the universe gets steadily

less ordered with the passage of time. A bedroom will naturally get untidy, but to restore it to a tidier state requires the input of some effort. The only way for molecules to align themselves in a more ordered state is for their neighbours to assume a greater amount of disorder. For example, once sugar is dissolved into water, the only way for the sugar molecules to reorganise themselves into a crystal is to boil off the water. Thus, the cost of ordering the crystal is met by the disorder entailed by transforming the water into chaotic steam. When it comes to mixing oil and water, there is no chaos to meet the cost of ordering the water molecules into clathrates, which is why they do not mix at room temperature and atmospheric pressure.

Particulars are missing from this explanation. Testosterone is very slightly polar because it has alcohol groups at either end, each consisting of an oxygen atom and a hydrogen atom. The arrangement of their electrons mimics that of a water molecule, which explains why testosterone is very weakly soluble in water. But the bulk of the molecule is non-polar, which explains how it can diffuse freely through the cell membrane. Another complication is that water molecules can, sparingly, pass through the cell membrane but only if they are headed for an area in which they are less populous, also in accordance with thermodynamic laws. In fact, it is larger, polar molecules, such as dopamine, or species with a high charge density, such as calcium ions (Ca^{2+}), that cannot traverse the cell membrane without the help of a transporter as the thermodynamic costs described above cannot be met (ref. 30, p. 474; for a more detailed explanation of the relevant thermodynamic principles, see the Appendix).

One kind of cell into which testosterone may roam is a neuron. Once inside the neuron, it teams up with molecules inside the cell before heading to the nucleus, where it may set in motion the production of nitric oxide synthase (NOS), an enzyme that does not just make us feel aroused but also primes our genitals to act on our urges.

Steroid hormones interact with cells in quite a different way to the other kinds of signalling biomolecules we have met so far. The receptors in taste cells and post-synaptic neurons were lodged into the cell membrane, enabling biomolecules that cannot easily enter the cell to transmit signals inside. In contrast, testosterone can enter the cell unhindered and, once inside, it pairs up with androgen receptors, which float around the cell untethered.[27]

There may be some other steps involved as it is still not clear exactly which hormone binds with the NOS-expressing receptor. Research led by Dr Elaine Hull at the University of Florida suggests that oestradiol is the steroid that drives NOS production in male rats.[31] A critical difference between rats and humans is that one cannot legally inject a needle into a human brain to extract samples in most places before dissecting the brain to make sure the samples came from the right spot. This is how microdialysis was used to gather the data on rodents, highlighting the difficulty of figuring out how the process works in humans. Just because rats use a certain hormone to express NOS is no guarantee that humans use the same one. Even if the hormone had been conclusively identified, the details of how it expresses the gene would still be pretty murky. The science of how androgens guide gene expression is still not well understood.[32]

That said, emerging evidence seems to suggest that androgen receptors behave like museum curators. Each of our cells contains a length of DNA that measures approximately two metres, but that is somehow crammed into a nucleus measuring about six millionths of a metre across. As impressive as this packing feat may be, it seriously curtails access. It is not possible to arrange our genome so that every section is constantly on show. Instead, its exhibits are constantly rotated between storage and display, with biomolecules such as hormones acting as decision-makers in this process.

These powers are exerted *via* histone modification. A histone is a special kind of protein that facilitates this extraordinary packing feat of DNA. When you buy string, it is usually coiled around a cardboard tube that helps it to keep its shape. In the same manner, DNA coils itself around these histones so all of our genes can be bunched up tight and packed into a small space. When a cell needs to access a particular stretch of DNA, it has to be unwound from the histone. This is where histone modification is involved. As we have seen, proteins act in different ways depending on which other substances they are bound to. Not only can such interactions make them change shape, they can also make them more or less attractive to certain neighbours. Therefore, the histones can be chemically altered to make them more or less attractive to the DNA coiled around them. Obviously, if the histone is less attractive, then the DNA can be uncoiled more easily and *vice versa*. Histones are usually modified with one of two chemical groups: acetyl groups make them less attractive to DNA so that it can be more easily unwound. Conversely, methyl groups can be used to make the histones cling more tightly, but they can also do the opposite depending on what other modifying groups have been added to the histone. The main point is that the histones can be chemically altered, either to make the DNA harder or easier to unspool,[27] as shown in Figure 6.4.

Based on this hypothesis, testosterone probably expresses NOS broadly, as shown in Figure 6.5. Having entered the cell, testosterone or one of its metabolites binds to an androgen receptor, producing an androgen receptor complex. This androgen receptor complex teams up with an identical neighbour to form a homodimer (ref. 33, p. 41). As a result of this bonding, the homodimer develops an affinity for a specific stretch of DNA located near the *NOS* gene called a promoter region (ref. 34, p. 135). A useful additional effect is that binding the androgens has also made the receptor pair more attractive to another protein called a coactivator. Consequently, the homodimer clamps onto the promoter region whilst simultaneously binding to a coactivator. Finally, the coactivator probably attaches an acetyl group to the histone, weakening its hold on the DNA and thus enabling the *NOS* gene to be unspooled for transcription.[27]

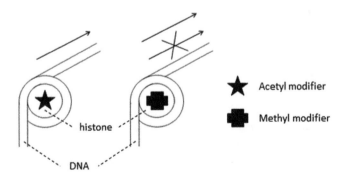

Figure 6.4 Acetyl modifiers allow DNA to be unwound from the histone. Methyl modifiers can prevent or promote unspooling, depending on what other modifiers are present.

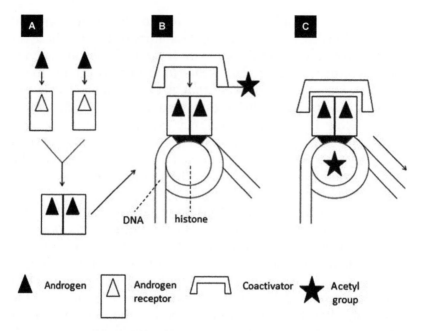

Figure 6.5 (A) Androgen molecules bind with receptors, which pair up as homodimers. (B) Acting as a transcription factor, the homodimer binds to the promoter region of the DNA (shaded). (C) Having bound the homodimer, the coactivator transfers an acetyl group onto the histone, enabling the *NOS* gene to be unspooled.

Expression of NOS starts about 40 minutes after the testosterone enters the cell and the process continues for a few hours.[34] This may partly explain another finding from Hull's research team. Three groups of rats were castrated with predictable effects on their inclination to mate. After being injected with testosterone, their libido was revived and they were once more able to copulate. Each group received daily injections of testosterone for either 2, 5 or 10 days. None of the two-dose rats copulated, just over half of the five-dose rats got frisky, but all six of the ten-dose rats got their mojo back.[35] The fact that a batch of enzymes takes several hours to assemble must surely be a factor in the failure of the two-dose group to copulate, although many other variables exist. Meanwhile, Hull's team discovered something else of interest: the ten-dose rats all showed elevated levels of dopamine in the medial preoptic area of the brain, suggesting that dopamine may also play a role in lust.

We have already seen evidence that dopamine may motivate sexual behaviour. In Chapter 3 the distinction between wanting and liking was considered in the context of the man with a compulsion to masturbate. He repeatedly activated electrodes implanted in his brain in positions thought to correspond to the motivational dopamine pathways. If this is the case, then what is the connection between NOS and dopamine?

6.3   A RADICAL TRANSMITTER

To answer this question, we need only interpret the name of the enzyme. Unlike some of the arcane alphanumerical tags we have seen, NOS has a name that tells you what it does: it synthesises nitric oxide. In order to do this, a ready supply of the amino acid arginine is required, along with a few other supplies, including oxygen. Various substances cooperate to tack an oxygen atom onto one of the nitrogen atoms in arginine and then another oxygen atom knocks them out of the amino acid in the form of nitric oxide, as shown in Figure 6.6.

Survivors of school chemistry may wonder why it is not simply called nitrogen oxide. This is because several such oxides exist, including N_2O, NO_2, N_2O_4 and the one that concerns us, NO. The two elements struggle to form compounds in which both gain a full outer shell of electrons. Sometimes, such as in the case of nitric oxide, this can result in an unpaired electron on one of the atoms (shown as a dot in Figure 6.6), putting the oxide in the category of a free radical. These are the biological bogeymen, which fuel demand for so-called superfoods that are said to be

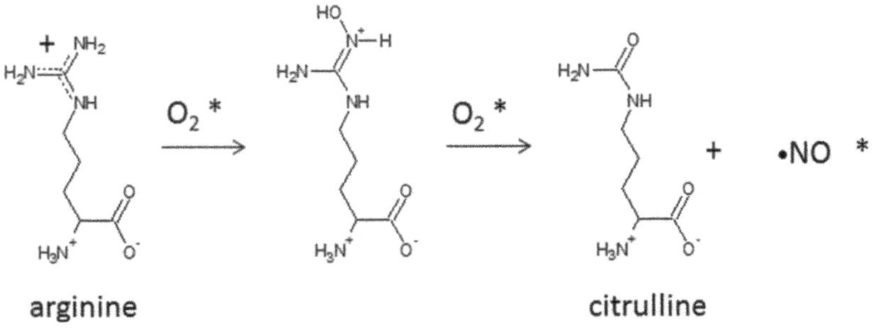

Figure 6.6   Arginine is converted into citrulline and nitric oxide (* *along with other substances*).

rich in radical-dispatching antioxidants. The fact that our bodies contain an enzyme whose sole purpose is to produce NO clearly demonstrates that radicals are not always harmful to health. Another oxide of nitrogen demonstrates that non-radicals can equally be harmful to health. Nitrous oxide, N_2O, was responsible for a tragedy that hit the family of Elaine Hull. Having devoted a significant proportion of her career to the elucidation of the biological role of the radical NO, it was the non-radical N_2O, better known as laughing gas, which caused the death of her son Geoffrey, who suffocated after inhaling it from a canister.

This is not to say that nitric oxide is completely harmless. On the contrary, it can cause mutations in DNA, react with oxygen to produce the acid-rain-causing NO_2 and may even promote the formation of tumour.[36] So, how on Earth can it be beneficial for our bodies to produce the stuff? The answer to this question is one to which the scientific community was initially quite resistant (ref. 37, p. 151): it is, in fact, a neurotransmitter. And it is not the only toxic neurotransmitter. Another one is hydrogen sulfide, the rotten egg smell used in stink bombs. Even more surprising was the discovery that carbon monoxide, the lethally toxic gas inhaled from car exhaust fumes by some suicide victims, is also a neurotransmitter.[37] Perhaps we should not be too surprised, scientists have long contemplated the paradox that oxygen, another unlikely radical, seems simultaneously to sustain us while slowly killing us. These findings show that context is equally important to dosage when it comes to gauging the toxicity of compounds.

Another obstacle to the acceptance of nitric oxide as a neurotransmitter must have been its unconventional behaviour. Compared to the dopamine-style of neurotransmitter we met in Chapter 5, it is different in almost every way imaginable. Dopamine is stored in vesicles that launch the transmitters out of the neuron so that they can activate the receptors on the neighbouring cell. Having achieved this, they are vacuumed up by the dopamine transporter and repackaged into vesicles for reuse as transmitters, to be released when the cell is next depolarised. Nitric oxide does none of the above. First of all, it cannot be stored. Its radical nature makes it highly unstable, giving it a half-life of 3 to 4 seconds. This also means that it cannot be gathered back up for reuse. And it does not activate receptors on neighbouring neurons.

This, of course, raises the question of what it does do. The expression *to fire a neuron* denotes the depolarisation of that neuron, culminating in the release of neurotransmitters from stored vesicles. Nitric oxide does not *fire neurons*, instead it adjusts their settings. Petrol and oxygen explosively combine to propel cars, but the engine might be prevented from working at all if the settings on the carburettor are wrong. Nitric oxide tinkers with the carburettor, manipulating the proportion of fuel and air. The radical transmitter makes alterations to neurons, such as how easily and for how long they fire. For example, it can influence ion channels, causing them to admit ions more slowly and hence making the neuron harder to fire.[38] Another of its techniques is similar to cocaine. Both substances reduce the speed of the dopamine transporter.[39] This is the membrane-straddling protein that vacuums up dopamine after it has been released from the neuron. By decelerating transport, nitric oxide enables the dopamine that has already been released to continue triggering the neighbouring neuron. In this way, nitric oxide amplifies a dopamine signal rather than triggering a signal of its own.

What happens next is open to speculation. It would be convenient to conclude that these dopamine currents in the medial preoptic area interact with those in the nucleus accumbens, thus triggering the "wanting" for sex. It is unlikely to be this simple. Furthermore, dopamine is not the only neurotransmitter with which nitric oxide interacts. One of its key collaborators is oxytocin,[40] which fits with the finding that rats that received injections of oxytocin directly into their brains immediately developed erections.[41] Chapter 8 will consider pathways of the neurotransmitter in more detail.

6.4   PRIMING GENITALS

Lust is the urge and sex is the goal. Assuming these motivational strategies have secured the attentions of a mate, the biology moves downstairs. Impulses hurtle from the central to the peripheral nervous system, triggering a much clearer-cut role for NOS in the genitals of both men and women. Tumescence is the state of being swollen and it happens the same way in the penis as in the clitoris.[42] Blood flows into the organ through blood vessels, the walls of which can be made to expand or contract by neighbouring muscle cells. When the muscle cells contract, the vessels get narrower, rather like when you pull a drawstring bag shut. When the cells relax, the vessels stretch out and swell up with blood, resulting in tumescence.[40]

Muscle cells are triggered by nerve signals. In this case, the cells receiving the signal have very similar chemistry to the lust-stoking brain cells described above. Both contain NOS and thus both produce nitric oxide. Just as in the brain, the nitric oxide diffuses from one cell to the next, in this case migrating into the muscle cells lining the blood vessels. Once inside, it starts to look like a familiar story told from a different perspective. Many of the biomolecules we met in Chapters 3 and 4 are involved, but they interact in different ways and, in some cases, the same characters seem to be played by different actors.

In Chapter 3 we saw how the two proteins actin and myosin cause skeletal muscle cells to contract. By repeatedly changing shape and alternately attracting and relinquishing its partner, the myosin hauls the actin, like a person pulling a rope. Each molecule of myosin was powered by adenosine triphosphate (ATP), but another chemical species is necessary to activate muscle contraction. Nerve impulses release calcium ions (Ca^{2+}) from an internal holding bay called the lumen. These ions activate another set of proteins called the troponin complex, which in turn activates another protein called tropomyosin.

Tropomyosin winds around the actin filaments like a vine around a tree, as shown in Figure 6.7. In its resting state it acts as a barrier, preventing the myosin and actin molecules from interacting with each other. As such, the nodding myosin cannot get any purchase on the actin filament and the muscle cell is, therefore, unable to contract. However, when calcium ions are released from the lumen, they bind to one of the proteins in the troponin complex, which consequently changes shape. This change ripples around the other proteins in the vicinity, triggering further changes of shape, including that of tropomyosin. Consequently, it moves out of the way of the myosin and actin so that the two can collaborate to contract the cell.[27]

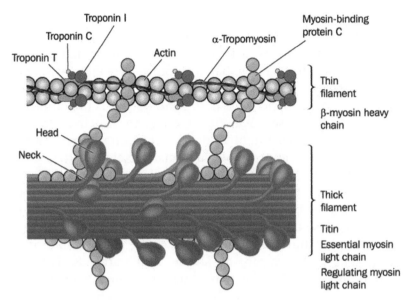

Figure 6.7 Tropomyosin acts as a barrier between myosin and actin. (Image from Shutterstock, artwork © Blamb).

Two different kinds of muscle cell exist, but they each function the same basic way. The muscles in the arm are striated, named for their striped appearance. The muscle cells in the genitals are called smooth muscle cells because they are not striped. Although the two cell types differ, the primary action of actin–myosin and tropomyosin are the same. Both cells require ATP to fuel the nodding myosin, and Ca^{2+} ions to move the tropomyosin "safety catch" out of the way of the actin.

In Chapter 4 we saw how taste cells also have to be flooded with Ca^{2+} ions in order to transmit signals. On that occasion, the receptor triggered a sequence of dominoes that ultimately released a molecule called IP_3, which unlocked the door to the cell's Ca^{2+} storage bay, the lumen. Getting an erection requires the relaxation of the smooth muscle cells in the penis. Rather than flooding the cell with Ca^{2+} ions, the goal in this case is to keep them locked up. This is achieved by the recruitment of a locksmith who changes the lock on the lumen.

When the nitric oxide enters the smooth muscle cell, it binds to, and thus activates, a protein called soluble guanylyl cyclase.[43]

Again, the name gives us an idea of the function. It takes a guanine-based compound and fashions it into a loop. Guanine is one of the components of DNA and is actually named from the Spanish *guano* for bird or bat droppings in which it is abundant.[44] Modern-day use of the chemical to give a pearly appearance[45] to synthetic pearls and shampoos was pre-empted by an ancient Japanese cosmetic practice, in which lucky individuals would smear nightingale droppings onto their faces, a recently revived trend dubbed a "geisha facial".[46] It is also a constituent of guanosine triphosphate (GTP), which is very similar to ATP. It consists of a molecule of guanine, with a tail formed of three phosphate groups and a ribose group. What the guanylyl cyclase does is lop off two of the phosphate groups and forge the remaining one into a cyclic ring on the side of the pentagonal ribose group, as shown in Figure 6.8. This new compound is called cyclic guanosine monophosphate (cGMP).

Figure 6.9 shows how cGMP causes tumescence. The freshly minted cGMP then activates another protein called a kinase. This family of enzymes is characterised by its ability to attach phosphate groups to other proteins.[30] The kinase in question is called protein kinase G (PKG), a reference to its activation by the guanine-containing cGMP. Our locksmith is now recruited. It drifts towards the lumen, where it attaches a phosphate group to the IP_3-gated ion channel.[47] This causes the ion channel to change shape so that the key no longer fits. Passing IP_3 molecules are now unable to bind with the ion channel, meaning that it keeps the Ca^{2+} ions safely locked up inside the lumen. In fact, PKG also effects other changes, all of which combine to reduce the concentration of Ca^{2+} ions.[43] Thus, the actin–myosin complex is rendered unable to contract, the smooth muscle cells

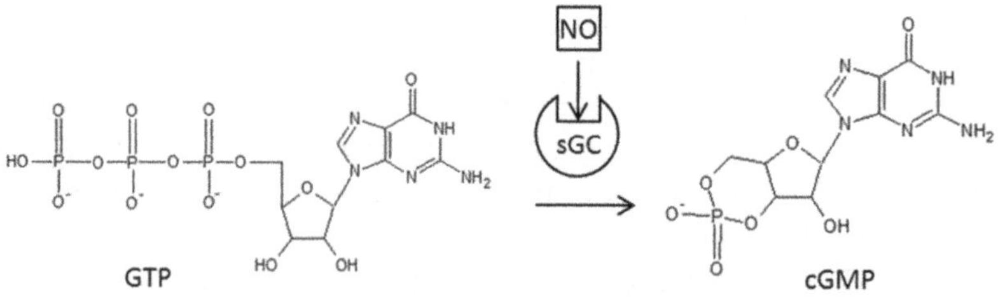

Figure 6.8   Nitric oxide activates the enzyme soluble guanylyl cyclase, which catalyses the conversion of GTP into cGMP.

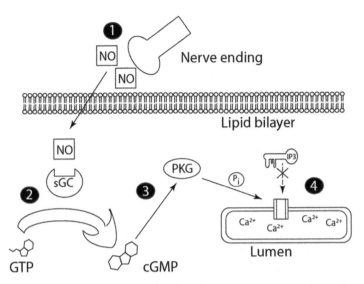

Figure 6.9 (1) Nitric oxide, released from nerves, diffuses into the cell. (2) Nitric oxide enables conversion of GTP to cGMP. (3) cGMP activates PKG. (4) PKG attaches a phosphate group (P_i) to the lumen ion channel, rendering IP$_3$ unable to release Ca^{2+} ions. Adapted from *Testosterone and Erection.*[43]

relax, the walls of the blood vessels slacken, blood gushes into the organ, and either penile or clitoral tumescence is achieved.

That takes care of getting it up, but what happens when you want it to go back down? To reverse tumescence, the smooth muscles need once again to constrict the arteries and stem the inflow of blood. This requires another protein called phosphodiesterase 5. This enzyme breaks down the cGMP so that it can no longer activate the PKG that ultimately increases the concentration of Ca^{2+} ions, meaning that the actin–myosin complex can once again contract. This is the arena in which Viagra works its miracles. Interestingly, it works not by producing erection-enabling chemicals but by preserving the ones that get rid of them. Sildenafil, the generic name for the wonder drug, is a phosphodiesterase inhibitor, meaning that it prevents phosphodiesterase from doing its job. It binds to the enzyme and causes it to change into a shape that can no longer break down cGMP. The cGMP sticks around, continuing to activate PKG and thus maintaining the relaxed state of the muscles that regulate blood flow into the penis.

6.5 THE LUST HYPOTHESIS

Testosterone plays a part in both directions. Just as in the brain, it promotes the expression of the NOS that provides the nitric oxide and it also interacts with other proteins to reduce the sensitivity of the smooth muscle cell to Ca^{2+},[43] meaning that greater concentrations of Ca^{2+} are required for the muscles to squeeze the erection-maintaining blood vessels shut. So, it is clear that testosterone plays a role not only in initiating lust but also in acting on it.

But questions remain. We have seen a possible mechanism by which testosterone and its fellow androgens underwrite the lust in phase 1 of Helen Fisher's theory of romantic lust. But testosterone is the engine oil rather than the petrol. The car will not run without it, but adding extra oil does not make it go any faster. This hints at the involvement of other biological processes and, if that is the case, what are they and what are the implications for Dr Fisher's theory? She argues that lust is an undiscriminating urge; it may select, but it is not selective. The commonly held notion has long been that surging hormones overwhelm us into bed. That no longer seems to be the case, but there is a strong case that they play a role in motivating us to mate.

REFERENCES

1. D. Campbell, "Flagging US Males Turn to Testosterone Treatment", *The Guardian*. Available at: http://www.theguardian.com/world/2003/nov/13/usa.health. [Accessed: 17-Feb-2015.]
2. R. McKie, "Robin McKie: Men Can Now Boost Their Testosterone Levels", *The Guardian*. Available at: http://www.theguardian.com/education/2002/jul/28/medicalscience.science. [Accessed: 16-Feb-2015.]
3. A. Morales, The Long and Tortuous History of the Discovery of Testosterone and its Clinical Application, *J. Sex. Med.*, 2013, **10**(4), 1178–1183.
4. M. J. Aminoff, MD, *Brown-Sequard: An Improbable Genius Who Transformed Medicine*, OUP, New York, USA, 2010.

5. Brown-Sequard, Note on the Effects Produced on Man by Subcutaneous Injections of a Liquid Obtained from the Testicles of Animals, *Lancet*, 1889, 105–107.
6. R. P. Maines, *The Technology of Orgasm: "Hysteria", the Vibrator, and Women's Sexual Satisfaction*, JHU Press, 2001.
7. A. J. Cussons, J. P. Walsh, C. I. Bhagat and S. J. Fletcher, Brown-Séquard Revisited: A Lesson from History on the Placebo Effect of Androgen Treatment, *Med. J. Aust.*, 2002, 177(11), 678–679.
8. L. M. Schwartz and S. Woloshin, Low "T" as in "Template" How to Sell Disease, *JAMA Intern. Med.*, 2013, 173(15), 1460–1462.
9. "Abbott 2012 Annual Report". Abbott Laboratories, 2013.
10. S. Tavernise, F.D.A. Panel Backs Limits on Testosterone Drugs, *N. Y. Times*, http://www.nytimes.com/2014/09/18/health/testosterone-drugs-fda.html. [Accessed: 16-Feb-2015.]
11. C. Roni Rabin for *Well*, "Weighing Testosterone's Benefits and Risks", *The New York Times*. Available at: http://well.blogs.nytimes.com/2014/02/03/weighing-testosterone-benefits-and-risks/. [Accessed: 16-Feb-2015.]
12. J. Emsley, *Vanity, Vitality, and Virility: The Chemistry behind the Products you Love to Buy*, Oxford University Press, Oxford, 2004.
13. E. Nieschlag, H. M. Behre and S. Nieschlag, *Testosterone: Action, Deficiency, Substitution*, Cambridge University Press, 2012.
14. J. Bancroft, "The Behavioral Correlates of Testosterone", in *Testosterone*, Cambridge University Press, 4th edn, 2012.
15. Tyler Vigen, "Divorce Rate in Maine Correlates with Per Capita Consumption of Margarine (US)". Available at: http://www.tylervigen.com/view_correlation?id=1703. [Accessed: 06-May-2016.]
16. S. F. Gilbert, *Developmental Biology*, Sinauer Associates, Sunderland, Mass, 9th edn, 2010.
17. J. R. White, D. A. Case, D. McWhirter and A. M. Mattison, Enhanced Sexual Behavior in Exercising Men, *Arch. Sex. Behav.*, 1990, 19(3), 193–209.
18. C. Jung-Hoffmann and H. Kuhl, Divergent Effects of Two Low-dose Oral Contraceptives on Sex Hormone-binding

Globulin and Free Testosterone, *Am. J. Obstet. Gynecol.*, 1987, **156**(1), 199–203.

19. C. A. Graham, J. Bancroft, H. A. Doll, T. Greco and A. Tanner, Does Oral Contraceptive-induced Reduction in Free Testosterone Adversely Affect the Sexuality or Mood of Women?, *Psychoneuroendocrinology*, 2007, **32**(3), 246–255.

20. J. Bancroft, D. Sanders, D. Davidson and P. Warner, Mood, Sexuality, Hormones, and the Menstrual Cycle. III. Sexuality and the Role of Androgens, *Psychosom. Med.*, 1983, **45**(6), 509–516.

21. S. Davis, M.-A. Papalia, R. J. Norman, S. O'Neill, M. Redelman, M. Williamson, B. G. A. Stuckey, J. Wlodarczyk, K. Gard'ner and A. Humberstone, Safety and Efficacy of a Testosterone Metered-dose Transdermal Spray for Treating Decreased Sexual Satisfaction in Premenopausal Women: A Randomized Trial, *Ann. Intern. Med.*, 2008, **148**(8), 569–577.

22. A. Tuiten, J. Van Honk, H. Koppeschaar, C. Bernaards, J. Thijssen and R. Verbaten, Time Course of Effects of Testosterone Administration on Sexual Arousal in Women, *Arch. Gen. Psychiatry*, 2000, **57**(2), 149–153, 155–156.

23. S. R. Davis, Androgen Therapy in Women, Beyond Libido, *Climacteric*, 2013, **16**(S1), 18–24.

24. W. Somboonporn, R. J. Bell and S. R. Davis, "Testosterone for Peri and Postmenopausal Women", in *Cochrane Database of Systematic Reviews*, John Wiley & Sons, Ltd, 1996.

25. S. R. Davis, 'Testosterone use in women', in *Testosterone*, Cambridge University Press, 4th edn, 2012.

26. A. W. Norman and H. L. Henry, *Hormones*, Academic Press, 3rd edn, 2014.

27. J. M. Berg, J. L. Tymoczko and L. Stryer, *Biochemistry: International Edition*, W. H. Freeman, New York, 7th edn, 2011.

28. C. Tanford, *Hydrophobic Effect: Formation of Micelles and Biological Membranes*, John Wiley & Sons Inc, New York, 1973.

29. L. Brown, "Clathrate", *The New Shorter Oxford English Dictionary on Historical Principles*, Oxford University Press, Oxford, 1993, vol. 1.

30. H. Lodish, A. Berk, C. A. Kaiser, M. Krieger, A. Bretscher, H. Ploegh, A. Amon and M. P. Scott, *Molecular Cell Biology: International Edition*, W. H. Freeman, 7th edn, 2012.

31. S. K. Putnam, S. Sato, J. V. Riolo and E. M. Hull, Effects of Testosterone Metabolites on Copulation, Medial Preoptic Dopamine, and NOS-immunoreactivity in Castrated Male Rats, *Horm. Behav.*, 2005, **47**(5), 513–522.

32. C. W. Hay, K. Watt, I. Hunter, D. N. Lavery, A. MacKenzie and I. J. McEwan, Negative Regulation of the Androgen Receptor Gene Through a Primate-Specific Androgen Response Element Present in the 5′ UTR, *Horm. Cancer*, 2014, **5**(5), 299–311.

33. O. Hiort, R. Werner and M. Zitzmann, "Pathophysiology of the Androgen Receptor", in *Testosterone*, Cambridge University Press, 4th edn, 2012.

34. W. H. Walker, Molecular Mechanisms of Testosterone Action in Spermatogenesis, *Steroids*, 2009, **74**(7), 602–607.

35. S. K. Putnam, J. Du, S. Sato and E. M. Hull, Testosterone Restoration of Copulatory Behavior Correlates with Medial Preoptic Dopamine Release in Castrated Male Rats, *Horm. Behav.*, 2001, **39**(3), 216–224.

36. B. Weinberger, D. L. Laskin, D. E. Heck and J. D. Laskin, The Toxicology of Inhaled Nitric Oxide, *Toxicol. Sci.*, 2001, **59**(1), 5–16.

37. Squire, *Fundamental Neuroscience*, Academic Press, Amsterdam; Boston, 4th edn, 2008.

38. E. R. Kandel, J. H. Schwartz, T. M. Jessell, S. A. Siegelbaum and A. J. Hudspeth, *Principles of Neural Science*, McGraw Hill, New York, 5th edn, 2013.

39. S. Pogun, M. H. Baumann and M. J. Kuhar, Nitric Oxide Inhibits [3H]Dopamine Uptake, *Brain Res.*, 1994, **641**(1), 83–91.

40. K.-E. Andersson, Mechanisms of Penile Erection and Basis for Pharmacological Treatment of Erectile Dysfunction, *Pharmacol. Rev.*, 2011, **63**(4), 811–859.

41. A. Argiolas, M. R. Melis and G. L. Gessa, Oxytocin: An Extremely Potent Inducer of Penile Erection and Yawning in Male Rats, *Eur. J. Pharmacol.*, 1986, **130**(3), 265–272.

42. F. S. Gragasin, E. D. Michelakis, A. Hogan, R. Moudgil, K. Hashimoto, X. Wu, S. Bonnet, A. Haromy and S. L. Archer, The Neurovascular Mechanism of Clitoral Erection: Nitric Oxide and cGMP-stimulated Activation of BKCa Channels, *FASEB J.*, 2004, **18**(12), 1382–1391.

43. M. Maggi and H. M. Behre, "Testosterone and Erection", in *Testosterone*, Cambridge University Press, 4th edn, 2012.
44. "Guanine", Wikipedia, the free encyclopedia. 08-Feb-2015.
45. "Animal Ingredients List", PETA.
46. C. Connell, "The Most Cringe Inducing Facial Ever: Made from Birds' Mess", *Mail Online*. Available at: http://www.dailymail.co.uk/femail/article-2641957/The-cringe-inducing-facial-The-good-news-beats-Botox-The-bad-news-birds-mess.html. [Accessed: 07-Apr-2015.]
47. P. Komalavilas and T. M. Lincoln, Phosphorylation of the Inositol 1,4,5-Trisphosphate Receptor Cyclic GMP-dependent Protein Kinase Mediates cAMP and cGMP Dependent Phosphorylation in the Intact Rat Aorta, *J. Biol. Chem.*, 1996, **271**(36), 21933–21938.

The Chemistry of Romantic Love

"I thought it was the worst pile of blubbery school-girl mush I've ever been compelled to endure."[1] So speaks Arnold Rimmer on the climax of a weepy movie, watched with his crewmates on the space ship Red Dwarf. In the UK sitcom, Rimmer's own death is just one of a long line of personal misfortunes that have distorted the vending machine repairman's youthful ambition into embittered cynicism. Resurrected as a hologram following his death in a radiation leak he is unable to experience any form of physical contact but sublimates his anguish in the form of petty, conniving cowardice. So, it is hardly surprising that he sneers at the romantic movie's male lead when he sacrifices his career for his paramour, even though they will not be together. That is until he is spirited onto a holoship, also the name of the episode,[2] which is like something out of Aldous Huxley's *Brave New World*. The arrogant crewmembers introduce themselves by name and IQ, and scorn romantic love as "a short-term hormonal distraction, which interferes with the pure pursuit of personal advancement."[1] Nevertheless, ship regulations require that they enjoy sexual congress at least twice a day to promote optimum health and they arrange their liaisons as casually as if chatting about the weekend. So, Rimmer finally gets to enjoy some long yearned

The Chemistry of Human Nature
By Tom Husband
© Tom Husband 2017
Published by the Royal Society of Chemistry, www.rsc.org

for physical contact and he does so with his attractive host Nirvanah Crane.

Rimmer is determined to join the ship's crew but can only do so by challenging one of the existing members to an intelligence test. He attempts to cheat his way on board by applying a mind patch, but it fails midway through the challenge. Crushed, he goes to admit defeat but is surprised to learn that his anonymous competitor has pulled out. He has won, but his fleeting success is destroyed by the discovery that his competitor was Nirvanah Crane. Her defeat means that she is effectively dead, as the ship's computer will run Rimmer's hologram in her place. Rimmer realises that she has fallen in love with him and explains this to the ship's captain as he resigns. Captain Platini cannot believe that Commander Crane would succumb to a "temporary hormone imbalance",[1] nor that Rimmer is willing to sacrifice his place on the holoship, especially when they will remain apart.

"You're wrong," replies Rimmer. "We won't be apart, we just ... won't be together." Finally, his old self returns as he disgustedly adds: "I cannot *believe* I just said that!"[1]

The idea of romantic love as a mere hormonal imbalance is a common theme in science fiction, but there are several points of interest in this plot. First of all, as we saw in Chapter 6, it is lust that surfs our surging hormones, yet sex is not an imbalance that the crew of the holoship shuns. Not only that, but, as we will see in Chapter 12, testosterone is also implicated in the elevation of status, such as would be entailed by the holoship crew's "pursuit of personal advancement." It might be more accurate to describe love as a biochemical imbalance because a raft of different substances, including hormones and neurotransmitters, are thought to pull our strings when we fall prey to the temporary insanity that has bewitched poets for millennia. But, of course, the term "imbalance" suggests a malfunction and, in evolutionary terms, love is anything but.

The intriguing point in the *Red Dwarf* episode is the idea that holograms could have biochemical imbalances. In the show, holograms are light-based replicas that think and feel as the deceased individuals for which they substitute. This projection is made by a computer program that has the person's every memory stored in a databank and, as such, holograms have access to all of their memories and react to events entirely as their real-life counterparts used to before they died. It is only make believe, of course, like the teleporter that beamed up Captain Kirk, but it highlights an interesting point with implications for artificial intelligence. How could romantic love be simulated? Semantics aside, romantic love is thought to result from increased activity of a particular brand of neurotransmitters called monoamines. So, for Rimmer to experience love, the hologram-beaming computer program would have to simulate the assault on our mindset that those bioactive molecules achieve.

The first biochemical perspective on love was put forward by a psychiatrist called Michael Liebowitz, who noticed many interesting parallels between mental health problems and romantic love. In a particularly illustrative case, there was a man who ran off with a woman as a rapidly burgeoning love affair took them to five different cities in as many days.[3] But their romance struck an iceberg in the form of the woman's sister, who coolly asked when she had stopped taking her bipolar medication. The unfortunate man had been swept away by a mania he mistook for blossoming love.

Dr Liebowitz is probably the first person to describe romantic love as an addictive state, noting the withdrawal-like symptoms that accompany its sudden ending. People in the first throes of romantic love think obsessively about their paramours and crave constantly to be with them. In his book *The Chemistry of Love*,[3] Liebowitz makes frequent comparisons between romantic euphoria and the high that accompanies amphetamine use. Takers of the drug also experience optimism during the high, as well as an urge to talk and be sociable. By contrast "speed come downs" are depressing withdrawal-like states that, Liebowitz argues, resemble break ups.

Building on these ideas, Liebowitz suggested that romantic love was composed of attraction and attachment. Attraction is the yearning component, whereas attachment is the pleasant feeling of being with your sweetheart. He also felt that attraction without attachment was *not* romantic love, as it doesn't discriminate between different partners.

This differs from Dr Helen Fisher's model considered in the introduction to this section, which includes lust, as well as attraction and attachment. Her addition of lust enables a valid distinction because, whilst we may lust after several people at once, it's rare to go Romeo-and-Juliet over two people at the same time. Moreover, the giddy thrill of blossoming romance is a different beast from the warm glow after years of marriage, which is how she characterises her version of the attachment phase.[4]

Liebowitz hypothesised that the chemical part of attraction was caused by stimulant neurotransmitters, such as noradrenaline, serotonin and dopamine, as well as another neurotransmitter called phenethylamine (PEA), which he termed "endogenous amphetamine".[3] These neurotransmitters are collectively known as the monoamines and are shown in Figure 7.1. Just as opioids may be produced inside the body or from poppies, so the stimulating effects of PEA closely resemble those of exogenous amphetamine. Exogenous means *from the outside*, while endogenous means *from the inside*. Similarly, Liebowitz proposed that endorphins cause the pleasurable sensation that promotes attachment.

7.1 MONOAMINES

Perhaps the most interesting thing about Liebowitz's research is his use of anti-depressants to help people with romantic problems. He felt that he was particularly able to help patients with a certain condition called hysteroid dysphoria. These unlucky souls are wont to fall in love with anyone. Their threshold for falling in love is too low, but their own mood soars and crashes with the climate of the relationship. They need constant stimulation and attention, making them struggle with the end of the honeymoon period. Often, they will leave their partners to seek novelty in new relationships. In other cases, the lack of novelty manifests as perceived rejection and they demand constant reassurance that their partner still loves them, frequently driving their partner away. Other patients tolerate woefully unreasonable abuse from their partners because they are so terrified of being alone. In many cases, Liebowitz was able to alleviate these problems by prescribing a kind of anti-depressant called a monoamine oxidase (MAO) inhibitor. This was a major step towards the contemporary biochemical model of romantic love.

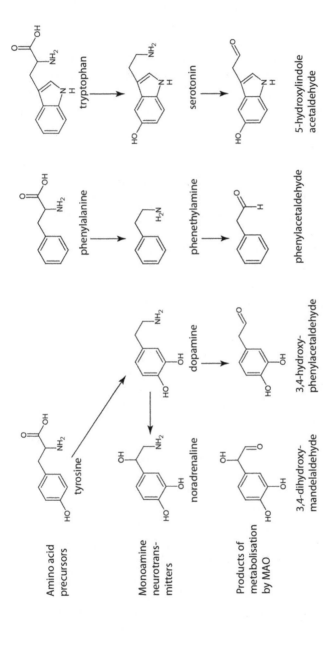

Figure 7.1 The amino acids tyrosine, tryptophan and phenylalanine are precursors for the monoamine neurotransmitters noradrenaline, dopamine, PEA and serotonin. All four are metabolised by monoamine oxidase (MAO) into aldehydes, which are then further metabolised into other products.

Liebowitz was a psychiatrist and his decision to prescribe MAO inhibitors was surely inspired by the fact that they were popular anti-depressants of the day. Unfortunately, side effects included sudden and massive spikes in blood pressure, which could cause stroke or even death. These were caused by a substance common in fermented foods called tyramine, meaning that blood pressure could be successfully regulated, providing patients were willing to give up tyramine-rich foods, including beer, wine and cheese (not a diet that I could stick to). Liebowitz's prediction that superior anti-depressants would replace MAO inhibitors fairly well pre-empted the rise to dominance of selective serotonin reuptake inhibitors (SSRIs), such as Prozac, which, in turn, confirms the complicity of serotonin in our sense of well-being. However, Dr Liebowitz maintains that no other variety of anti-depressants, including SSRIs, can replicate the effect of MAO inhibitors.

On the subject of serotonin, chocolate has more to show us about chemistry. Like dopamine, noradrenaline and PEA, serotonin is a monoamine. This family of compounds can be defined as aromatic biochemicals that are synthesised from aromatic amino acids (ref. 5, p. 1040). All of the compounds in Figure 7.1 share the same hexagon motif. The term "aromatic" refers to the delocalisation of electrons denoted by the alternating single and double bonds introduced in the context of dopamine in Chapter 3. Each of the transformations represented by the arrows in Figure 7.1 requires at least one enzyme. When we eat foods rich in tyrosine, such as cheese (the Greek word for which—*tyros*—inspired the amino acid's name), the body uses enzymes to convert the amino acid first into dopamine and then into noradrenaline (tyrosine is also the precursor for the blood-pressure-increasing tyramine). PEA is synthesised from phenylalanine, and serotonin is synthesised from tryptophan,[6] which is abundant in chocolate. Both serotonin and chocolate are linked with well-being. Is there more to the connection than coincidence?

There have been two significant theories about the effects of chocolate ingredients on brain chemistry, including one from Dr Liebowitz himself. He noted that chocolate was rich in PEA, the amphetamine doppelganger he had linked with love, and also that many of his heart-broken patients liked to guzzle the brunette confection. Was there a connection? No. A group of researchers ingesting copious quantities of the stuff succeeded only in giving themselves headaches, while failing to measure elevated levels of PEA in their urine.[3] In a similar vein, it has often been suggested that chocolate boosts serotonin levels—explaining its comforting appeal—because it is rich in tryptophan, the precursor from which our bodies manufacture the neurotransmitter. Ironically, chocolate contains *too much* protein to affect neural serotonin levels.[7] This may seem at odds with the fact that tyramine can increase blood pressure. The explanation for this is that tyramine acts *via* the interface between the gut and the *peripheral* nervous system, but our *central* nervous system is not so easily hijacked by nutrients in our diet. The brain uses its own stash of neurotransmitters to reward us for fuelling it with high energy foods, *including* chocolate, as we saw in Chapter 5, and this is the best explanation for the thrills of chocoholism.

7.1.1 Getting Rid of Monoamines

So much for how monoamines are made, but the body also needs a way to destroy them in order to maintain a suitable balance. In fact, this is true of most biochemicals. In Chapter 6 we saw how Viagra maintains erections not by *producing* the chemical that fuels them but by *preserving* it. Cyclic guanosine monophosphate (cGMP) helps relax the muscles that admit blood to the penis, but having laboured to produce the stuff the body immediately deploys the enzyme phosphodiesterase to get rid of it. In order to sustain the erection, the man's brain has to keep signalling for more cGMP to be produced. This, of course, has pros and cons, as anyone for whom floppiness has wrecked play will attest. On the plus side, it means that a man's first erection doesn't last the rest of his life.

When the body needs to dispose of monoamines, it recruits the enzyme MAO. The name indicates that the enzyme oxidises the monoamines, which, in this case, means that the amine (NH_2) groups of noradrenaline, dopamine, PEA and serotonin are switched for double bonded oxygen atoms, as shown in Figure 7.1. The aldehydes created in this first stage of metabolism are transformed into various other metabolites, which may be flushed from the body in our urine. Equally, the

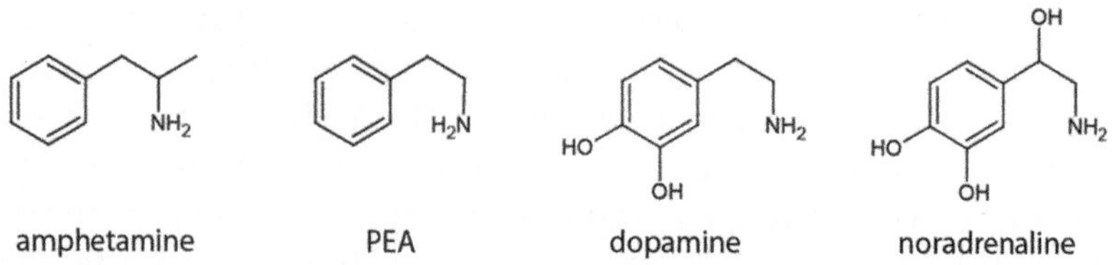

Figure 7.2   The relationship between amphetamine and selected monoamine neurotransmitters.

neurotransmitters may be transformed into each other, thanks to the similarity between their structures, as shown in Figure 7.2. The addition of two –OH groups transforms PEA into dopamine, and a third –OH produces noradrenaline. Note also the similarity to amphetamine. This was an important factor in Liebowitz's hypothesis.

Finally, MAO *inhibitors* do the same basic thing as Viagra— they inhibit enzymes from getting rid of useful substances. By stopping MAO from oxidising monoamines, stocks of serotonin, dopamine, PEA and noradrenaline are kept at higher levels, making them more available to trigger neurons. Since all four are implicated in the maintenance of mood, it follows that keeping them around may lift spirits.

This was just what Liebowitz found with some of his hysteroid dysphoric patients.[3] One benefit was that patients were no longer so desperate to fall in love that they grabbed the first unsuitable urchin that crossed their path. Alternatively, having found love, his patients were no longer racked by a sense of rejection simply because their partner wanted to spend an evening without them. In short, the intense craving for fireworks had been dampened to levels that fostered rather than sabotaged true love. Liebowitz was quick to note that his ideas merely formed a hypothesis. He first noticed parallels between amphetamine-induced euphoria and romantic love. Next, he allowed this fledgling theory to guide the prescription choices he made for those patients struggling with love. Finally, he took the success of MAO inhibitors with hysteroid dysphoric patients in support of his hypothesis that love is a kind of addiction. Evolutionarily, our species benefits when individuals are motivated to find love, but the point is made yet again that you can have too much of a good thing. Heightened sensitivity to romantic cravings can wreck a person's love life. At least that seems to be the case from Liebowitz's anecdotal evidence.

7.2 THE HYPOTHESIS DEVELOPS

Helen Fisher was the next person to write the book on love. She built on Michael Liebowitz's work, both by seeing if she could find incidental support from the work of other scientists and by conducting her own research. Her research began in earnest during the 1990s, a time when PEA had been written off as a mere supporting actor in the neural machinations of human nature. In fact, it was considered to be a secondary messenger, like the nitric oxide that we met in Chapter 6, meaning that it was held to adjust the settings of neurons rather than actually firing them. This conclusion was reached partly because PEA existed at much lower levels than its counterparts, dopamine, noradrenaline and serotonin,[8] and Fisher focused her attentions on the other monoamines.

One thing that seized her attention was the role of serotonin in obsessive-compulsive disorder (OCD). SSRIs, the anti-depressants mentioned earlier, extend the window in which serotonin molecules can activate the receptors of post-synaptic neurons. Furthermore, they have been used to treat OCD.

Dr Fisher wondered whether the same neurotransmitter might account for the obsessive notes of romantic love. She found support for the idea in a study that directly compared the two states.[9] There were three groups in the study: 20 people in love, 20 people with OCD and a control group of 20 members. Members of the love group were selected only if they had been together for less than six months and, curiously, if they had not yet had sex. The results were that both the OCD and the love groups had lower levels of serotonin in their bloodstream than the control group. Of course, it is very dicey to conclude that serotonin levels must be lower in the brain just because they are lower in the bloodstream, as Fisher herself noted. With an associate by the name of J. Anderson Thomson Jr, they went on to hypothesise that SSRIs might dampen romantic love. They found a collection of case studies in which subjects had either complained that SSRIs had cooled their passion, or had noted that coming off the pills had revived their amorous feelings.[10] At first, one participant claimed she had not experienced any side effects from the pills. When probed specifically about *sexual* side effects, she revealed that she had been having less sex, but she had chalked that up to feeling less in love with her boyfriend. Again, as SSRIs *increase* the opportunity for serotonin to activate post-synaptic neurons, the story does support the idea that lower serotonin levels characterise love. On the other hand, a collection of anecdotes may be compelling but is not real evidence at all.

The other obvious problem with the serotonin evidence is the clash with Liebowitz's theory. MAO inhibitors elevate the levels of all the monoamines, including serotonin. Fisher was arguing that love blossoms when the brain is stingy with serotonin, but Liebowitz had based his theory on a brand of medication that increases opportunities for the serotonin to tweak receptors. On the other hand, it was precisely the obsessive traits of his patients that the psychiatrist relieved with the MAO inhibitors. Perhaps his hysteroid dysphorics had low levels of serotonin even before they fell in love, plummeting to such depths when they were in love as to make their levels of obsession unsustainable.

Evidence supporting the role of noradrenaline came from animal studies. Various experiments had linked the neurotransmitter with sexual motivation and sexual arousal. When exposed to male urine, female prairie voles released the monoamine in their olfactory bulbs[11] (the part of the brain that processes information from nasal receptors). In another study female rats released noradrenaline during mating.[12] This lesser-known biochemical is linked with another trademark of romantic love, as we shall see later.

The main focus of Dr Fisher's development of the love hypothesis was our old friend dopamine. This should be no surprise. The idea that love is like an addictive state is gaining traction all the time[13,14] and we have already seen that dopamine is the *capo di tutti i capi* of addiction.

In Chapter 6 we saw that testosterone is thought indirectly to increase the flow of dopamine traffic *via* the secondary messenger nitric oxide. Incidentally, Dr Fisher had a friend who had recently started taking a kind of anti-depressant that elevates dopamine levels. Not only did the friend start feeling hornier, but she also started having regular multiple orgasms.[4]

7.3 DUPING DOPAMINE RECEPTORS

Other studies suggest that dopamine triggers more than mere lust. Like humans, the prairie vole is monogamous and scientists have been able to manipulate their affections using a drug with Shakespearean powers. In *A Midsummer Night's Dream*, the mischievous Puck spreads the juice of a magical flower onto the eyes of the fairy queen Titania and the mortal Lysander. Actually, Puck has mistaken Lysander for Demetrius, who is after Lysander's lover Hermia. Upon waking, Titania and Lysander fall in love with the first person they see, which, in Titania's case, is the horse-headed Bottom, while Lysander falls for Helena, Demetrius' jilted ex-lover. Chaos ensues until Puck uses more flower juice to make the right people fall in love. Not only have scientists pioneered a real-life version of the flower potion, they have developed another that switches love off, although, perhaps fortunately, these effects have only been demonstrated in voles.

The two kinds of drugs are called agonists and antagonists. Both work by mimicking ligands, the name given to any biochemical that binds to a receptor. In the vole studies, the ligand in question was dopamine. In fact, such drugs have far wider applications than turning voles promiscuous. The dopamine agonist apomorphine has previously been used to treat erectile dysfunction, morphine addiction, Parkinson's and, ineffectively, homosexuality.[15] Meanwhile, the dopamine antagonist haloperidol is a potent antipsychotic, used to treat conditions such as schizophrenia.

Both drugs work by duping receptors. Figure 7.3 shows how the structures of dopamine and apomorphine are quite similar so the agonist can bind to the dopamine receptor and cause it to change shape, just as dopamine would. Conversely, antagonists are saboteurs. Although the similarity in structure is not immediately obvious, haloperidol can also slip inside the dopamine receptor, but the bulky chain poking out of the nitrogen atom acts like a stick jammed between the spokes of a wheel.[16] Extending beyond the usual binding site,[17] it jams the receptor and prevents it from changing shape. In both cases, actual dopamine is blocked from entering the receptors, as shown in Figure 7.4, but while agonists activate the associated processes inside the cell, antagonists effectively take the receptor out of play. Agonists mirror the effects of their mimicked ligands, while antagonists block them.

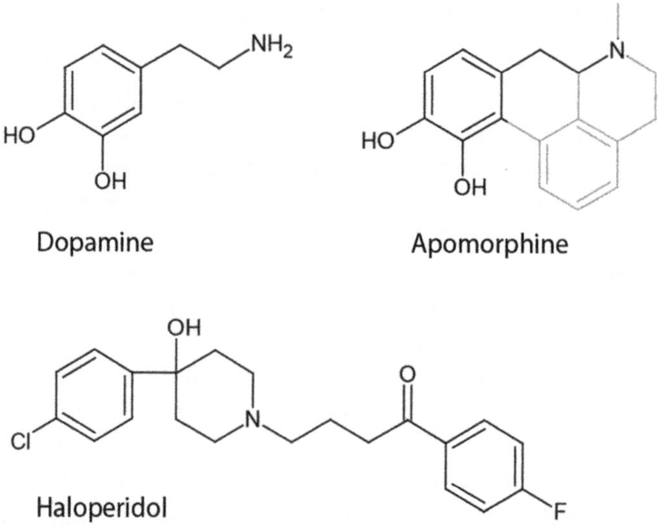

Figure 7.3 The structures of apomorphine and haloperidol sufficiently resemble that of dopamine to bind with the dopamine receptor. (The bold section of apomorphine emphasises the dopamine resemblance.)

In a study led by Brandon Aragona,[18] antagonists made the voles lose interest in their significant others, whilst agonists rekindled the flame. Both drugs were injected directly into the nucleus accumbens, the brain's pleasure centre that we met in Chapter 3. When antagonists were administered, the male voles continued mating, but showed no preference for the female they got jiggy with. Next, a group of sexually naïve voles were kept in cages with females for six hours but were stopped from mating. After the six hours, those that had been injected with agonists mated exclusively with these female companions, falling "in love" with the first mate they saw after taking the potion, just like the Shakespeare characters. Since the drugs in question were dopamine agonists and antagonists, the study strongly suggests that dopamine plays a key role in mate selection *for prairie voles*.

Is there no end to the abilities of dopamine? So far, it has addicted us to drugs, food and potentially other practices, like sex and even social media. In the medial preoptic area of the brain it seems to trigger lust in rats, while dopamine in the nucleus accumbens makes voles fall in love.

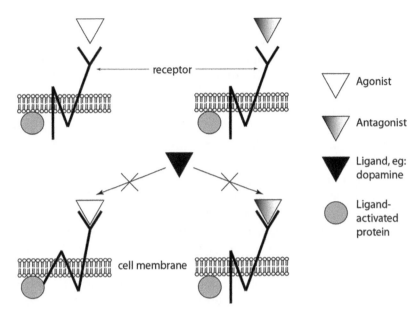

Figure 7.4 The action of agonists and antagonists. (Left) An agonist binds
with the receptor, causing it to change shape and hence activate
the protein. (Right) An antagonist binds with the receptor but does
not cause its shape to change. Both molecules prevent the receptor
from binding with its normal ligand.

7.4   WHAT ABOUT HUMANS?

Humans are not voles and "love" is probably not the best word to
describe the interactions between these burrowing rodents.
Having found further evidence in support of Liebowitz's
hypothesis, Fisher set out to find human data. Teaming up with
the neuroimaging expert Dr Arthur Aron and his colleagues,
she scanned the brains of 17 people who had recently fallen in
love. This was not the first such study, and there have been
others since, but all of these scans found the same thing.

Romantic love activates the brain's reward centre. Different
parts of the brain have lit up in several different studies, but
there are two that are almost always activated: the ventral tegmental area (VTA) and the caudate. Both of these are known to
play key roles in the brain's reward system as they are frequently
activated in response to rewards,[19] such as food, financial gain,
cocaine and alcohol.[13] Furthermore, these are areas in which
dopamine is known to have impressive clout. Consequently,

Fisher and Aron took their findings in support of the role of dopamine in romantic love not only because they had identified the involvement of a dopamine-rich area of the brain but because that part of the brain is associated with the addictive craving hypothesised to characterise love.[20]

This hypothesis provides another link to the suggested overlap with OCD. Neurons from the VTA are known to project to another brain region associated with reward called the orbitofrontal cortex (OFC),[21] part of whose job it is to keep up to date with what cues are still paying out.[22] For example, in a study called the Iowa gambling test, participants win money by drawing cards from four different decks. Two decks issue weak cards, while the others issue strong cards and the point is to figure out which are the strong decks. People with impaired function of the OFC struggle to make this deduction.[23] Damage to this region also seems to reduce the person's ability to defer gratitude as they will choose smaller short-term rewards over larger long-term rewards.[24] The OFC has also been shown to behave unusually in people with OCD, for example, devouring glucose much more rapidly than is common, suggesting over-activity of the region.[25] Could the blossoming of romantic love mimic that over-activity to foster the obsessive notes of human attraction identified by Fisher and Liebowitz?

Just how sure can we be that dopamine is tugging our heart strings? The most likely kind of brain scan to be referenced in popular culture is functional magnetic resonance imaging (fMRI), which is the variety used by Fisher's team. Participants are inserted into a chunky, claustrophobic cylinder, the kind in which news presenter Jon Snow freaked out in during a televised study into the effects of cannabis.[26] The information such scans can provide is a strong suggestion for the part of the brain that is active at particular times. So, if someone is given a spatial reasoning task, such as mentally rotating a three-dimensional image, the scan is *likely* to highlight the part of the brain responsible for spatial reasoning.

fMRI uses quantum mechanics to monitor the rate at which cells are metabolising oxygen. Cells need oxygen to generate energy and they get it from a protein in red blood cells, more technically known as haemoglobin. When haemoglobin is carrying oxygen, it is described as oxygenated. Once it hands over the oxygen to a cell, it becomes deoxygenated. Since fMRI is able to detect the proportion of oxygenated to deoxygenated haemo-globin, researchers can compare the rates at which different parts of the brain are using oxygen. The more oxygen a region consumes, the more important it is deemed to be to the cognitive process under investigation.

fMRI is an aptly named process because magnetism plays three different roles. The chunkiness of an fMRI scanner is down to a colossal electromagnet, which generates a magnetic field of 1.5 to 3 Teslas, more than 30 000 times the strength of the Earth's magnetic field. Unpaired protons in hydrogen atoms create mini-magnetic fields as they spin on the spot. Electrons, neutrons and protons all spin on the spot, and the movement of electrical charge is what generates the mini-magnetic field. Pairs of particles will cancel out each other's magnetic field, which is why MRI requires nuclei with an odd total number of protons and neutrons. In the same way, unpaired electrons also generate magnetic fields. The crucial ingredient for fMRI is the difference between oxygenated and deoxygenated haemoglobin. The former does *not* have un-paired electrons, whereas the latter does. As a result, deoxygenated haemoglobin is described as *paramagnetic*. The chunky electro-magnet provides a constant magnetic field, but powerful sensors can detect minute fluctuations in the total field strength caused by the hydrogen nuclei, as well as the deoxyhaemoglobin electrons. The scanners can detect when paramagnetic deoxyhaemoglobin is flushed out by oxygenated haemoglobin, as it brings fresh oxygen supplies to the cells that need them.[27]

How can you be sure that a scanned brain is occupied with being in love, rather than some other exploit? Our brains always have some scheme on the go, so researchers use contrasting stimuli to firm up their findings. Fisher's team directed their participants to gaze at photos of their romantic partners, then, as the scan continued, they asked participants to count backwards in increments of seven from a given number.[28] By identifying the brain regions that were *only* active while participants looked at the photos, the team established the neural correlates of love.

7.5 WHICH NEUROTRANSMITTER?

A brain scanner can tell you which part of the brain is devouring oxygen, but it cannot tell you the neurotransmitters being released. Dr Fisher's team were especially struck by the clear role of the VTA, described as *dopamine rich*.[20] We know this region makes frequent use of the neurotransmitter, but even single neurons release more than one variety of neurotransmitter,[29] let alone entire brain regions. The study strongly suggests that dopamine is involved, but it does not confirm it.

Studies in better-funded fields back up their fMRI findings with data from other instruments. One such method is positron emission tomography (PET), actually discovered before fMRI. Unfortunately, PET is cash ravenous thanks to its dependence on a nearby cyclotron. An early variety of particle accelerator, the cyclotron is used to pelt a suitable element with high-speed protons. The long tubes and powerful magnetic fields required to do this make cyclotrons expensive, likely explaining the underuse of PET in studies of romantic love.

The purpose of the cyclotron is to make a radioactive tracer by using a form of modern-day alchemy. The dream of alchemists was to convert cheap metals like lead into gold and modern-day science has developed the means to do this. Elements are defined by the number of protons contained in each atom. Any atom can be classified as lead if it contains 82 protons, while atoms of gold are obliged to contain 79 protons. Sadly, transmuting lead into gold in a particle accelerator turns out to be woefully unprofitable, but the same science can be used to make radioactive tracers. For example, if boron-11 is pelted with high-speed protons, the element is transformed into carbon-11.[30] Boron has five protons, while carbon has six. As the high-speed proton is absorbed into the transforming nucleus, a neutron is knocked out, producing atoms with the same mass but a different atomic number. The resulting carbon-11 is radioactive, with a half-life of roughly 20 minutes, meaning that researchers must race to use the radionuclide before it decays beyond use.

The next stage is to incorporate the carbon-11 into some form of drug that can be administered to research subjects. These radionuclides are frequently incorporated into antagonists such as those we met above. Carbon-11 is a great choice because virtually all drugs contain carbon and it is commonly used to produce a dopamine antagonist called raclopride. This freshly minted radiotracer is then administered to the patient. Before long, the unstable carbon-11 nucleus will decay and, when it does, it produces a neutron and a positron (a positively charged electron and an example of the famous antimatter). This positron flies off at the speed of light (ref. 5, p. 428) before colliding with an electron. When this happens, two beams of gamma rays fly off in opposite directions and it is the *simultaneous* detection of such rays by which the PET scanner determines the location of the radiolabelled antagonist.[5] Basically, a PET scanner works by causing antimatter explosions inside your brain.

Using some clever maths, this tool can even show the proportion of antagonists that are bound to receptors.[31] How is that useful? It basically indicates the availability of receptors, suggesting how busy those receptors are with their usual ligands. Back in the realms of dopamine and food, a PET study was carried out using raclopride.[32] The scan showed that fewer molecules of raclopride were bound after a meal than when the participants were fasting. Earlier, we saw how antagonists can block neurotransmitters, but the same thing can happen in reverse, as shown in Figure 7.5. What the findings suggest is that the meal triggered the release of dopamine, which consequently hogged the dopamine receptors. When the raclopride molecules turned up, all the receptors were busy, so the PET scan showed that few of them were able to bind receptors.

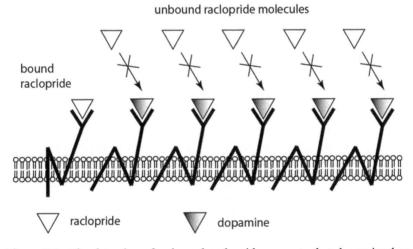

Figure 7.5 The detection of unbound raclopride suggests that dopamine has already bound the corresponding receptors.

PET could be used in conjunction with fMRI in order to confirm which neurotransmitters make us fall in love. Radiotracer-implanted raclopride could be used to gauge the role of dopamine, while other antagonists would have to be doctored to test the involvement of serotonin and noradrenaline. The hugely prohibitive costs of the research make it unlikely, especially as research into love is less of a priority than the prevention of neurological diseases like Parkinson's, Alzheimer's and so on.

7.6   WHY DOES MY HEART GO BOOM?

If PET is prohibitively expensive then what other options are available? Is there some other aspect of being in love that fits with our existing knowledge of neuroscience? The words *emotion* and *feeling* are often used interchangeably, but they are not the same. One way to distinguish them is as follows: an emotion is the physiological changes that happen in your body, while feelings are the mental processes that accompany these changes.[5] Romantic love is not just a giddy feeling, it's the sweaty palms, faster beating heart and the enhanced ability to cope without sleep.[20]

What is it that makes the heart beat faster? When it comes to the accelerated rhythms of our pounding hearts, noradrenaline strikes again. We have seen how noradrenaline is implicated in sexual activity in rodents. As a monoamine, its levels would have been affected by the MAO inhibitors prescribed by Liebowitz. But, in love, it seems that noradrenaline acts beyond the borders of the brain, venturing into the peripheral nervous system to accelerate the heart rate. What's really interesting is that love may tap into a primeval system adapted to protect us from harm: the *fight-or-flight* response.

This adaptation is often referred to as the adrenaline rush, a simplification as adrenaline is not the only biochemical involved, but it is a useful moment to consider its relationship with noradrenaline. Figure 7.6 shows how closely related the biochemicals are. The prefix *nor-* means demethylated because, where adrenaline has a methyl- group (CH_3), noradrenaline has merely a hydrogen atom.

But what is the adrenaline rush or, more accurately, the *fight-or-flight* response? This adaptation enables the body to release a

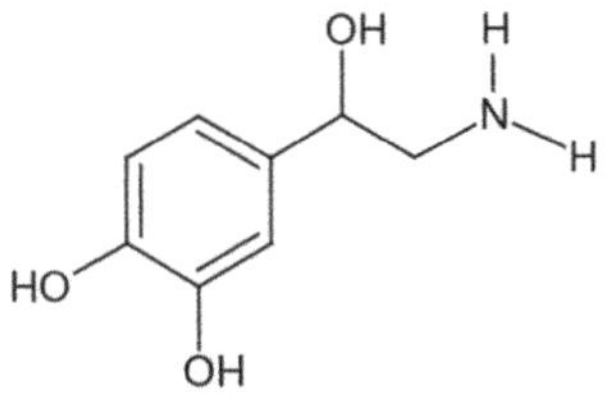

UK: Noradrenaline
US: Norepinephrine

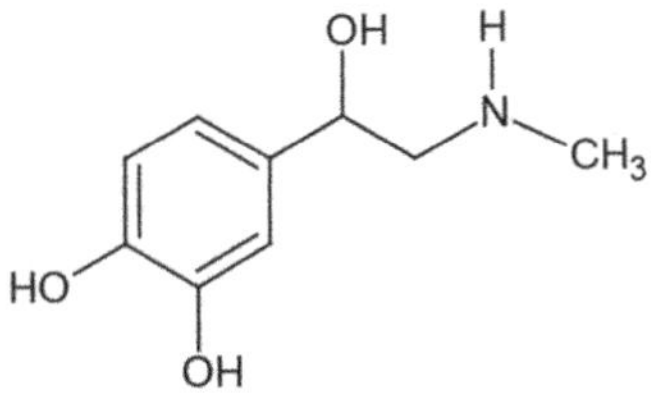

Adrenaline
Epinephrine

Figure 7.6 The structures of adrenaline and noradrenaline (note that the US terms are epinephrine and norepinephrine, respectively).

sudden burst of energy when a fright demands fight or flight. When a hungry predator or a marauding rival clan ambushed our ancestors, an arsenal of biochemicals would prime the body to take rapid and energetic action. Adrenaline, released from the adrenal glands of the kidney, would constrict arteries causing increased blood pressure and also redirecting blood from non-vital processes, such as digestion, to those organs that would be needed for fleeing or fighting—namely, muscles. In addition, it would jettison fuels from storage, releasing glucose from the liver and fatty acids from fat cells. The response also promotes sweating to maintain body temperature during this period of sudden physical exertion. Meanwhile, noradrenaline, acting *via* the peripheral nervous system, makes the heart beat harder and faster so that blood is pumped more rapidly to the necessary organs.

What role could the fight-or-flight response play in love? Even an unusually difficult in-law would be unlikely to elicit such an intense response, but it does seem that the body activates certain components of the response when we fall for someone. While fighting or fleeing a partner must surely indicate malfunction, the fright aspect is easier to accept. From experience, many of us will vouch that love can be terrifying. The stability of a long-term relationship is hard won after months of fretting. And as stability crystallises, the giddy thrill of romance dissipates; the excitement dies with the fear.

If fear is a crucial ingredient of love, perhaps we should be unsurprised that one part of the brain seems to juggle both emotions, as technically speaking it is two parts. Only the left-hand amygdala can be seen in Figure 7.7, but we all have a pair

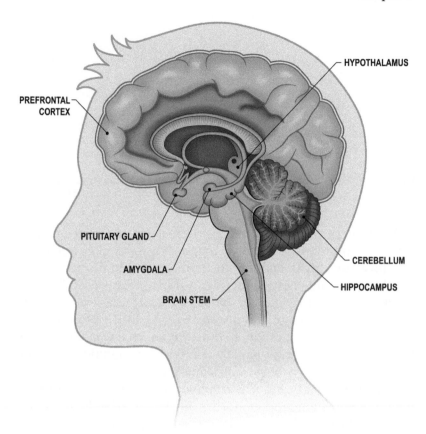

Figure 7.7   The hypothalamus, pituitary gland and amygdalae play a part in the fight-or-flight response.
(Image from iStock, artwork © James Kopp).

of the almond-shaped nuclei, one in each temporal lobe. The diagram also shows the hypothalamus and the pituitary gland, which work alongside the amygdalae to initiate the fight-or-flight response *via* the adrenal glands and the sympathetic nervous system.[5] Activity in the amygdalae was indicated by an fMRI scan during a study into romantic love.

The study investigated the pain-killing effects of love.[33] The research, led by Jarred Younger and including Helen Fisher's associate Arthur Aron, scanned participants while pressing heat blocks into their left hands. Each participant compared three activities for pain relief: looking at a photo of their partner; looking at a photo of a close friend or family member; or a distraction task (similar to the counting down task of Fisher and Aron's earlier study). Gazing at their lover and the distraction task proved to be the most effective forms of pain relief, but only while looking at the photo of their partner did participants show activation of their amygdalae during the fMRI scan. This suggests that the amygdalae may play a role in romantic love. Furthermore, researchers Tobias Esch and George Stefano have suggested that activation of the amygdalae could account for the increased heart rate that thumps the beat to the first notes of love.[34]

Whether or not romantic love can be considered to trigger the fight-or-flight response is a matter of contention. In fact, the body is able to activate individual components of the response separately,[5] so it is more likely that love activates specific pathways that happen to form part of the danger response. Nevertheless, the fabled adrenaline rush is a good match for the breath-taking delirium of burgeoning romance; increased energy, sweaty palms and a thudding heart are just a few symptoms common to both states. These can all be effected by the sympathetic nervous system,[20] part of the automated mental apparatus that takes care of the things over which we have little or no conscious control, such as maintaining a constant temperature of 37 °C. The likelihood is that both love and the fight-or-flight response use the same machinery in slightly different ways.

7.7 ANOTHER ROLE FOR NORADRENALINE

So, how do our lovers make our hearts race? Noradrenaline causes the heart to beat harder and faster by triggering the production of cyclic adenosine monophosphate (cAMP). The challenge is to achieve this in thousands of different cells in no more than an instant so as to get the sudden rush of energy associated with the "adrenaline rush". This is achieved by an elegant signal cascade in which a single molecule of noradrenaline can generate billions of molecules of cAMP across the heart within moments.

The first step is for noradrenaline to bind a G-protein coupled receptor. We have already met this variety of receptor in Chapter 2 in the form of the sugar-detecting protein that straddles the membrane of the taste cell. As the name indicates, it is coupled with a G-protein, which is lodged into the membrane just next to the receptor. It is called a G-protein because it interacts with guanosine triphosphate (GTP), the energy-providing molecule we met in Chapter 6. Notice that both *adenosine* triphosphate (ATP) and *guanosine* triphosphate play roles in this process. In Chapter 2 we saw how ATP fuels muscle contraction when it is hydrolysed to adenosine *di*phosphate (ADP). Meanwhile, in Chapter 6 we saw how the conversion of GTP into cGMP plays a pivotal role in the process that enables erections of the penis or clitoris. In this signal cascade the two swap roles and, as shown in Figure 7.8, this time GTP is hydrolysed into guanosine *di*phosphate (GDP), whilst ATP is converted into cAMP.

The chemistry between the G-protein and GTP is what gives this elegant process its pizazz. We have already seen how ATP provides energy by lopping off one of its three phosphate groups to produce ADP. Not only does GTP share this ability, but the hydrolysis acts as a timer in the noradrenaline signal cascade, dictating the window of time available for the production of cAMP. This is achieved through its interaction with the G-protein composed of three different subunits denoted alpha α, beta β and gamma γ (in other words, A, B and C).[†] It is the G-protein that hydrolyses the GTP, but this provokes a bout of shapeshifting that makes the whole signal cascade possible. The overall signal cascade is shown in Figure 7.9.[35] It starts with a single molecule of noradrenaline binding to the receptor, transforming it into the right shape to bind to the G-protein coupled receptor. Notice that the G-protein is already bound to GDP, the product of a previous cycle of hydrolysis. Binding the receptor prompts the G-protein to spit out the GDP, only for a molecule of GTP to slip into its still warm grave. This causes the α-subunit to change shape and hence to detach from the β and γ subunits. In its new conformation it is now the perfect shape to bind with adenylyl cyclase, which has an analogous function to the similarly named guanylyl cyclase from Chapter 6. Adenylyl cyclase is the enzyme that converts ATP into cAMP. At the same time, the α-subunit hydrolyses the GTP back into GDP. Once the hydrolysis is complete, the α-subunit resumes its original shape, causing it to disengage from—and hence deactivate—the adenylyl cyclase, and enabling it to reunite with the β and γ subunits.

[†]Before the Greek community sentences me to a horrific Sisyphean eternity, I should clarify that c and γ are equivalent in terms of alphabetical position rather than sound.

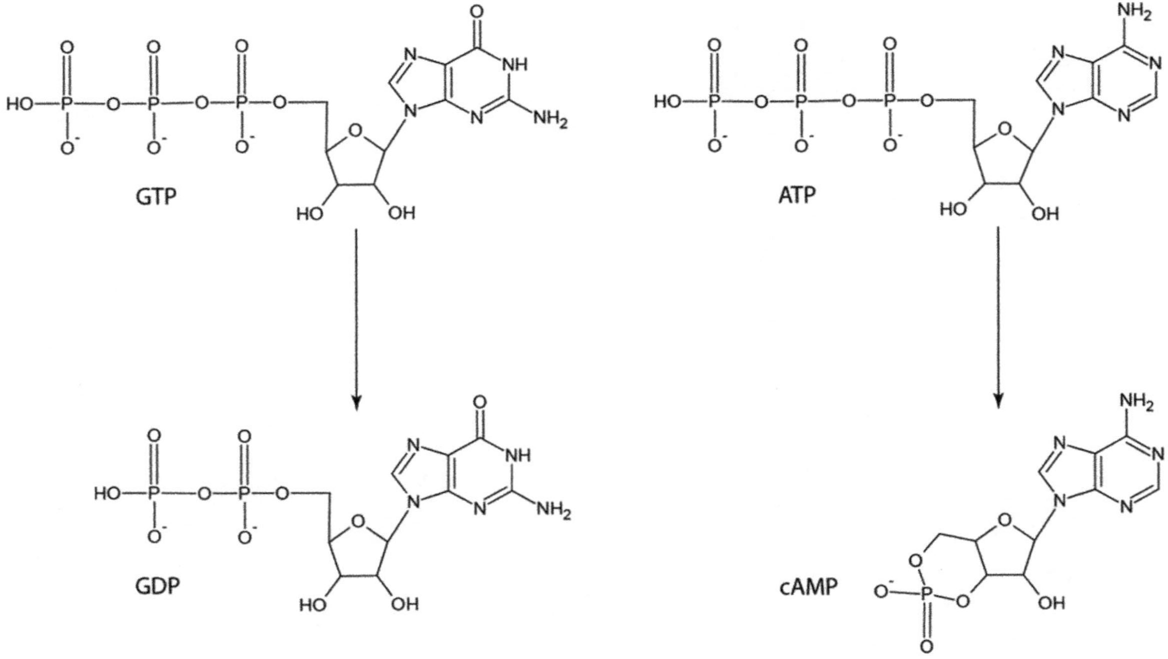

Figure 7.8 GTP is converted into GDP, while ATP is converted to cAMP.

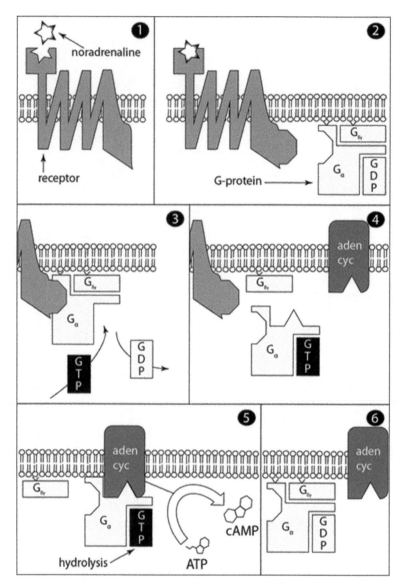

Figure 7.9   Noradrenaline signal cascade in the heart. (1) Noradrenaline binds
the receptor..., (2) ...causing it to change into a shape suitable to
bind G-protein. (3) Binding the receptor causes the G-protein to
eject GDP and bind GTP. (4) Binding GTP causes the α-subunit of
the G-protein (G$_\alpha$) to change shape and hence detach from the βγ-
subunits (G$_{\beta\gamma}$). (5) The α-subunit binds adenylyl cyclase (aden cyc),
which converts ATP into cAMP. Simultaneously, the α-subunit
hydrolyses GTP back to GDP. (6) When hydrolysis is complete,
the subunits recombine and adenylyl cyclase is deactivated.

So, the signal cascades as follows: a single molecule of noradrenaline tweaks the receptor, toggling the α-subunit from its GDP-bound to its GTP-bound state. The α-subunit is a slow mover compared to the adenylyl cyclase. In the same time it takes to hydrolyse its one molecule of GTP, the enzyme can convert myriad molecules of ATP into cAMP. But once the hydrolysis is complete, the timer alarm rings and the conversion comes promptly to a stop. In that same window of time the noradrenaline has already rushed off and triggered the same process in thousands of other cells.

But how does all that cAMP rev up the heart? This is where the above-mentioned role reversal takes place. In Chapter 6 we saw how the cGMP effectively "locks the doors" of the cellular lumen, preventing the release of the Ca^{2+} that would otherwise cause the erection-regulating muscles to contract. This is achieved through its interaction with protein kinase G (PKG), which phosphorylates the relevant ion channels, *reducing* the rate at which Ca^{2+} is released.[36] In this case, the opposite occurs; cAMP *unlocks* calcium ion channels, allowing Ca^{2+} ions to gush in from outside the cell. Appropriately, it does this by interacting with protein kinase A (PKA), and both kinases phosphorylate L-type calcium channels but with contrary effects,[37] as shown in Figure 7.10. So, the PKA triggers the influx of Ca^{2+}, which causes the cardiac cell to contract. This is how noradrenaline makes your heart go *boom*!

What can we say for certain about love? Liebowitz originated the hypothesis that the monoamines whip up the emotion of love. Helen Fisher gathered further support for the hypothesis, while adding lust to her predecessor's model of attraction and attachment. She also contributed to a growing evidence base that seems to implicate the brain's reward system, providing further credence to Liebowitz's suggestion that love is an addiction-like state. Further research is needed to confirm Fisher's hypothesis that love is characterised by elevated levels of noradrenaline and dopamine but reduced levels of serotonin. The science lies somewhere between a sturdy hypothesis and a fledgling theory. Although, the strongest part of the hypothesis is perhaps the notion that dopamine drives the craving to be near our sweethearts in those early days of romance. While fMRI scans can only suggest which areas of the brain are active, this data can be interpreted with the more comprehensive research described in

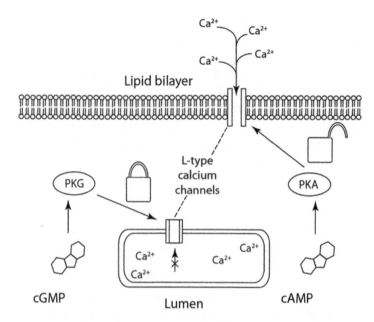

Figure 7.10 The contrasting effects of cGMP and cAMP. cGMP reduces the rate at which Ca^{2+} is released from the lumen, while cAMP accelerates its flow into the cell.

Chapter 3. PET scans and other instruments strongly support the role of dopamine in guiding us towards food. Now there is a growing consensus that romantic love activates the same areas of the brain.

In evolutionary terms, this makes perfect sense. The self-replicating molecules that brought life into existence need fuel and building supplies in order to keep replicating, and dopamine is emerging as a key player. Not only does it guide us towards energy- and protein-rich foods, it also pushes us to create a playground in which for the self-replicators to go on replicating after our own bodies expire. As we are about to find out, romantic love is the ideal way to bait us into raising the neediest of all offspring—*Homo sapiens* junior.

REFERENCES

1. With thanks to Grant Naylor Productions for permitting reproduction of these quotes, excerpted from the Red Dwarf series 5 "Holoship" episode. Grant Naylor Productions, 1992, written by Rob Grant and Doug Naylor.

2. J. May, "Red Dwarf", 20-Feb-1992.

3. M. Liebowitz, *Chemistry of Love*, Berkley, New York, 1984.

4. H. Fisher, *Why We Love: The Nature and Chemistry of Romantic Love*, Holt Paperbacks, New York, Reprint edn, 2004.

5. E. R. Kandel, J. H. Schwartz, T. M. Jessell, S. A. Siegelbaum, and A. J. Hudspeth, *Principles of Neural Science*, McGraw Hill, New York, 5th edn, 2013.

6. D. Voet and J. G. Voet, *Biochemistry*, Wiley, 4th edn, 2011.

7. G. Parker, I. Parker and H. Brotchie, Mood State Effects of Chocolate, *J. Affective Disord.*, 2006, **92**(2), 149–159.

8. D. Narang, S. Tomlinson, A. Holt, D. D. Mousseau and G. B. Baker, Trace Amines and their Relevance to Psychiatry and Neurology: A Brief Overview, *Bull. Clin. Psychopharmacol.*, 2011, **21**, 73–79.

9. D. Marazziti, H. S. Akiskal, A. Rossi and G. B. Cassano, Alteration of the Platelet Serotonin Transporter in Romantic Love, *Psychol. Med.*, 1999, **29**(3), 741–745.

10. H. E. Fisher and J. A. Thomson, Jr., "Lust, Romance, Attachment: Do the Side Effects of Serotonin-Enhancing Antidepressants Jeopardize Romantic Love, Marriage, and Fertility", in *Evolutionary Cognitive Neuroscience*, The MIT Press, Cambridge, 2006.

11. D. E. Dluzen, V. D. Ramirez, C. S. Carter and L. L. Getz, Male vole Urine Changes Luteinizing Hormone-releasing Hormone and Norepinephrine in Female Olfactory Bulb, *Science*, 1981, **212**(4494), 573–575.

12. A. M. Etgen and J. C. Morales, Somatosensory Stimuli Evoke Norepinephrine Release in the Anterior Ventromedial Hypothalamus of Sexually Receptive Female Rats, *J. Neuroendocrinol.*, 2002, **14**(3), 213–218.

13. B. Acevedo and A. Aron, "Romantic Love, Pair-bonding, and the Dopaminergic Reward System", *Nat. Dev. Soc. Connect. Brain Group Wash. DC Am. Psychol. Assoc.*, 2014.

14. M. Reynaud, L. Karila, L. Blecha and A. Benyamina, Is Love Passion an Addictive Disorder?, *Am. J. Drug Alcohol Abuse*, 2010, **36**(5), 261–267.

15. G. Smith, A. Bartlett and M. King, Treatments of Homosexuality in Britain Since the 1950s—An Oral History: The Experience of Patients, *BMJ*, 2004, **328**(7437), 427.

16. M. Y. S. Kalani, N. Vaidehi, S. E. Hall, R. J. Trabanino, P. L. Freddolino, M. A. Kalani, W. B. Floriano, V. W. T. Kam and W. A. Goddard, The Predicted 3D Structure of the Human D2 Dopamine Receptor and the Binding Site and Binding Affinities for Agonists and Antagonists, *Proc. Natl. Acad. Sci. U. S. A.*, 2004, **101**(11), 3815–3820.

17. J. M. Berg, J. L. Tymoczko and L. Stryer, *Biochemistry: International Edition*, W. H. Freeman, New York, 7th edn, 2011.

18. B. J. Aragona, Y. Liu, J. T. Curtis, F. K. Stephan and Z. Wang, A Critical Role for Nucleus Accumbens Dopamine in Partner-preference Formation in Male Prairie Voles, *J. Neurosci.*, 2003, **23**(8), 3483–3490.

19. H. A. Harsay, M. X. Cohen, N. N. Oosterhof, B. U. Forstmann, R. B. Mars and K. R. Ridderinkhof, Functional Connectivity of the Striatum Links Motivation to Action Control in Humans, *J. Neurosci.*, 2011, **31**(29), 10701–10711.

20. H. Fisher, A. Aron and L. L. Brown, "Romantic Love: An fMRI Study of a Neural Mechanism for Mate Choice, *J. Comp. Neurol.*, 2005, **493**(1), 58–62.

21. R. D. Oades and G. M. Halliday, Ventral Tegmental (A10) System: Neurobiology. 1. Anatomy and Connectivity, *Brain Res. Rev.*, 1987, **12**(2), 117–165.

22. Y. K. Takahashi, M. R. Roesch, T. A. Stalnaker, R. Z. Haney, D. J. Calu, A. R. Taylor, K. A. Burke and G. Schoenbaum, The Orbitofrontal Cortex and Ventral Tegmental Area are Necessary for Learning from Unexpected Outcomes, *Neuron*, 2009, **62**(2), 269–280.

23. A. Bechara, A. R. Damasio, H. Damasio and S. W. Anderson, Insensitivity to Future Consequences Following Damage to Human Prefrontal Cortex, *Cognition*, 1994, **50**(1–3), 7–15.

24. N. D. Volkow, G.-J. Wang, J. S. Fowler and D. Tomasi, Addiction Circuitry in the Human Brain, *Annu. Rev. Pharmacol. Toxicol.*, 2012, **52**, 321.

25. L. Menzies, S. R. Chamberlain, A. R. Laird, S. M. Thelen, B. J. Sahakian and E. T. Bullmore, Integrating Evidence from Neuroimaging and Neuropsychological Studies of Obsessive-compulsive Disorder: The Orbitofronto-Striatal Model Revisited, *Neurosci. Biobehav. Rev.*, 2008, **32**(3), 525–549.

26. V. Curran, "Jon Snow's Negative Experience on Skunk is Very Understandable", *The Guardian*. Available at: http://www.the-guardian.com/science/blog/2015/feb/23/jon-snows-on-skunk-the-cannabis-trial-channel-4. [Accessed: 23-May-2015.]

27. R. Cabeza and A. Kingstone, *Handbook of Functional Neuro-imaging of Cognition*, MIT Press, Cambridge, Mass, 2nd revised edn, 2006.

28. A. Aron, H. Fisher, D. J. Mashek, G. Strong, H. Li and L. L. Brown, Reward, Motivation, and Emotion Systems Associated with Early-stage Intense Romantic Love, *J. Neurophysiol.*, 2005, **94**(1), 327–337.

29. Squire, *Fundamental Neuroscience*, Academic Press, Amsterdam; Boston, 4th edn, 2008.

30. International Atomic Energy Agency, *Cyclotron Produced Radionuclides: Principles and Practice*. 2009.

31. A. A. Lammertsma and S. P. Hume, Simplified Reference Tissue Model for PET Receptor Studies, *Neuroimage*, 1996, **4**(3), 153–158.

32. D. M. Small, M. Jones-Gotman and A. Dagher, Feeding-induced Dopamine Release in Dorsal Striatum Correlates with Meal Pleasantness Ratings in Healthy Human Volunteers, *NeuroImage*, 2003, **19**(4), 1709–1715.

33. J. Younger, A. Aron, S. Parke, N. Chatterjee and S. Mackey, Viewing Pictures of a Romantic Partner Reduces Experimental Pain: Involvement of Neural Reward Systems, *PLoS ONE*, 2010, **5**(10), e13309.

34. T. Esch and G. B. Stefano, The Neurobiology of Love, *Neuroendocrinol. Lett.*, 2005, **26**(3), 175–192.

35. H. Lodish, A. Berk, C. A. Kaiser, M. Krieger, A. Bretscher, H. Ploegh, A. Amon and M. P. Scott, *Molecular Cell Biology: International Edition*, W. H. Freeman, 7th edn, 2012.

36. R. E. Klabunde, *Cardiovascular Physiology Concepts*, Lippincott Williams and Wilkins, Philadelphia, 2005.

37. F. Schröder, G. Klein, B. Fiedler, M. Bastein, N. Schnasse, A. Hillmer, S. Ames, S. Gambaryan, H. Drexler, U. Walter, S. M. Lohmann and K. C. Wollert, Single L-type Ca(2+) Channel Regulation by cGMP-dependent Protein Kinase Type I in Adult Cardiomyocytes from PKG I Transgenic Mice, *Cardiovasc. Res.*, 2003, **60**(2), 268–277.

The Chemistry of Attachment

At 1.15 in the morning, on April 4th, 2013, the Maine state police arrested a burglar emerging from a camp site dining hall with $280 worth of stolen food.[1] This was not a one-off hit; they had captured a persistent offender who had committed more than a thousand burglaries over nearly three decades. Initially, the man refused to answer questions. When the interrogating officer softened her approach, he said he was reluctant to speak because he felt ashamed. There was almost certainly another reason why conversation was not forthcoming—he had only spoken one word to another person in the last 27 years: "Hi." The authorities had captured the legendary North Pond Hermit.

Christopher Thomas Knight vanished into the woods aged 20. He would later tell his biographer Michael Finkel[2] that he was never happy at school or in work. One day he simply drove a car into the middle of the woods and remained there until he was caught 27 years later. Knight says that he initially attempted to forage for food but was unable to survive on roadkill. Although stealing made him feel scared and guilty, it was the only way he could survive in total isolation, which was all he wanted. Although fierce winters drove him to prayer and, Finkel speculates, contemplation of suicide, his divorce from society made him content. He was happiest alone.

Following his incarceration, a mental health evaluation suggested Knight may have Asperger's syndrome. People with this autism spectrum disorder struggle with social interaction and non-verbal cues. One telling note of Finkel's biography is the moment when Knight explains his aversion to eye contact, saying the face broadcasts too much information. Assuming the diagnosis was correct, there is a strong possibility that Knight's brain had some kind of difficulty with a substance called oxytocin. Researchers believe that autism disorders may result from mutations in the gene that codes for the oxytocin receptor.[3] The hypothesis is hardly radical. Decades of research confirm that this neuropeptide plays a critical role in the usually social nature of humans and other species. This could explain Knight's aversion to human interaction.

What is so remarkable about Knight's story is that he found discomfort in something that most of us consider a comforting necessity—social attachment. This is broadly defined as the pleasure we take in association with other people and most especially our loved ones. Both Fisher and Liebowitz have a place for attachment in their theories of romantic love. In Liebowitz's model it is the pleasurable sensation of spending time with our partners and particularly during physical contact.[4] Fisher extends this idea to characterise the companionate love that endures after the flame of our romance has dwindled, when dizzy excitement mellows into contented security.[5] In *Captain Corelli's Mandolin*, author Louis de Bernières paints a beautiful distinction between two kinds of love.[6] Being *in love*—which any fool can manage—is like the eruption of a volcano. But the heroine's father counsels her to seek a deeper variety of love in which the lovers are like trees, their roots mingling underground until they have become so entwined that two trees have become one. In academic terms one interpretation would be that being *in love* is the attraction phase, while *love* is the attachment phase. In our brains the crackle of our neural circuitry quietens to a reassuring hum. The neurotransmitters that had us pining for our beloved return to normal levels and a new blend of biomolecules emerges to stoke the embers of the romantic afterglow.

First comes love, then comes marriage, then comes a little baby crying in the carriage. In the developed world, people are increasingly choosing to dishonour the playground ditty, but it remains true that the majority of women in the US and UK have had children by their forties.[7] Most women cannot ignore their biological clock and that's the way evolution adapted us. Romantic love has proven strikingly effective at perpetuating the ancient process of self-replication.

This is the purpose of mating drives. Just as evolution has equipped us with cellular machinery that drives us to eat and breathe, so powerful processes compel us to mate. Eating provides the energy and building supplies necessary for self-replication; reproduction creates the vehicles in which for it to take place. The breeds of self-replicator that still exist are the ones whose ancestors evolved the most effective means by which to compete for resources. Enough humans currently roam the globe to show what a boon romantic love has been for the self-replicators.

This does not explain why we have romantic love. It is a non-essential ingredient of the mating drive. Sea turtles do things very differently from us.[8] Inveterate loners, they only spend time together when they are born and when they are mating. Their courtship rituals are almost comically direct. The male might nibble the female's neck and if she does not swim off then he starts having sex with her. Next, the female finds a beach on which to bury the eggs, which she then deserts forever. When the eggs hatch, the offspring dig their way out and then dash *en masse* to the sea, dashing through a danger zone of hungry predators like soldiers going over the top in trench warfare. This is a chilling vision of childbirth for a social creature like *Homo sapiens*, but it gets results. Roughly a tenth of the young turtles make it to adulthood. An average female sea turtle might lay about 5000 eggs in her life, of which 500 would be likely to survive to adulthood. Even the most prolific human mother on record comes nowhere close to this figure. Mrs Vassilyev gave birth to sixteen pairs of twins, seven sets of triplets and four sets of quadruplets, all but two survived to adulthood.[9]

Turtles are winning one numbers game but losing another. Female for female, turtles outbreed *Homo sapiens* by hundreds to one, but we are trouncing them in the population game. Romantic love may have played a critical role in the evolution of this edge.

Helen Fisher's model of romantic love includes lust, attraction and attachment. She argues that the three can act independently or in tandem.[10] Lust is indiscriminate and may lead to reproduction in isolation from the other two mating drives. In pre-civilisation times, attraction evolved because it enabled individuals to focus their efforts on one mating partner and thereby conserve energy, suggests Fisher. Essentially, we are adapted for attachment because it fosters selflessness not only towards mating partners but also the offspring they produce.

This compassionate tendency may partly explain our success in the population game. Humans show significantly more compassion to their children than do turtles. Evolutionary theory suggests this post-natal care may have helped develop the adaptation that effectively removed us from the food chain: intelligence. Unlike other species, the brains of human children continue to grow after they are born, explaining how our cranial growth sidestepped the limiting confines of the birth canal. It is highly advantageous that our species has adapted to develop this way, but the trade-off is that the human infant is utterly dependent on its parents for much longer than any other species.[11] This is in stark contrast to the sea turtles who can expect precisely zero hours of parenting. They enjoy fleeting camaraderie with their siblings, running together to the sea before separating forever. So, while turtles produce more young, it is the care provided to *Homo sapiens* children after they are born that may have helped us to get so smart. The tragedy is how we sometimes use that intelligence; another important reason our population dwarfs that of turtles is because we hunt them for food, to make leather or even to stuff them as curios.[12]

It is ironic that the caring streak of parenting humanity produced an intelligence that can be so careless of others, especially considering that the evolution of *Homo sapiens'* bigger and better brain is strongly linked with the development of social intelligence.[13]

Why do sea turtles and humans do things differently? We have a common ancestor with the reptiles but our self-replicating genes diverged. Random glitches in the replication process would have produced mutations, leading to the formation of new proteins and new assortments of proteins. Those developments that enhanced the ability of the organism to bid for fuel and building materials would have been retained and the amalgamation of all such modifications is what we see in each species. Mutations are offered at random and evolution selects the ones that produce the best adaptations. Your species can only have a particular feature if your ancestors chanced to have the corresponding mutations.

A huge complication is that we share the spoils of many relevant mutations with the turtles. It seems that long before *Homo sapiens* branched off from his hominid ancestors, various mutations had created a family of biomolecules that drive social functions in all vertebrates and even some invertebrates.[14] This set of peptides is known as the oxytocin/vasopressin family. Certain fish employ isotocin and vasotocin, while amphibians, birds and reptiles adopt the latter in conjunction with mesotocin. Mammals get the title track, oxytocin and vasopressin, which are specific biomolecules that also christened the family. As reptiles, sea turtles express the versions called mesotocin and vasotocin, complicating the matter of how they evolved differently to humans.

While these peptides invariably promote social processes, the particular process in question varies by species. Vasotocin makes sparrows aggressively territorial, but administering the same peptide to zebra finches does not cause them to start angrily defending their property. But what use could solitary sea turtles have for social hormones? Although they generally eschew one another's company, once a year they meet to mate and this is the likely role for vasotocin. In fellow reptiles *Cnemidophorus inornatus*, commonly known as little striped whiptail lizards, vasotocin is released in the brain of the male when it mounts the female. Meanwhile, in *Homo sapiens* oxytocin has been linked with a variety of behaviours and emotions, including the feelings of compassion of parents for their children.

Oxytocin and vasopressin are very similar. Both act as hormones as well as neurotransmitters. Both are peptides and both are produced in the same seemingly inefficient way; the body makes a huge protein, then chops it up and throws most of it away. The tiny fragment left over will be either oxytocin or vasopressin, depending on which gene produced the protein.[15] The process is shown in Figure 8.1.

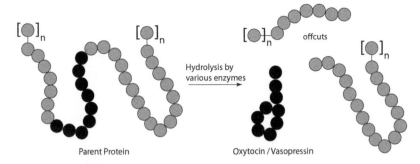

Figure 8.1 Synthesis of oxytocin and vasopressin begins with the production of a parent protein. The protein is cleaved by enzymes, producing vasopressin or oxytocin depending on which gene was translated. The square bracket notation indicates the presence of an unspecified number of additional amino acids (n) in the parent protein.

The peptides differ by just two amino acids. Ten different amino acids, displayed in Figure 8.2, can be used to make either vasopressin or oxytocin. Seven appear in both but where vasopressin contains arginine and phenylalanine, oxytocin has leucine and isoleucine. The striking similarity in their structures, shown in Figure 8.3, explains the observation that each may activate the other's receptors,[16] albeit with a weaker affinity.

They are also similar in terms of the jobs they do around the body. Both oxytocin and vasopressin have social as well as non-social functions. Vasopressin signals the kidney to reduce the flow or urine to the bladder. Likewise, oxytocin has various mechanical roles, such as signalling for a mother's breasts to release milk and causing the uterus to contract during childbirth. Both neuropeptides also have clear roles in social behaviour, helping individuals to forge bonds and then to stay bonded. These biomolecules enable us to love in the broadest sense. Not only do they drive the afterglow of romantic love for our sweethearts but also for our children, family and friends.

8.1 THE PUPPET MASTERS OF MAMMALIAN ATTACHMENT

In Chapter 7 we saw evidence that the monogamy of prairie voles seems to rely on dopamine in some way. Dopamine agonists, the drugs that mimic the effect of dopamine, cause the rodents to shack up with the first available mate, while antagonists, the

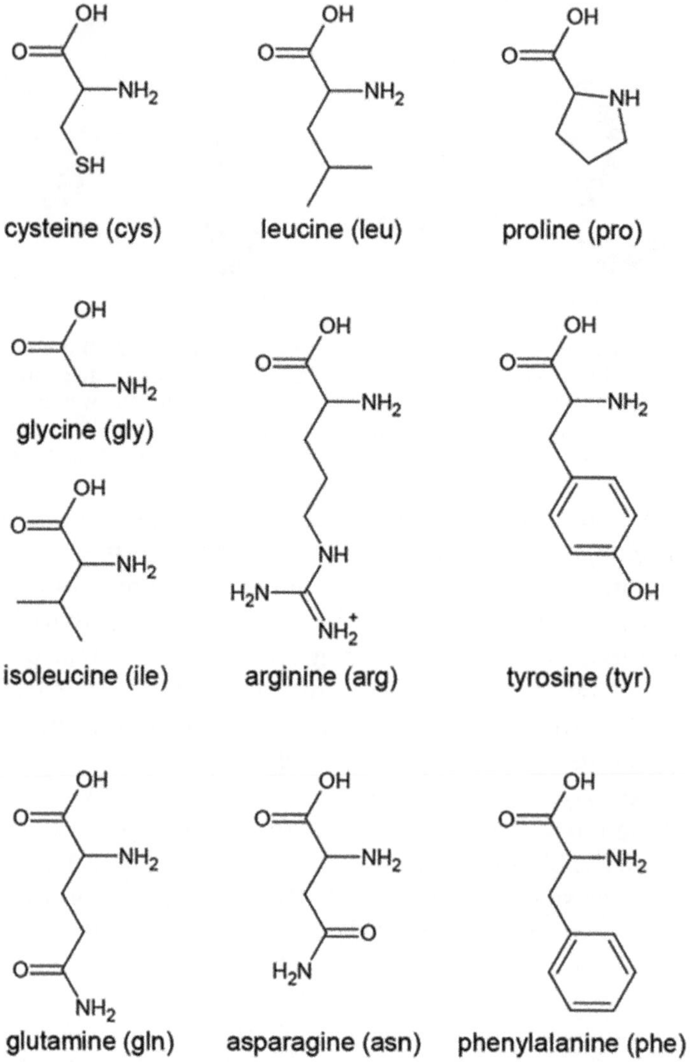

Figure 8.2   The ten amino acids used to construct vasopressin or oxytocin.

variety that blocks dopamine, turned them promiscuous. But dopamine is not the only neurotransmitter with these powers.

A very similar set of experiments suggests that oxytocin and vasopressin also encourage prairie voles to settle down with one significant other. After male prairie voles mate, they turn into perfect fathers and husbands. They stay close to the mother, lash out at male intruders who might be after their women and help

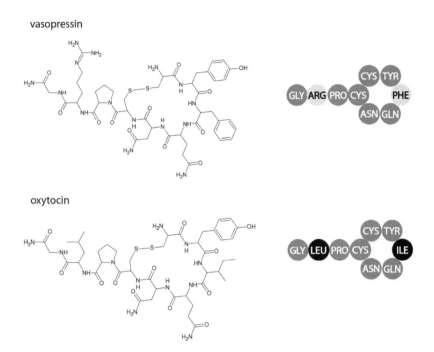

Figure 8.3 The structures of vasopressin and oxytocin.

to raise the children. A team of researchers led by James Winslow and Thomas Insel suspected the involvement of vasopressin, so they designed an experiment to switch off the effects of the neuropeptide.[17] First, they observed the transformation in voles after mating. Virgin voles would react with curiosity when intruders were placed in their cage, but that all changed after they had been intimate with a lady. Now the rodents would threaten and even viciously attack intruders and, after mating, they spent all of their time with their mate.

What the researchers did next was to measure the effect of an antagonist. Just as we saw in Chapter 7, vasopressin antagonists blocked the corresponding receptors without activating them. When these antagonists were injected into the voles their behaviour was completely different. Firstly, they no longer attacked intruders and, secondly, when they were given the option to enter the cage occupied by their mate or another cage with a different female in it, they spent the same amount of time with both voles. The team concluded that vasopressin played a clear role in pair formation.

Female voles are wired slightly differently. In their case, it was oxytocin that made them stick with their partner. In another experiment carried out by Insel, this time with Terrence Hulihan, injections of oxytocin encouraged the females to spend more time with their mate, while oxytocin antagonists made them less choosy about which males they spent time with.[18] The oxytocin injections were so effective that females even developed attachments to males they had not had sex with.

8.2   OXYTOCIN—THE CUDDLE HORMONE?

Oxytocin has overtaken vasopressin in the media hall of fame, where it has been crowned the "cuddle hormone". Just as we saw that dopamine was originally miscast as Dionysus in Chapter 5, so oxytocin's moniker has proved misleading. Headlines such as 'New Oxytocin Neuroscience Counters "Cuddle Hormone" Claims' in *Scientific American*[19] seem to suggest that the neuropeptide is nothing to do with cuddles. Actually, it is a cuddle hormone but also so much more.

Just as vasopressin brings out the parent in male voles, so oxytocin makes female rodents "cuddle" their pups. Female virgin rats have no maternal instinct towards pups and will even attack them. But once they produce their own offspring they start crouching over, grooming and licking their litter. The role of oxytocin has been demonstrated by two methods. First of all, virgin pups have been transformed into doting mothers following injections of oxytocin. Secondly, actual mothers have cast aside their parenting duties when injected with oxytocin-blocking antagonists. There is a clear role for oxytocin, as well as vasopressin, in the maternal behaviour of rats, mice and prairie voles.[16]

Oxytocin seems to encourage rodents to cuddle—or at least groom—their pups, but do our brains dole out the neuropeptide when we receive cuddles? Scientists are in the business of doing experiments to prove their hunches. So, have any of them done a study in which cuddles caused oxytocin to spike? Yes. Quite a lot, in fact, have found exactly that. Julianne Holt-Lunstad, Wendy Birmingham and Kathleen Light divided married couples into experimental and control groups. The experimental group engaged in regular "warm touch", which the researchers defined as behaviours such as holding hands, hugging or sitting or lying "cuddled up". They found not only that cuddling increases blood levels of oxytocin but also that it reduces stress and blood pressure in men.[20] Another study also recorded increased oxytocin after warm touch.[21] Still another experiment found that both mothers and their children had higher levels of oxytocin in their urine after a play session involving "warm, physical contact".[22] Unfortunately, all these studies suffer from the same drawback as the one linking serotonin with obsessive-compulsive disorder in Chapter 7: blood or urine levels of any substance are not necessarily indicative of their activity in the brain.

Additional evidence for the hypothesis was gathered with an imaginative sequence of experiments. Kerstin Uvnäs Moberg at the Karolinska Institute carried out much of the seminal research into oxytocin in the 1980s and 1990s. In 1988 she conducted an experiment with Solveig Stock, which found that stroking rats on their backs elevated the levels of oxytocin in their blood.[23] They wanted to find out whether the increased levels of oxytocin in the blood resulted from the neuropeptide's release in the brain. Her team tested the hypothesis by stepping back to view the role of oxytocin in the broader context of stress relief.

In Chapter 7, opioids were identified as the neurological agents behind the feel-good factor of attachment. Drugs like heroine and morphine, plant-extracted imitators of the body's own stash of endorphins, are prescribed for pain relief but abused for euphoria. They make us forget our worries, as well as our aches and pains. Having established that opioids underwrite the comfort mammals find in contact with their mothers, researchers asked the next obvious question: how does skin contact trigger the release of opioids? This is when a starring role opened up for oxytocin.

Uvnäs Moberg's team conceived an ingenious experiment to test whether soothing skin contact would increase a rat's tolerance to pain.[23] An experimental group of rats was tickled with vibrating devices, gently warmed at a temperature of 40 °C or given electro-acupuncture, which is basically acupuncture with electrified needles. (It sounds pretty traumatic but apparently acupuncturists swear by it even with human patients.) Blood samples confirmed that the techniques had elevated oxytocin levels.

The experimental and control groups were each subdivided into two or more groups. Within each group, one sub-group was given saline solution while another sub-group was given oxytocin antagonists. Finally, the rats had their tails placed in water at a temperature of 50 °C, which would be uncomfortably hot. The researchers timed how long it took for the rats to register their discomfort by flicking their tails. All of the groups flicked their tails very rapidly, with one exception: the sub-group that had received the soothing skin treatments and the saline solution. By contrast, those that had received skin contact and oxytocin antagonists flicked their tails as quickly as the rats that had no skin contact at all. So, the skin treatment increased their tolerance to pain in the absence of oxytocin antagonists, strongly suggesting that oxytocin played a key part in the mediation of the pain relief. Uvnäs Moberg contributed to another experiment that recorded similar observations when rats were gently massaged.[24] We will consider the relationship between oxytocin and pain-relieving opioids later.

8.3   THE ROLE OF TEMPERATURE

It is interesting that researchers have taken to using the term "warm touch" as a catch all reference to different kinds of physical contact. Although the expression seems to indicate the warmth of the emotion conveyed by the touch rather than literal temperature increase, we have already seen how oxytocin levels were elevated in rats that were gently heated. Does this mean that warm activities, such as taking a bath or sunbathing on a beach, on some level simulate a cuddle? This poses questions about exactly how cuddles are detected.

Simulated attachment can be very rewarding to orphaned or isolated animals. One researcher managed to make ducklings develop an attachment to his boot,[25] but that was nothing compared to a series of experiments that turned out to be seminal in the theory of attachment.

In 1955 a researcher named Harry Harlow began breeding rhesus monkeys in order to observe their problem-solving abilities. The newly born offspring were raised in isolation to stop them from catching infections, but Harlow was disappointed to find that, unlike previous monkeys he had worked with, these had no inclination or aptitude for problem solving. Then, his team noticed an odd fixation of the monkeys. Their cages were lined with unused diapers to use as bedding, but the monkeys clung to them all the time, even taking them when they were moved to new cages. Harlow hatched the idea that the soft diapers were a kind of mother substitute.

The team gave the monkeys a choice of two fake mothers. They fashioned the substitutes to be similar in size to an adult female rhesus and they even installed milk bottles, from which the monkeys could drink *via* tubes protruding from the dummy's chests. One of the dummies was made from cloth and the other was made out of wire. What they found was that all of the monkeys clung to the cloth mother, not only while drinking milk but simply for the comfort they found in it.[26] The likelihood is that the warm feeling of the soft fabric triggered the release of oxytocin, which in turn caused a soothing pay-out of opioids.

Might this explain the attachment children form to cuddly toys? In the *Rosebud* episode of *The Simpsons*, Mr Burns is disconsolate at the absence of his cherished childhood toy Bobo. Fate delivers the teddy bear to Maggie Simpson, much to the delight of Homer, who sees the opportunity to sell Bobo back to Mr Burns. But the doting father turns down $1 million and three Hawaiian Islands when he sees that a tearfully reluctant Maggie has developed her own attachment to the bear. The chords this strikes with reality may draw on oxytocinergic activity.

8.4   CHEATING THE SENSORS

If the idea that a fluffy rag could really dupe the body into thinking it's getting hugged still feels too fishy, there is another way to view the issue. How exactly would your body detect "cuddles"? Through a cuddle receptor? Of course not. The body basically has to gather sensory input and deduce when it is being embraced. All sensory data is delivered *via* four different families of receptor: pressure-detecting, heat-detecting, chemical-detecting and light-detecting. Obviously, the latter is used exclusively in our eyes. We have already met the chemoreceptors that detect different nutrients on our tongues. The same variety is embedded in our skin to detect substances like histamine, which prompts us to scratch the itch that likely indicates the presence of a blood-sucking insect to be dispatched. The skin also has a huge array of thermoreceptors to detect temperature changes and mechanoreceptors to detect pressure changes. It is these last two that are most likely to register skin-to-skin contact, which raises the interesting question of how they work.

Both the pressure and temperature receptors are actually nothing more than ion channels. These are the membrane-straddling proteins that ferry charged ions, such as positively charged calcium (Ca^{2+}) or sodium (Na^+), across cell membranes. Once enough positive ions have been scooped into the cell, it will depolarise, triggering the release of neurotransmitters and the transmission of a signal along the nerves to the brain.

Mechanoreceptors are the simpler variety to understand. Although their precise mechanism is not completely understood, their basic function is fairly straightforward. When pressure is applied to the skin, the cells are squeezed and that creates tension in parts of the cell membrane. When you squeeze a balloon, the parts you are not holding bulge out, which actually stretches the rubber. A cell behaves the same way. The simplest mechanoreceptors open up when the cell membrane on either side of them is stretched apart. More complex versions are moored to the structural scaffolding that gives the cell its shape.[27] Even so, they operate by the same basic principle. When pressure deforms the cell, the membrane moves relative to the scaffolding, which causes the mooring proteins to yank open the ion channel. Both varieties are shown in Figure 8.4.

Temperature-detecting ion channels are much less straightforward. At the time of writing, no one knows exactly how they work, but what is known demands a working understanding of various thermodynamic laws. (See the Appendix for a detailed explanation.)

Some interesting things are known about heat-detecting ion channels. On the subject of duping our neural system, for our brains the question "hot hot or spicy hot?" has only one answer. This is a typical question to ask someone who has just described freshly served food as hot, but it is these same heat-detecting ion channels that detect rising temperature, as well as the presence of capsaicin, the spicy ingredient in chilli peppers. Again, no one is too sure why this happens, but menthol compounds have an analogous effect on cold-detecting ion channels. I was once

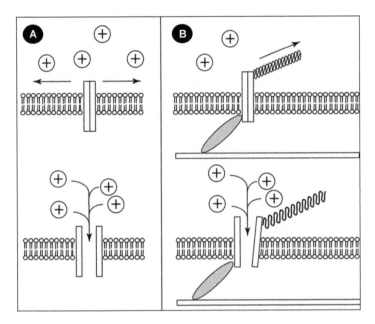

Figure 8.4 Pressure-detecting mechanoreceptors. Type A opens in response to tension in the membrane. Type B is moored to structural proteins inside and outside of the cell. Tension resulting from pressure applied to the cell pulls the ion channel open *via* these structural proteins.[28]

forced to use tiger balm as mosquito repellent on holiday. In spite of the scorching heat, I was racked with the shivers once I had covered myself in the ointment. These compounds inadvertently bind to a particular part of the ion channel and trigger a conformational change that forces them open. The resulting signal indicates to our brains that the temperature has changed and our brain responds as if the temperature had changed, which is why chilli makes us sweat and menthol makes us shiver.

There are also some fascinating discrepancies between different species of animal. Capsaicin has no effect on birds, whereas it is painful or poisonous to mammals. The researchers Joshua Tewksbury and Gary Nabhan have shown that the spicy compound puts off consumers that destroy its seeds, while remaining neutral to those that distribute them unharmed.[29] They rationalised this as follows: mammals chew seeds, whereas birds do not. This means that seeds pass intact through a bird's digestive system but are destroyed by mammals. Sure enough, Tewksbury and Nabhan showed that fewer seeds germinated having passed through a rodent's guts compared to a bird's. Capsaicin is not the only self-defence mechanism employed by plants. Uncooked kidney beans are lethally dangerous because they contain compounds called proteases, which deactivate our digestive enzymes and render us unable to metabolise protein. We may hypothesise that another genetic mutation may have produced a compound that stung the mouths of both mammals and birds. But this mutation was unlikely to replicate itself in future generations, because it puts off all consumers, regardless of which might help it produce offspring. However, the genes that created capsaicin guaranteed their place in future generations because the compound simultaneously repels consumers that destroy the seeds, while remaining palatable to those that help them grow. This also suggests that birds do not share our heat-detecting ion channels.

This comparison between the ion channels of different species is relevant. Many species *do* have the same temperature-detecting proteins but with a curious twist: the same proteins that detect heat in one species detect cold in other species. The ion channel named transient receptor potential cation channel subfamily V member 1 (TRPV1) opens in response to heat in flies and snakes but detects cold in mice.[30] This has led to suggestions that the proteins may act as subordinates to some yet-to-be-discovered temperature-detecting module, but recent theoretical investigations refute the idea. David Clapham and Christopher Miller have suggested that the same proteins may open in response to rising or falling temperatures.[30] As if that was not odd enough, they also theorise that the proteins operate by doing the impossible.

These heat receptors seem to work by doing what oil and water cannot. In Chapter 6 we considered why hydrophobic cooking oil does not mix with hydrophilic water. Water is polar, owing to its uneven distribution of electrons between positive nuclei, and is hence able to form hydrogen bonds. This forces them to weave clathrates (cages) around the hydrophobic oil molecules, which cannot hydrogen bond since they are non-polar. The more ordered nature of the clathrates makes the process thermodynamically unfavourable.

Proteins cannot be described as hydrophilic or hydrophobic—they are both. They are made of amino acids, which may have polar or non-polar sidechains. Of the amino acids we met earlier, hydrophobic acids include isoleucine and phenylalanine, while hydrophilic varieties include asparagine and glutamine, as shown in Figure 8.5. Proteins invariably contain polar and non-polar amino acids and, for the same reason that oil and water do not usually mix, they tend to tuck their hydrophobic sections into their core, while wearing their hydrophilic sections on the surface, as shown in Figure 8.6. The resulting hydrophilic exterior enables the proteins to mix with water.

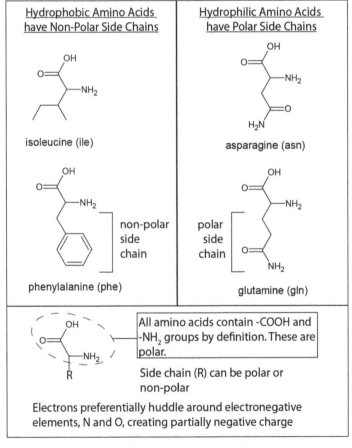

Figure 8.5 Examples of polar and non-polar amino acids.

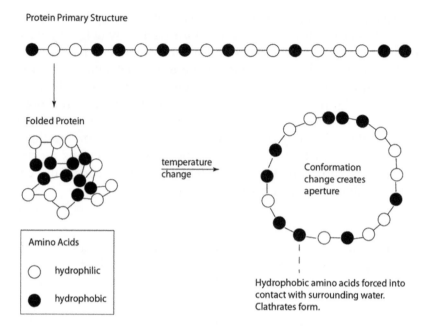

Figure 8.6 Proteins usually tuck their hydrophobic sections into their core to keep them out of contact with the surrounding water. When the temperature changes, the hydrophobic core is thrust into contact with the surrounding water.

The crazy thing is that when these heat receptors get hot, they seem to thrust their hydrophobic sections into the water, also shown in Figure 8.6. Even more crazy is that they do the same whether you make them hotter or colder. Whether the temperature increases or decreases it becomes thermodynamically favourable to force their "water-hating" non-polar sections into contact with their accursed foe.

Thermodynamics, as the name suggests, relates the temperature of particles to their other forms of energy. For example, transferring heat to a collection of particles will increase their average speed, just as increasing their average speed, such as by crushing them into a smaller space, will cause their collective temperature to increase. Thus, it is the changes in temperature that cause heat-detecting ion channels to arrange themselves in different ways.

Baron Chanda, Sandipan Chowdhury and Brian Jarecki have put Clapham and Miller's theory to the test. Clapham and Miller suggested that as the number of a protein's hydrophobic amino acids increases, the more thermodynamically favourable it will become for these hydrophobic sections to mix with water in response to temperature changes.[30] Putting this theory to the test, Chanda, Chowdhury and Jarecki built their own heat receptor. They started with a voltage-gated ion channel, like the ones we met in Chapter 3, and then spliced in some extra hydrophobic amino acids. Then they increased the temperature and compared the rate at which ions were transferred into the cell. As predicted, they found that additional hydrophobic regions inclined the ion channel to open up in response to temperature changes.[31]

Let's review the situation. Oil and water do not mix. This is because oil is hydrophobic and water is hydrophilic. Plunging hydrophobic molecules into water forces the water to form clathrates around the hydrophobic molecule, which is a thermodynamically unfavourable process. But when *extra* hydrophobic regions are added to an ion channel it suddenly becomes thermodynamically favourable for those hydrophobic regions to mix with water. Is this the impossible made possible? No. Here are some problems with this reasoning. First, oil and water *do* mix. Second, hydrophobic amino acids are not oil. Third, extra hydrophobic regions do not, on their own, cause the ion channel to open up.

Do oil and water mix or do they not? A curious thing about thermodynamics is that while its laws are immutable, they can only predict the behaviour of the crowd. Analysts have to predict the behaviour of people when they decide how much electricity needs to be available during, say, the World Cup. They look at past data, consider other factors and eventually make a decision of the sort that: in the UK there are X televisions and Y% of them will be on during the match between Germany and England, meaning that we need Z watts of electricity. They cannot say: Mrs Horowitz of 22 Acacia Avenue in Bournemouth will watch the start of the match but switch it off after exactly 36 minutes. Similarly, when we say that oil and water do not mix, what we really mean is that the *majority* of the oil and water molecules will not mix. The reason for this limitation of thermodynamics is that it is based on probability. The probability that *all* of the water and oil molecules will mix is so low that it is effectively impossible, but the probability that individual molecules will mix is far less prohibitive. The reality is that a small proportion of oil molecules will stray into the water, which has ramifications for our mysterious ion channel.

Hydrophobic amino acids are not oil. In Chapter 6 we compared the ability of oil and water molecules to form hydrogen bonds. Each water molecule can form four hydrogen bonds with neighbouring water molecules and, having done so, every single atom in each water molecule will be involved in a hydrogen bond. Conversely, the typical molecule you would find in a bottle of cooking oil has roughly 150 atoms, of which a maximum of six can form hydrogen bonds. We can infer two important points from this: firstly, molecules in cooking oil have only a very limited ability to form hydrogen bonds and, secondly, the part of the molecule that cannot form hydrogen bonds has a very large surface area. Now we need to compare hydrophobic amino acids to our molecule from the cooking oil. From Figure 8.5 we can see that any amino acid has the ability to form hydrogen bonds. As such, rather than describing the whole amino acid, we should really say that its sidechain is either hydrophobic or hydrophilic. Even then an amino acid's hydrophobic sidechain, composed of fewer than 20 atoms, will have a much smaller surface area than the hydrophobic region of the molecule from the cooking oil. In the simplest terms even hydrophobic amino acids are *much* less hydrophobic than the molecules that compose cooking oil.

A temperature change, as well as extra hydrophobic amino acids, is needed to make the heat receptor open up. We could add as many extra hydrophobic amino acids as we wanted, but if the heat receptor stays the same temperature it will not open up. The point of this distinction is that when the heat receptor is at its normal temperature it is still very much thermodynamically unfavourable for the hydrophobic regions to mix with the water. In spite of the differences with cooking oil the same principles apply: mixing the hydrophobic regions with water will force the water molecules to form the cage-like clathrates around the exposed hydrophobic sidechains, which is thermodynamically unfavourable. The difference is that the smaller surface area changes the threshold at which this thermodynamic barrier can be overcome. Dodecane, a hydrophobic compound, and water have to be heated to 374 °C under a pressure of 22.1 megapascals before they will completely mix.[32] At this temperature the water molecules have gained enough kinetic energy to break free from their clathrates. But even hydrophobic amino acids are much less hydrophobic than dodecane or cooking oil, meaning the same result can be achieved at much lower temperatures. According to Dr Chanda, a modest increase in temperature is enough to dispatch the clathrates that form around the hydrophobic regions as they are thrust into the water.[33] This reduces the thermodynamic penalty of mixing hydrophobic with hydrophilic.

There is also a snowball effect. Heat-sensing cells will not contain just one heat receptor but hundreds or even thousands. They are sensitive not only to changes in temperature but also to changes in the cell's potential, and thus the voltage across its membrane. If one heat receptor starts to open up it will transfer positive ions inside the cell, increasing the cell's potential. This, in turn, will cause more heat receptors to open up, transferring more ions inside and increasing the potential further, causing still more heat receptors to open up. In other words, not all of the heat receptors open up at the same time, but each one that does open encourages its neighbours to do likewise.[33]

If the explanation feels unsatisfying, there is experimental proof to back it up. Clapham and Miller's theory has been tested by another group of scientists. When Sandipan Chowdhury, Brian Jarecki and Baron Chanda built their own heat receptor using the other group's theory, they started with a voltage-gated ion channel and then spliced extra hydrophobic amino acids into its structure. As predicted, they found that additional hydrophobic regions inclined the ion channel to open up in response to temperature changes.[31] Even if the theory is incomprehensible, evidence suggests it is right.

A caveat is needed here. There is more than one type of heat receptor and it is doubtful that the one we have considered is linked to oxytocin release. The TRPV1 ion channel that we have looked at signals the detection of painfully hot stimuli at temperatures of 45 °C or more. A more likely contender for interaction with oxytocin is TRPV3, which detects warm temperatures above 35 °C.[34] Just as capsaicin randomly activates TRPV1, so camphor, the extract of the camphor tree, activates TRPV3, which likely explains its long-running appeal as a folk remedy to relieve itchiness and other ailments.[35] This distinction between the different ion channels explains why eating chilli pepper is painful because it masquerades as something dangerously hot, while a warm hug is soothing. The structure of all TRP channels is very similar, suggesting that their mechanisms of heat detection may also be similar.

8.5   NOT A CUDDLE HORMONE

If cuddles elevate levels of oxytocin, then why should it not be called the cuddle hormone? Because it is so much more than *just* a cuddle hormone. Even before it was nicknamed it was known to play a role in childbirth, as well as breast feeding. In its guise as a social hormone, cuddles are just the beginning.

Recent research suggests that oxytocin may tune our attention to social cues. This resulted from the neuropeptide's discovery in the most unexpected of places—the brain's auditory cortex. When rodent mothers move around, their pups cling to their underbellies. A drawback of this mode of transport is that pups can fall off, but nature has equipped them with a backup: the pups squeal until the mother comes back and collects them. The researchers suggest that the auditory oxytocin may help the mothers to zero in on the distress calls.[36] Other work suggests that the biomolecule may help people to recognise positive facial expressions.[37]

Another problem with the cuddle crown is that oxytocin has a dark side. The parable of the Good Samaritan encourages us to look after anyone in need, even if they are not members of our clan or biological in-group. Neurology shows that oxytocin is an unlikely puppet master for the boundless compassion of the biblical do-gooder. Several studies have indicated that another way the neuropeptide zeroes in on social cues is by categorising people into in-group and out-group, after which it promotes favouritism towards the in-group and non-cooperation with outsiders.[38] The scourge of nepotism may thrive on the so-called cuddle hormone.

8.6   THE NEUROMODULATOR

No matter what signal triggers it, ultimately oxytocin will do something in the brain. It primarily acts as a neuromodulator, a concept we met in Chapter 6. Neuromodulation differs from neurotransmission in various ways. Neurons transmit signals in both your nerves and your brain, but there are differences in the way they do so. The neurons in nerves have little need to interpret information; their role is merely to pass on sensory data gathered from receptors. They typically use fast-acting ionotropic receptors. But there is another, slower-acting variety called a metabotropic receptor, which is more likely to be found in neurons that contemplate complex information, such as social cues. The metabotropic variety characterises neuromodulation.

The speed of the action depends on the mechanism. Ionotropic receptors are simply ion channels that are opened by neurotransmitters, as shown in Figure 8.7. When the neurotransmitter

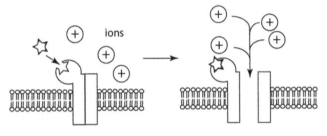

A Ionotropic Receptor

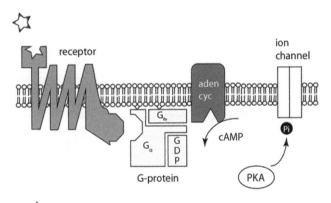

B Metabotropic Receptor

Figure 8.7 The different actions of ionotropic (A) and metabotropic (B) receptors. Ionotropic receptors are ion channels that open as soon as they bind to the corresponding neurotransmitter. Conversely, metabotropic receptors recruit secondary messengers to indirectly modify remote ion channels.

binds to the ion channel it immediately opens up and starts transferring ions across the membrane. The chain of command is much longer for metabotropic receptors. In this case, the neurotransmitter binds the receptor, which then dispatches a secondary messenger. This messenger may interact with several other proteins before one of them finally does something to an ion channel. The good news is that typical metabotropic mechanisms look very similar to other processes we have seen, such as noradrenaline making our hearts beat faster. Figure 8.7 also shows an example involving a G-protein. Acting *via* adenylyl cyclase and cyclic adenosine monophosphate (cAMP), protein kinase A (PKA) is recruited to phosphorylate the ion channel, which causes it to behave differently.[34] The channel may transfer ions more or less rapidly as a result, which, in turn, may have the effect of enhancing or decreasing the neuron's sensitivity to depolarisation by other neurotransmitters, such as dopamine.

Oxytocin modulates neuron function in a number of ways, not all of which are completely understood. A particularly extreme example is when it activates a kind of self-destruct mechanism. When there is too much oxytocin sloshing around your brain, your neurons switch themselves off to its wiles. They do this by scooping all of the receptors up in a bubble of membrane, which is drawn inside the cell, putting them out of reach from other oxytocin molecules, as shown in Figure 8.8. The bubble is really a

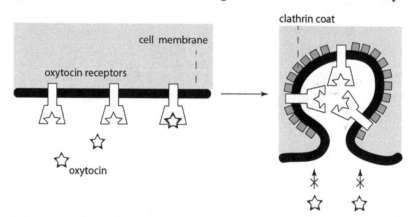

Figure 8.8 Oxytocin-triggered endocytosis of oxytocin receptors. When the receptor is triggered, the membrane curves inwards, drawing the receptors inside the cell. The nascent vesicle is stabilised by a clathrin coating.

vesicle and the process, endocytosis, is the reverse of exocytosis. A biological polymer called clathryn forms a coating around the vesicle to stabilise it. The endocytosis mechanism safeguards your brain from getting addled by excessive oxytocin, which is sometimes released in stressful situations.[39] Obviously, if you are running from a sabre-toothed tiger you do not want to get too relaxed!

Two other mechanisms are very familiar. Figure 8.9 shows how oxytocin triggers a process similar to the one observed when the tongue tasted glucose in Chapter 2. Acting *via* a particular G-protein called q/11, oxytocin triggers the release of inositol trisphosphate (IP$_3$), which in turn releases Ca^{2+} from the lumen. This has a snowballing effect on the positivity of the inside of the cell, firstly because the Ca^{2+} ions are positive, and secondly because they deactivate the potassium channels. The job of these specific channels is constantly to bail potassium ions *out* of the cell, usually to restore the neuron to normal levels of positivity after it has been depolarised. The other mechanism is almost identical to the one shown in Figure 8.7. In this case, the ion channel is programmed to bring sodium ions into the cell more quickly, although it should be noted that this is achieved *via* some unknown mechanism that does not involve PKG. At any rate, all of these mechanisms have the effect of making the inside of the cell more positive, which means it can be depolarised much more easily by any other neurotransmitters it recognises. Oxytocin is also able to make cells *less* sensitive to depolarisation by setting the ion channels to eject or keep out positive ions.

Oxytocin has a huge impact as a modulator. It can speed up or slow down various ion channels, giving it direct influence over the charge of the inside of the cell, making neurons easier or harder to depolarise.

8.7   REWARD

The next obvious thing to consider is what kinds of cells oxytocin modulates and where. This key-shaped peptide helps us zero in on social cues. It drives us towards our mates and then drives us to cherish the children we produce with them. The oxytocin gene guarantees its replication in future generations by motivating us

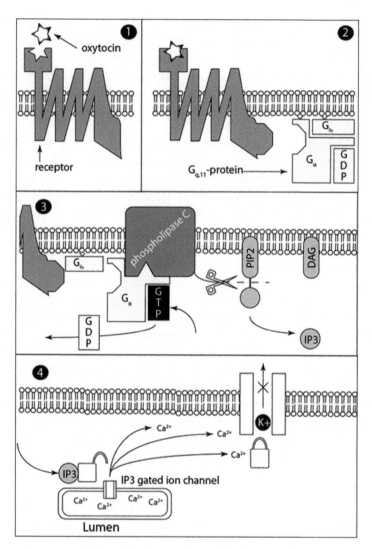

Figure 8.9 An oxytocin neuromodulation pathway. (1) Oxytocin binds the receptor. (2) The receptor activates the $G_{q/11}$-protein, which releases guanosine diphosphate (GDP). (3) While hydrolysing newly bound guanosine triphosphate (GTP), the G-protein activates phospholipase C, which causes phosphatidylinositol 4,5-bisphosphate (PIP2) to release IP_3. (4) IP_3 unlocks the lumen ion channel. The resultant increase in the cellular concentration of Ca^{2+} deactivates the potassium ion channel.

to look after the future generation. Our reward for perpetuating the self-replication of this gene is the delight we find in cuddling our partners, nurturing our children or laughing with our friends. In short, oxytocin twiddles the dials of our motivation and reward circuitry, swelling the currents of our old friends, dopamine and the opioids.

This is hardly surprising. The involvement of dopamine was already hinted at by the strikingly similar love potions in this chapter and the last. Female voles shack up with a partner whether they are given dopamine agonists or oxytocin. Whilst Liebowitz reported the association between opioids and attachment before the prosocial inclinations of oxytocin had been unearthed, it is clear that oxytocin modulates dopamine- and opioid-peddling neurons.

The location of these neurons should not be too surprising either. Oxytocin receptors can be found in the nucleus accumbens and ventral tegmental area, as well as the amygdala and hippocampus. The nucleus accumbens is the pleasure centre thought to dole out opioids after we have followed the dopamine trail to some delicious but vein-clogging snack. The ventral tegmental area is another landmark in the reward circuitry, which lit up in just about all magnetic resonance imaging (MRI) scans of couples in love (Chapter 7).

When lactating rodent mothers are suckled by their pups, neural activity increases in several areas, including the nucleus accumbens and ventral tegmental area. When a female rat licks or grooms her children, dopamine is released in the nucleus accumbens. Moreover, dopamine reuptake inhibitors, drugs that mimic the ability of cocaine to switch off dopamine transporters (Chapter 3), make the rodents more attentive mothers, whilst dopamine antagonists do the opposite.

There is a clear crossover between oxytocin and dopamine during sex as well. Earlier, we saw how most oxytocin is produced in the paraventricular nucleus. These oxytocinergic neurons also have dopamine receptors, which, when stimulated, cause rats to get erections. Another way to give them erections is to inject oxytocin into the ventral tegmental area, which also causes a dopamine spike in the nucleus accumbens. Both the erection and the dopamine hike can also be increased by dishing out oxytocin in the amygdala and hippocampus.[40]

If dopamine is for wanting and opioids are for liking, sex should be a cocktail of both with a twist of oxytocin. A study with human participants[41] saw the blood levels of oxytocin shoot up to four times their normal amount immediately after orgasm but only in the control group. In the treatment group, whose participants had been given opioid antagonists, not only was there no oxytocin spike, but they reported being less aroused and enjoying their orgasm less.

Opioids do not only reward us during sex. It has long been agreed that they mediate social behaviour[42] and there is a growing consensus that they keep prairie voles faithful to their partners.[43] We have already seen how oxytocin and dopamine antagonists can make the rodents stray from their significant others. Opioid antagonists have been added to the list of potions that can derail their domestic bliss.[44] Opioids also seem to foster relationships between offspring and their parents. Earlier, we saw how baby rats scream when separated from their mothers so that the parent goes back to fetch them. But mice that were genetically modified to lack opioid receptors were indifferent to both the presence and the absence of their mothers.[45]

There are many other studies linking oxytocin to the reward circuitry. In a way, it would be more surprising if the reward circuitry weren't involved. No one can yet clearly explain how opioids deliver their soothing euphoria, but our species' long-running fascination with the spoils of the poppy confirms that ability. Dopamine guides us towards opportunities to mate, as well as to eat, both of which are behaviours that promote the optimum conditions in which for the self-replication of our genes to persist. Eating provides the twin streams of fuel and building materials, while reproduction guarantees that self-replication can carry on after we die.

Oxytocin and vasopressin seem to play important roles in social attachment. More is known about the former, explaining its dominance of this chapter. The likelihood is that vasopressin also alerts us to social cues *via* interaction with our reward system, whether to promote beneficial interactions or to avoid hazardous ones. In either case, it is highly believable that opioids are the culprit for the wonderful feeling we get when we spend time with the people we love, whether cuddling, laughing, lavishing affection on our progeny or even celebrating a diamond wedding anniversary.

REFERENCES

1. Craig Cros Kennebec, 'After 27 Years of Burglaries, "North Pond Hermit" is Arrested', *The Portland Press Herald / Maine Sunday Telegram*.
2. M. Finkel, "The Strange Tale of the North Pond Hermit", *GQ*. Available at: http://www.gq.com/story/the-last-true-hermit. [Accessed: 26-Jul-2015.]
3. E. Lerer, S. Levi, S. Salomon, A. Darvasi, N. Yirmiya and R. P. Ebstein, Association Between the Oxytocin Receptor (OXTR) Gene and Autism: Relationship to Vineland Adaptive Behavior Scales and Cognition, *Mol. Psychiatry*, 2007, **13**(10), 980–988.
4. M. Liebowitz, *Chemistry of Love*, New York, Berkley, 1984.
5. H. Fisher, *Why We Love: The Nature and Chemistry of Romantic Love*, Henry Holt & Company, New York, 1st edn, 2004.
6. L. De Bernières, *Captain Corelli's Mandolin*, Secker & Warburg, London, 1994.
7. F. Hardy, "Sad and Deluded – Or Just Honest? Meet the Couple Who Say the Secret of a Perfect Marriage is NOT Having Children", *Mail Online*. Available at: http://www.dailymail.co.uk/femail/article-2399338/Meet-couple-say-secret-perfect-marriage-NOT-having-children.html. [Accessed: 27-Jun-2015.]
8. "Sea Turtle Conservancy: General Behavior of Sea Turtles". Available at: http://www.conserveturtles.org/seaturtleinformation.php?page=behavior. [Accessed: 19-Jun-2015.]
9. "Most Prolific Mother Ever", *Guinness World Records*. Available at: http://www.guinnessworldrecords.com/world-records/most-prolific-mother-ever. [Accessed: 20-Jun-2015.]
10. H. E. Fisher, Lust, Attraction, and Attachment in Mammalian Reproduction, *Hum. Nat.*, 1998, **9**(1), 23–52.
11. W. Trevathan, *Human Birth: An Evolutionary Perspective*, Transaction Publishers, 1987.
12. "Sea Turtle Conservancy: Green Sea Turtle". Available at: http://www.conserveturtles.org/seaturtleinformation.php?page=green. [Accessed: 20-Jun-2015.]
13. R. I. Dunbar, Coevolution of Neocortical Size, Group Size and Language in Humans, *Behav. Brain Sci.*, 1993, **16**(4), 681–694.

14. T. R. Insel and L. J. Young, Neuropeptides and the Evolution of Social Behavior, *Curr. Opin. Neurobiol.*, 2000, **10**(6), 784–789.
15. A. W. Norman and H. L. Henry, *Hormones*, Academic Press, 3rd edn, 2014.
16. O. J. Bosch and I. D. Neumann, Both Oxytocin and Vasopressin are Mediators of Maternal Care and Aggression in Rodents: From Central Release to Sites of Action, *Horm. Behav.*, 2012, **61**(3), 293–303.
17. J. T. Winslow, N. Hastings, C. S. Carter, C. R. Harbaugh and T. R. Insel, A Role for Central Vasopressin in Pair Bonding in Monogamous Prairie Voles, *Nature*, 1993, **365**(6446), 545–548.
18. T. R. Insel and T. J. Hulihan, A Gender-specific Mechanism for Pair Bonding: Oxytocin and Partner Preference Formation in Monogamous Voles, *Behav. Neurosci.*, 1995, **109**(4), 782–789.
19. H. Shen, 'New Oxytocin Neuroscience Counters "Cuddle Hormone" Claims', *Scientific American*, 25-Jun-2015. Available at: http://www.scientificamerican.com/article/new-oxytocin-neuroscience-counters-cuddle-hormone-claims/?utm_source=twitterfeed&utm_medium=twitter&utm_campaign=Feed%3A+ScientificAmerican-Twitter + %28Content%3A+Global+Twitter+Feed%29. [Accessed: 01-Jul-2015.]
20. J. Holt-Lunstad, W. A. Birmingham and K. C. Light, Influence of a "Warm Touch" Support Enhancement Intervention Among Married Couples on Ambulatory Blood Pressure, Oxytocin, Alpha Amylase, and Cortisol, *Psychosom. Med.*, 2008, **70**(9), 976–985.
21. K. M. Grewen, S. S. Girdler, J. Amico and K. C. Light, Effects of Partner Support on Resting Oxytocin, Cortisol, Norepinephrine, and Blood Pressure Before and After Warm Partner Contact, *Psychosom. Med.*, 2005, **67**(4), 531–538.
22. J. Bick and M. Dozier, Mothers' and Children's Concentrations of Oxytocin Following Close, Physical Interactions with Biological and Non-biological Children, *Dev. Psychobiol.*, 2010, **52**(1), 100–107.
23. K. Uvnäs-Moberg, G. Bruzelius, P. Alster and T. Lundeberg, The Antinociceptive Effect of Non-noxious Sensory Stimulation is Mediated Partly Through Oxytocinergic Mechanisms, *Acta Physiol. Scand.*, 1993, **149**(2), 199–204.

24. G. Agren, T. Lundeberg, K. Uvnäs-Moberg and A. Sato, The Oxytocin Antagonist 1-Deamino-2-D-Tyr-(Oet)-4-Thr-8-Orn-Oxytocin Reverses the Increase in the Withdrawal Response Latency to Thermal, But Not Mechanical Nociceptive Stimuli Following Oxytocin Administration or Massage-like Stroking in Rats, *Neurosci. Lett.*, 1995, **187**(1), 49–52.

25. K. Lorenz, Der Kumpan in der Umwelt des Vogels, *J. Für Ornithol*, 1935, **83**(2), 137–213.

26. J. Haidt, *The Happiness Hypothesis: Putting Ancient Wisdom to the Test of Modern Science*, Arrow, London, 1st edn, 2007.

27. S.-Y. Lin and D. P. Corey, TRP Channels in Mechanosensation, *Curr. Opin. Neurobiol.*, 2005, **15**(3), 350–357.

28. 'MECHANORECEPTOR CITATION TEXT'.

29. J. J. Tewksbury and G. P. Nabhan, Seed Dispersal. Directed Deterrence by Capsaicin in Chilies, *Nature*, 2001, **412**(6845), 403–404.

30. D. E. Clapham and C. Miller, A Thermodynamic Framework for Understanding Temperature Sensing by Transient Receptor Potential (TRP) Channels, *Proc. Natl. Acad. Sci.*, 2011, **108**(49), 19492–19497.

31. S. Chowdhury, B. W. Jarecki and B. Chanda, A Molecular Framework for Temperature-dependent Gating of Ion Channels, *Cell*, 2014, **158**(5), 1148–1158.

32. S. Deguchi and N. Ifuku, Bottom-up Formation of Dodecane-in-Water Nanoemulsions from Hydrothermal Homogeneous Solutions, *Angew. Chem. Int. Ed.*, 2013, **52**(25), 6409–6412.

33. B. Chanda, Personal correspondence, 05-Aug-2015.

34. E. R. Kandel, J. H. Schwartz, T. M. Jessell, S. A. Siegelbaum, and A. J. Hudspeth, *Principles of Neural Science*, McGraw Hill, New York, 5th edn, 2013.

35. "Camphor: Uses, Side Effects, Interactions and Warnings", *WebMD*. Available at: http://www.webmd.com/vitamins-supplements/ingredientmono-709-camphor.aspx?activeingredientid=709&activeingredientname=camphor. [Accessed: 21-Jul-2015.]

36. B. J. Marlin, M. Mitre, J. A. D'amour, M. V. Chao and R. C. Froemke, Oxytocin Enables Maternal Behaviour by Balancing Cortical Inhibition, *Nature*, 2015, **520**(7548), 499–504.

37. A. A. Marsh, H. H. Yu, D. S. Pine and R. J. R. Blair, Oxytocin Improves Specific Recognition of Positive Facial Expressions, *Psychopharmacology (Berl.)*, 2010, **209**(3), 225–232.
38. C. K. W. De Dreu, Oxytocin Modulates Cooperation Within and Competition Between Groups: An Integrative Review and Research Agenda, *Horm. Behav.*, 2012, **61**(3), 419–428.
39. A. S. Smith and Z. Wang, Hypothalamic Oxytocin Mediates Social Buffering of the Stress Response, *Biol. Psychiatry*, 2014, **76**(4), 281–288.
40. T. M. Love, Oxytocin, Motivation and the Role of Dopamine, *Pharmacol. Biochem. Behav.*, 2014, **119**, 49–60.
41. M. R. Murphy, S. A. Checkley, J. R. Seckl and S. L. Lightman, Naloxone Inhibits Oxytocin Release at Orgasm in Man, *J. Clin. Endocrinol. Metab.*, 1990, **71**(4), 1056–1058.
42. S. L. Resendez, M. Kuhnmuench, T. Krzywosinski and B. J. Aragona, κ-Opioid Receptors within the Nucleus Accumbens Shell Mediate Pair Bond Maintenance, *J. Neurosci.*, 2012, **32**(20), 6771–6784.
43. L. J. Young, M. M. Lim, B. Gingrich and T. R. Insel, Cellular Mechanisms of Social Attachment, *Horm. Behav.*, 2001, **40**(2), 133–138.
44. J. P. Burkett, L. L. Spiegel, K. Inoue, A. Z. Murphy and L. J. Young, Activation of μ-Opioid Receptors in the Dorsal Striatum is Necessary for Adult Social Attachment in Monogamous Prairie Voles, *Neuropsychopharmacology*, 2011, **36**(11), 2200–2210.
45. A. Moles, B. L. Kieffer and F. R. D'Amato, Deficit in Attachment Behavior in Mice Lacking the μ-Opioid Receptor Gene, *Science*, 2004, **304**(5679), 1983–1986.

CHAPTER 9

The Chemistry of Baby Making

One of the great technological advances of the 20th century was to take the copulation out of population. Contraception allowed us to copulate without populating, while certain fertility treatments allowed us to populate without copulating. It is true that an increasing number of couples in the US and the UK are choosing not to have children, but there is still a sizeable majority of humans producing babies the old-fashioned way. Even if fertilisation takes place without having sex, a sperm and ovum still need to fuse in order for the resulting zygote to develop into a baby. This chapter will explore the chemistry of that process.

I find it absolutely fascinating that a single fertilised egg can grow into a baby. A more familiar part of the mystery is that one cell multiplies into roughly a trillion, which happens by cell division. As described, the self-replicators make copies of themselves, which in this context means that each of our 23 chromosome pairs produces an entire copy of itself, all of which are then packaged into a new daughter cell. But what is so curious is that these cells are not identical. Babies need eye cells, brain cells, liver cells, heart cells, *etc.*, all of which serve unique functions. How is it that one kind of cell diversifies into other kinds of cells? And what systems ensure that those diverse cells are grouped together in clumps so that our brains only make brain cells and our eyes only make eye cells? If kidneys started

The Chemistry of Human Nature
By Tom Husband
© Tom Husband 2017
Published by the Royal Society of Chemistry, www.rsc.org

growing teeth and big toes started growing eyeballs, it would be a real concern. The burgeoning field of epigenetics has been answering these questions.

9.1   EPIGENETICS

Genetics is about what proteins we can make, epigenetics is about which proteins get made. The convenient analogy is the cookbook. A cookbook contains recipes for different dishes, just as our genes act as recipes for different proteins we can make. Each chapter represents a chromosome, the individual recipes are genes and the paper on which they are printed is the DNA.

The analogy can be extended to include the chefs who use the recipes. Take all of the restaurants in a city. In these days of globalisation, a great many cuisines will be available: Chinese, Indian, Mexican, Lebanese, French, Italian and so on. Imagine each cuisine has its own set of recipes, for example, all of the Italian restaurants could share the same recipes for Italian dishes. Each restaurant needs a reference manual but rather than bankrolling the production of different editions for each cuisine, a single, giant manual is published that contains every recipe for all of the various cuisines. Next, the head chefs modify the books to keep their staff from getting confused. Each restaurant marks the recipes they need by folding the corners of the relevant pages. Also, pages containing unwanted recipes are glued to the previous page so that they can no longer be viewed. In each restaurant only the recipes for the relevant cuisine can now be accessed.

This is how our bodies use DNA to produce different types of cells. Broadly speaking, all of our cells contain all of the recipes for all of the proteins we as individuals have the capacity to make, but different genes can be switched on or off by epigenetic processes. This is how our body is able to host different kinds of cells. Each of them can be modified to produce different sets of proteins. Cells in the pancreas will need to produce insulin, whereas cells in the liver will make alcohol dehydrogenase, one of the enzymes that enable us to metabolise alcohol. The process by which cells develop their characteristic specialisms is known as differentiation.

Differentiation is a continuum, not unlike a person's education. We start off learning a wide range of subjects, which expands up to the age of roughly 14, before narrowing down to a single subject at university. Architects and medics train for a similar length of time, but if one member of each profession switched roles they would perform badly. Nevertheless, there were key stages in the development of each professional when they could have trained in the other direction. Cells also undergo a gradual specialisation, transforming from stem cells to differentiated cells. Stem cells have received much coverage in the media, thanks in part to their controversial harvesting from aborted foetuses. These are said to be cells that can be induced to become any other kind of cell, whereas differentiated cells are those that have taken on a certain role, such as a neuron, and can no longer produce different breeds. In fact, these terms fall at either end of the continuum. Stem cells differ in their flexibility, so they are said to have different degrees of potency. Totipotent (as in totally potent) cells have the ability to become any other kind of cell. Pluripotent cells can become any kind of cell *other than* the placenta[1] and so it goes on. As cells differentiate, it becomes more common to refer to them by what they *can* become rather than what they cannot. For example, mesenchymal stem cells have the ability to make cell types such as muscle and skin, which is arguably more informative than their descriptor: *multipotent*. Note that mesenchymal cells are differentiated but they still have the capacity to form other varieties. A cell that has completely specialised and can no longer be induced to become any other type of cell is said to be terminally differentiated.

Cells differentiate *via* epigenetics. Genetics concerns the proteins that our bodies have the ability to make, whereas epigenetics concerns the processes that decide which of those proteins get made, when, how often and for how long. In the cookbook analogy, epigenetic markings were represented by the corner folding of pages to highlight desired recipes and the gluing together of unwanted pages. We have already met some of the methods by which our bodies achieve this; in Chapter 6 we saw how testosterone helps to coordinate production of the protein nitric oxide synthase (NOS) by drawing the parent gene out of storage. The key players in the regulation of the epigenome are shown in Figure 9.1.

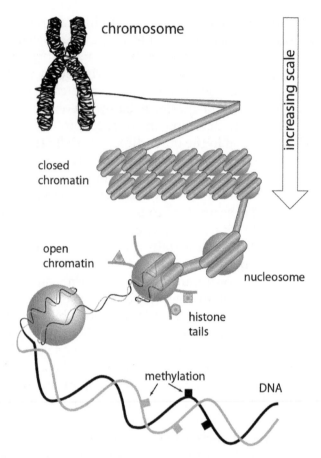

Figure 9.1 The chromosome, nucleosomes and chromatin.

Epigenetic modification regulates the accessibility of genes. Figure 9.2 shows that, when they are not in use, genes are coiled around units called histone octamers. As the name suggests, the octamer contains eight histones, each of which has a tail. A stretch of DNA coiled around an octamer is called a nucleosome and these nucleosomes, in turn, bunch together to form chromatin. Depending on what modification has taken place, the chromatin can take the form of an open house or an impenetrable fortress.

Most epigenetic modification is achieved with just two little chemical groups called acetyl and methyl. These small groups, shown in Figure 9.3, contain just six and four atoms,

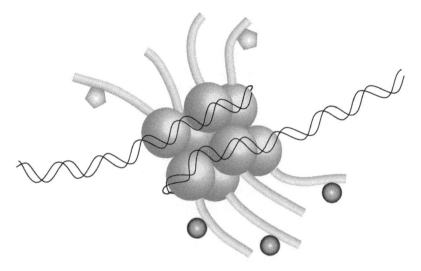

Figure 9.2 Histone octamers with methyl (black-outlined circles) and acetyl (pentagons) histone tail modifications.

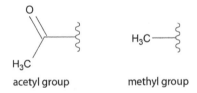

Figure 9.3 Acetyl and methyl groups: the currency of epigenetic markings.

respectively, and have the power to free up genes for transcription, as well as to consign vast swathes of DNA to deep storage. These are the main methods of differentiation. As cells specialise, they use methyl and acetyl modification to earmark the essential genes, whilst retiring the unwanted ones. At the highest level, methyl groups can tighten the packing of the chromatin, putting huge sections of DNA off limits. At the lower level, modification can make it harder or easier for individual genes to be unspooled from their histones.

As far as we know, acetyl modification always *promotes* gene expression. For example, in Chapter 6 we saw how testosterone, acting with other biomolecules, helps to unspool the *NOS* gene by coordinating the acetylation of its histone. The acetyl group is attached to a histone tail, which causes a redistribution of electrons as the strongly electronegative oxygen atom in the acetyl group draws electrons towards itself with consequences that ripple through the histone, just as we have seen mere calcium ions cause entire proteins to change shape. These electronic changes reduce histone's affinity for the DNA, making the NOS gene easier to unspool. This is how acetylation promotes gene expression, by making it possible to unwind the genes so that the transcriptional machinery can access them.

Methylation is the more versatile means of modification as it has contrary effects in different places. Like acetyl groups, methyl groups can be attached to the histone tails, but the effects are harder to predict. Methyl modification of histone tails can make DNA harder or easier to unspool, depending on what other histone modifications are present. Not only that, but there are several different positions on the tail where modifications can take place. In short, histone tails can be methylated so as to increase or decrease the accessibility of genes.

There is another variety of epigenetic marking in which methyl modification plays a much clearer role. Accessibility of genes is dictated by how tightly they are coiled up around the histone octamers and this can be controlled in two different ways. Not only can changes be effected by modifying the histone tails, but also *via* modification of the DNA itself. As DNA is composed of varying sequences of the nucleobases cytosine, guanosine, adenine and thymine, there are certain regions of DNA where cytosine neighbours guanosine much more often than usual, known as CpG islands, as shown in Figure 9.4. Just like the histone tails, the cytosine bases in these prevalent C–G sequences can also be methylated and the effects are dramatic; this is how long stretches of DNA are consigned to deep storage.

Rather than changing the "stickiness" of the histones, methylation of CpG islands alters the flexibility of the DNA. When the cytosine bases are methylated, the CpG islands are more flexible and hence more amenable to being coiled around the histones. By contrast, unmethylated CpG islands are stiffer, meaning that more energy is needed to wind them into nucleosomes (ref. 2, p. 296). It is the methylation of the CpG islands that transforms the chromatin into the impregnable fortress, putting genes out of reach of the transcriptional machinery. In other words,

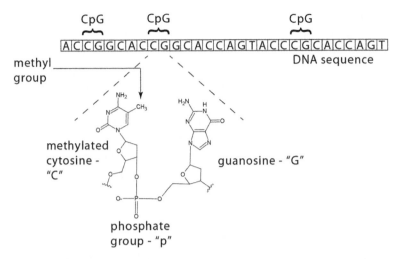

Figure 9.4 CpG islands are nucleotide pairings in which the nucleotides cytosine and guanosine occur side by side in the DNA sequence. The "p" refers to the linking phosphate group. A methyl group can be attached to the cytosine in the position shown.

methylation of DNA *reduces* gene expression, and not just of one gene but of many.

What controls this important process? Cells differentiate by switching different genes on or off and the tools of choice are methylation and acetylation. But there must be some kind of agency to operate the tools. This is the role of transcription factors. These are proteins that recognise and then attach themselves to specific stretches of DNA. Having clamped to these promoter regions, they usher in squads of enzymes to modify the histones, making genes more or less accessible as necessary. In Chapter 6 we saw how testosterone activates an androgen receptor to act as a transcription factor. Having bound the hormone, the receptor enters the nucleus, clamps to the DNA and then drafts in enzymes to acetylate the histone. Now the *NOS* gene can be unspooled and construction of the protein begins. This is just one example of a huge array of transcription factors that cooperate to switch on different combinations of genes in each different kind of cell.

There is another dimension to the definition of epigenetics with astonishing ramifications. Imagine a neural stem cell. It has the ability to become a neuron but not a liver cell or a pancreatic cell.

Like most cells it also has the ability to produce daughter cells by undergoing cell division. What will be the potency of the daughter cell? Will it be totipotent, like a freshly fertilised ovum, or multipotent, like its parent? The answer is multipotent. The daughter of a neural stem cell will be another neural stem cell, which has interesting implications. It means that when a cell divides, its current state of differentiation is inherited by the daughter cell, meaning that whichever epigenetic markings were present in the parent are passed on to the daughter. As such, division of a differentiated cell produces a daughter cell that is differentiated for the same job.

9.2 FOETAL DEVELOPMENT

The next stage of the mystery is how cells know what to become. How does a cluster of cells organise itself so that some of the cells grow into a brain, while others grow into an eyeball? Furthermore, how do they know where to grow? Why do we not sprout eyeballs where we usually expect to find our big toes?

The answer is cellular cooperation. The reason why an eyeball did not grow in place of your big toe is that the development of your eyeball was shaped by its immediate neighbours. This has been well researched in the African clawed frog. The development of a frog's eyeball begins with the formation of a cluster of cells called an optic vesicle. Next, the cells in the optic vesicle send biomolecular signals to the cells in the neighbouring ectoderm, which is the outer layer of the early embryo. The ectoderm responds by differentiating into the cells that will form the lens of the eye, but the cooperation does not stop there. Having differentiated, the lens tissues then beam signals back to the optic vesicle, causing it to develop into the retina. Basically, the optic vesicle and the neighbouring ectoderm co-construct the eyeball *via* the exchange of biomolecular signals (ref. 3, p. 80), as shown in Figure 9.5.

Throughout the development of the foetus, neighbouring cells co-direct their development. Wherever two different kinds of cells exist side by side, each emits different signals, which means that each *receives* different signals, with the result that each shapes the other's development. Going back to the cookbook analogy, it is as if the Mexican restaurateurs mark the Italian

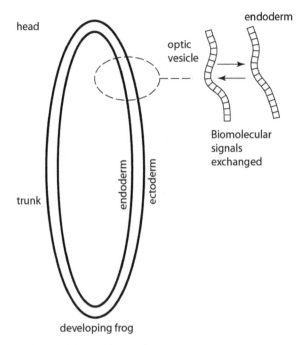

Figure 9.5 Development of a frog's eye. Biomolecular signals are exchanged between the optic vesicle and the neighbouring region of the ectoderm, which is induced to develop into lens tissue.

recipes for their neighbours and *vice versa*. Meanwhile, feet cells will not beam signals for their neighbours to become eyeballs. But if the optic vesicle brought the lens into existence, what brought the optic vesicle into existence?

There is another way to view this question. The development of an embryo is characterised by an ongoing diversification of cell types, such that a single cell—the fertilised ovum—multiplies into a colossal number, not just of cells but of *different kinds* of cells. In the earliest stages of the embryo, there is no optic vesicle but then it suddenly appears, begat from the assortment of cells that existed before, just as their diversity was somehow contrived from that singular brand: the fertilised ovum or zygote, as it is more correctly known. The real question is this: how is it that one kind of cell can produce different kinds of cells?

When a couple makes a baby, the sperm fertilises the ovum, or egg, to create a zygote. From that single cell a whole baby will grow. The building materials and fuel will be provided *via* the mother's placenta and the environment of the uterus will also have an impact on foetal development. But a huge proportion of the process will be directed from within the foetus. Somehow, these cells organise themselves into a screaming, gooey baby, which is pretty incredible. Returning to the analogy, it is as if an entire city's restaurant industry is seeded from a single chef. He turns up with his single manual, none of the pages of which have yet been marked, and then troops from building to building appointing head chefs and assigning them cuisines in which to specialise. These new recruits dutifully mark the relevant pages of their manual and then hire staff to prepare the appropriate dishes. Suddenly, this purpose-built town has a thriving restaurant community, serving Lebanese, French and Chinese cuisine, and everything else, all thanks to the organisational aplomb of the first chef.

The question of how one cell begets a plethora is easy to answer *in principle*. That is, it is easy to come up with plausible suggestions for how it *could* happen and, unfortunately, the actual answer has been much harder to elucidate. Back in 1972 the authors of a seminal research paper referenced three competing hypotheses.[4] Half a century later there are still three rival hypotheses,[5] all of which line up along another continuum. At one end is the idea that if one cell produces two different kinds of cells, the first cell must have been asymmetrical to start with. At the other end is the notion that the similar behaviour of identical cells necessarily leads to asymmetry, as we shall see. Many studies have been completed to improve our understanding of these processes. Unless otherwise stated, those referenced below were carried out on mice, which have been shown to provide a very suitable model for human embryonic development.

To consider these ideas in more detail it will be useful to consider the first stages in embryonic development. As shown in Figure 9.6, after the ovum is fertilised, cell division transforms the resulting zygote into two cells, then into four, then eight, sixteen and so on. Once the embryo has amassed 100 cells, it is referred to as the blastocyst and it is at this stage that it implants itself into the lining of the uterus. Not only do cells multiply during this pre-implantation stage, they also differentiate. By the time the blastocyst has formed, three distinct cell types have

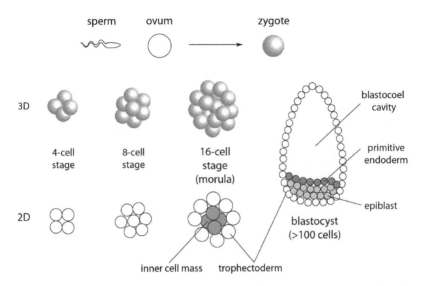

sperm   ovum   zygote

3D

4-cell
stage

8-cell
stage

16-cell
stage
(morula)

blastocoel
cavity

primitive
endoderm

epiblast

2D

blastocyst
(>100 cells)

inner cell mass   trophectoderm

Figure 9.6 Development of the zygote before it is implanted in the wall of the
uterus.

emerged: the epiblast, which will grow into the foetus, the
trophectoderm, which forms the placenta, and the primitive
endoderm, which will become the yolk sac that encapsulates
the foetus.[5] They are like senior school students—there is more
development to take place, but their subjects have now been
selected and the vista of their opportunities narrowed. The cells
that will grow into a baby have already lost their totipotency.

At what point does differentiation first occur? By the end of the
embryo's 16-cell stage, known as the morula, there are two cell
types: the trophectoderm and the inner cell mass. Researchers
say the first decision has taken place.[6] Meanwhile, the second
decision occurs when the inner cell mass divides into the epiblast and the primitive endoderm. There is a colossal amount of
work to be done before these processes are fully understood.

9.2.1 The Three Hypotheses

And so to the three hypotheses describing early development.
Hypothesis number one suggests that the zygote is asymmetrical
and, as a result, when it divides, it creates two slightly different
cells. The word "asymmetrical" should not be taken too literally
in this case because, mathematically, no cell could ever be truly

symmetrical. The nucleus especially would confound any such aspirations unless we had two *identical* copies of every single one of our genes. Instead, it refers to the distribution of cellular material outside the nucleus. The woman's ovum does not only consist of DNA and nutrients; it also contains proteins and other useful molecules that help guide the zygote's development. This hypothesis suggests that proteins may be distributed unevenly, such that one side of the cell has more of some varieties than the other and *vice versa*. Consequently, when the zygote divides, the resulting daughters have distinct populations, which enable differentiation. As ever, there is evidence for and against the hypothesis. Some studies have detected different epigenetic markings, or different gene expression in embryonic cells as early as the three- or four-cell stage,[7,8] while another found that all cells between the five- to eight-cell stages were expressing all the same proteins.[9] Having said that, the first two cited studies investigated mouse embryos, while the latter looked at human embryos.

Hypothesis number two, originally proposed by Martin Johnson and Carol Ziomek, suggests that cell fate is decided by the polarisation of cells.[10] Polarisation is when a cell assigns a front and back, as shown in Figure 9.7. Inside, the internal cargo is directed towards the clawed end along strands woven from the protein actin, which we met in Chapter 2. Later on we will see how this process enables nascent neurons to reach out and make connections with their neighbours. Meanwhile, the polarisation

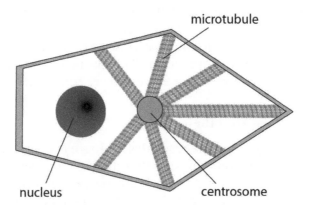

Figure 9.7   A polarised cell.

of cells gives us a means to understand how embryos can become asymmetrical. Suppose that all of the cells in the eight-cell embryo polarise, such that they are all pointing outwards. Next, all eight cells divide along the axis separating front from back. When a polarised cell divides along this axis, one of the two daughters remains polar, while the other ends up non-polar.[5] When all eight cells divide this way, the resulting 16-cell cluster will have eight non-polar cells in the centre and eight polar cells around the outside, as shown in Figure 9.8. The like behaviour of identical cells has produced non-identical cells. We already know that the outer cells of the morula are polarised and the inner cells are nonpolar.[11] The question is whether this *causes* their subsequent differentiation or results from it.

The third hypothesis concerns the way the identical cells interact with what's around them. As the cells proliferate, it quickly transpires that, although they may be identical, their environments are not. By the time the embryo has 16 cells, the

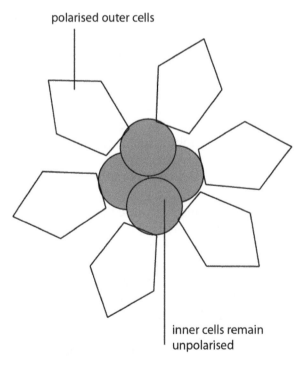

polarised outer cells

inner cells remain
unpolarised

Figure 9.8 The outer cells of the morula are polarised, while the inner cells remain unpolarised.

inner cells are completely surrounded on all sides by neighbours, while the outer shells have a portion of their membrane exposed. This basic fact gives its name to the *inside–outside* hypothesis, originated by Andrzej Tarkowski and Joanna Wróblewska[12] in 1967. Five years later, three researchers named Hillman, Sherman and Graham reported findings in support of the hypothesis.[4] Using mice, they soaked embryos containing two to four cells in a radioactive tracer called thymidine and then separated the cells. Next, each individual cell was combined with another embryo containing four to eight cells. The radioactive cells were added either to the outside or the inside of the unlabelled embryo. By tracking the radioactive signal, they found that the position of the labelled cell decided its fate. Left on the outside, the cells tended to develop into trophectoderm, whereas cells placed in the middle of the embryo were likely to develop into the inner cell mass.

9.2.2 Mechanism

The three hypotheses show how the like behaviour of identical cells can produce asymmetry, but they do not explain how cells differentiate. This remains a subject of fierce research today. While the matter is far from settled, many key players have been identified.

One potential mechanism for the inside–outside hypothesis involves a family of proteins called cadherins. This is a contraction of the longer title: calcium-dependent adhesion.[3] These are adhesive molecules that bind neighbouring cells together, as shown in Figure 9.9. As well as playing an interesting role in foetal development, they are also good candidates to explain how identical twins develop. Later, we will consider the identical twins June and Jennifer Gibbon. What was it that caused them to develop into monozygotic twins. It is impossible to know for sure, but it is possible that cadherins played a role. As the name shows, these proteins are dependent on calcium ions to bind together the cells of the embryo. If the concentration of calcium ions should fall, several of the cadherins will be deactivated with the result that cells in the embryo drift apart. This rationale does not explain why the embryo divides into two clusters of cells, rather than completely separating into a group of individual cells.

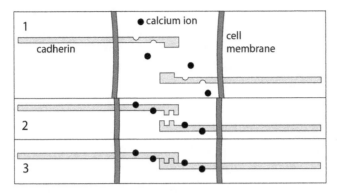

Figure 9.9 Cadherins depend on calcium to activate their cell-binding function.

Much more is known about the potential role of cadherins in foetal development. Like so many things in the body, cadherins have more than one thing that they do. As well as binding cells together, they also facilitate signalling between cells. For example, cadherins can increase the activity of the proteins Rac1 and Cdc42, whilst decreasing the activity of RhoA.[13] The cadherins could explain why the differing environments of the inner and outer cells cause them to differentiate into different cell types. The inner cells of the morula will be connected to neighbours on all sides by cadherins, whereas the outer cells will have no connection on their exposed surface. This means the inner cells would receive cadherin-transmitted signals from all directions, while the outer cells would receive them only from the unexposed sections of their membrane. All of the cells might be predisposed to develop a certain way *in the absence* of cadherin signalling, meaning those outer cells would behave differently than the inner cells.

In fact, studies have confirmed a role for cadherins in the first decision. One experiment found that when the gene for cadherin was deleted, most cells started to develop into trophectoderm,[14] the part of the embryo that the *outer* cells of the morula become. This finding supports the inside–outside hypothesis because it is those outer cells that would experience less interaction with cadherins. But the same findings show that the cadherins are not the head honchos of the first decision. Researchers noted that these cadherin-deleted cells had a much more trophectoderm character, but their ability to polarise was unaffected. Now other studies have confirmed that cells polarise without the help of cadherins,[15] suggesting that these binding proteins obviously do not head the chain of command. They are more captain than colonel.

9.2.3 Merging Hypotheses

Where competing hypotheses each have supporting data, sometimes the best option is to amalgamate them. This is what happened when the scientific world was rocked by the discovery of quantum physics. Beams of light had been shown to have both wave-like and particle-like properties. Eventually, it made sense to embrace the nonsensical and the wave–particle duality model was born. Similarly, what seems to be happening in developmental biology is an amalgamation of hypotheses two and three. The degree of cell contact and the polarity of cells have each been shown to affect development and are probably interlinked. The precise mechanism is yet to be uncovered, but the answer seems to involve something called the HIPPO pathway. The HIPPO pathway is another example of lighter-hearted nomenclature. Mutations in its starring protein, also called HIPPO, cause fruit flies to develop overgrown heads that look like a hippo's. Now, that strikes me as a bit of a contradiction, because the insects only come out like hippos when the protein is not working properly. The pathway's eponymous biomolecule is essentially named for the effects of its mutant alter-ego.

Misleading monikers aside, the effects of this malfunction point to the HIPPO pathway's best known role as a regulator of cell populations. Throughout our lives, cells in our bodies are constantly created and destroyed, but adulthood marks an important shift in the comparative rates. From the moment the ovum is fertilised, we would expect cells to be created more often than they are destroyed in order for that single zygote to multiply into the trillions of cells necessary to make a baby. But once we enter adulthood things slow down. Now we would hope that cells are created only as frequently as they are destroyed so that the total number of cells stays roughly constant. At this stage if one part of our body starts accumulating extra cells, a tumour might be developing, which is something to worry about. For this reason, cancer researchers have taken a keen interest in the HIPPO pathway, with the result that a family of the component proteins has been christened large tumour suppressors (LATS). Basically, the HIPPO pathway is famous for its regulation of cell proliferation; it is like a dial that can be twiddled either to speed up or slow down cell proliferation.

How does the HIPPO pathway help to cast the first decision in the human embryo? Given the contents of the last paragraph, would it not make perfect sense if the pathway's role in baby making were to coordinate this very same function, to ensure that the cells of each nascent organ proliferate at a proportionate rate? Unhelpfully, this does not seem to be the case. Although a clear role for the pathway has been identified, it does not seem to regulate the rate at which cells multiply.[16,17] Earlier, we identified transcription factors as the agents that use the tools of methylation and acetylation to epigenetically modify developing cells. The HIPPO pathway unleashes transcription factors in the outer cells of the morula, while restraining them in the inner cells.

Let's review the chain of command. The transcription factors operate the tools of methylation and acetylation to dictate which genes are expressed. In charge of these particular transcription factors is the HIPPO pathway. So who tells the HIPPO pathway to behave differently in the inner and outer cells of the morula?

This is where the hypotheses merge. We have seen how a cell's position in the morula seems to inform its fate, but also that a cell's polarity weighs into the decision. Enter angiomotin (AMOT), the most senior protein yet identified in this battalion.[11] Working alongside other biomolecules, AMOT acts as a trigger for the HIPPO pathway. Not only must the AMOT squad be activated, but it also needs to be in the right place to receive the signals. This is where the emerging asymmetry designates cellular fate. All cells are in the business of producing proteins and the cargo is subsequently shipped out to their membranes along molecular highways known as actin filaments. In polarised cells, such as those lining the outer edge of the morula, these highways are rerouted so that specific proteins are delivered only to certain parts of the cell (ref. 2, p. 1005).

In this case, freshly-minted AMOT proteins cluster in the front end of the cell, meaning that they are distributed along the outer edge of the morula.[18] This is the wrong place to receive the signals. Meanwhile, in the inner non-polar cells of the morula, AMOT is evenly distributed along the perimeter of each cell. Re-enter the adherens. Recall that these "sticky molecules", including the cadherins, bind cells together and beam signals between them. These transmissions are strongest among the inner cells of the morula, since neighbours surround them on all sides, and this is one of the signals to which AMOT is sensitive.[18] Not only do the inner cells receive stronger signals, but these are the cells where AMOT is available to receive them. Thus, we see how both polarity and cell position negotiate the first decision.[5,17] AMOT is exclusively activated in the inner cells of the morula, as shown in Figure 9.10.

Once AMOT is activated, the HIPPO's trigger is pulled. In the inner cells, the signalled AMOT goes on to activate the LATS battery of kinases, which in turn phosphorylates yes-associated protein (YAP).[16] This is another example of a protein being trapped beyond the reach of its associates, a process called sequestration. Phosphorylating YAP is like saddling it with a ball

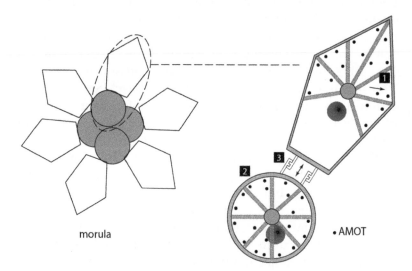

Figure 9.10 Polarised and non-polarised cells respond different to adherens signalling. 1) Molecules of AMOT are directed along microtubules to the cell membrane so that they reach only the front of polarised cells but... 2) ...are evenly distributed around the perimeter of non-polarised cells. 3) Non-polar cells are hence more sensitive to signalling from the adherens junctions.

and chain. In its unadulterated form, YAP can freely enter the nucleus, where it binds with and activates a transcription factor called TEA domain transcription factor 4 (TEAD4), but with the ball and chain attached, it can no longer fit through the holes leading into the nucleus. Thus does the HIPPO strike, a blizzard of activity that obscures the path through this linguistic minefield. When the HIPPO pathway is switched *on*, YAP is muzzled and the transcription factor TEAD4 is switched *off*. Conversely, switching off the pathway in the outer cells, switches on the transcription factor that differentiates them.[5]

Finally, the differential activation of TEAD4 will brand each cell type with a unique epigenetic signature. In the outer cells the activated transcription factor will use the tools of methylation and acetylation to bring a particular set of genes out of storage, namely *CDX2* and *GATA3*.[11] These are also transcription factors, so they may also go off and modify the histone tails or CpG islands of other strips of DNA, unearthing the genes to make still more new proteins. Meanwhile, in the inner cells where TEAD4 is switched off, a different transcription factor called *SOX2* is expressed,[11] producing proteins that will brand the DNA with their own epigenetic mark. These contrasting units of transcription factors will direct the inner and outer cells to express different sets of proteins, which shapes their development into the distinct cell types of trophectoderm and the inner cell mass.

This is the current best guess for how the first decision is made, although there is plenty more to find out. Certain data suggest the HIPPO pathway can be turned on in the absence of adherens signalling,[5] which poses further questions for the validity of the inside–outside aspect of the hypothesis. In considering cause and effect, philosophers ponder whether different factors are either necessary or sufficient for downstream events to unfold. Neither case is yet conclusively made for adherens signalling. However, this emerging hypothesis does plausibly link many disparate observations. As we have seen, identical behaviour of cells ultimately produces non-identical daughters, such that the outer cells polarise, while the inner cells experience higher degrees of adherens signalling. These subtle differences are amplified by a diverse squad of proteins, which interact to direct the production of different sets of genes in the inner and outer cells. From one variety of cells has emerged two in a process that snowballs until a myriad of cell types are woven into the fabric of a beautiful little baby.

9.3 EPIGENETIC INHERITANCE

The epigenetic revolution is likely to have far-reaching consequences for many disciplines. The epigenome is an important regulator of the genetic activity that enables the foetus to develop, which makes it a major focus. Parents want their children to grow up happy and healthy, and efforts to ensure that outcome start from the moment of conception. What makes the emerging field so tantalising is that it offers an arena in which for nature to continue its long-running battle with nurture because the epigenome is subject to modification not only by the complement of genes we happen to inherit but also *via* our interaction with the world around us.

A controversial claim is that we may inherit more than just genes from our parents. There is clear evidence in many species of plants and animals that parts of a parent's epigenome are inherited alongside their genes. If this is true for humans it means we inherit not only a selection of each parent's genes but also some of the settings regulating which are switched on and which are switched off. This has been a surprising discovery because the production of sex cells by meiosis is characterised by a complete reprogramming of the inherited genes, which amounts to removal of all the methylation and other epigenetic markings. This is how the earliest cells in the zygote come to be totipotent. Only there is some doubt as to just how complete the process of reprogramming is.

Many studies in mice and certain invertebrates have found that epigenetic markings can be inherited over multiple generations. Mice that are fed a high-fat diet can pass on a susceptibility to obesity and diabetes.[19] Similarly, when male mouse pups were raised in isolation from their mothers, they developed an enhanced susceptibility to stress, which was likewise transmitted down several generations *via* the epigenome.[20] In a more positive example of the phenomenon the *Caenorhabditis elegans* worm can epigenetically inherit a "nose" for a certain smell. When the nematodes are exposed to the olfactory cue early in their development, generations of offspring demonstrate an enhanced ability to move towards that same cue, with the added boon that the trait also seems to boost fertility.[21]

If similar mechanisms were confirmed in humans the implications could be huge. There have already been suggestions that when mothers are undernourished during pregnancy, their children can inherit a greater susceptibility to obesity, for which evidence has been provided by the Dutch famine at the end of World War II. However, these effects are not passed on beyond the second generation,[21] which poses questions for the notion of true epigenetic heritability. Nevertheless, it may yet transpire not only that our environment has the power to switch our genes on and off but that those settings echo through future generations.

9.4 CONCLUSION

This chapter has described the importance of epigenetic factors in the development of the foetus. These factors do not cease to influence our bodies the moment we are born but continue to affect us throughout our lives. The following chapters will explore how epigenetic activity could impact the development of complex traits, including violence and creativity. Currently, there is scant evidence for their propagation *via* the above-described mechanisms of epigenetic inheritance. But there is much more evidence that our interaction with our environment has the power to shape our characters by switching certain genes on or off.

REFERENCES

1. J. A. Lynch, *What Are Stem Cells?: Definitions at the Intersection of Science and Politics,* University of Alabama Press, 2011.
2. H. Lodish, A. Berk, C. A. Kaiser, M. Krieger, A. Bretscher, H. Ploegh, A. Amon, and M. P. Scott, *Molecular Cell Biology: International Edition,* W. H. Freeman, 7th edn, 2012.
3. S. F. Gilbert, *Developmental Biology,* Sinauer Associates, Sunderland, Mass, 9th edn, 2010.
4. N. Hillman, M. I. Sherman and C. Graham, The Effect of Spatial Arrangement on Cell Determination During Mouse Development, *J. Embryol. Exp. Morphol.,* 1972, **28**(2), 263–278.

5. C. Lorthongpanich and S. Issaragrisil, Emerging Role of the Hippo Signaling Pathway in Position Sensing and Lineage Specification in Mammalian Preimplantation Embryos, *Biol. Reprod.*, 2015, **92**(6), 143.

6. A. Jedrusik, Making the First Decision: Lessons From the Mouse, *Reprod. Med. Biol.*, 2015, 1–16.

7. M.-E. Torres-Padilla, D.-E. Parfitt, T. Kouzarides and M. Zernicka-Goetz, Histone Arginine Methylation Regulates Pluripotency in the Early Mouse Embryo, *Nature*, 2007, **445**(7124), 214–218.

8. A. Burton, J. Muller, S. Tu, P. Padilla-Longoria, E. Guccione and M.-E. Torres-Padilla, Single-cell Profiling of Epigenetic Modifiers Identifies PRDM14 as an Inducer of Cell Fate in the Mammalian Embryo, *Cell Rep.*, 2013, **5**(3), 687–701.

9. A. Galán, D. Montaner, M. E. Póo, D. Valbuena, V. Ruiz, C. Aguilar, J. Dopazo and C. Simón, Functional Genomics of 5- to 8-Cell Stage Human Embryos by Blastomere Single-Cell cDNA Analysis, *PLoS ONE*, 2010, **5**(10), e13615.

10. M. H. Johnson and C. A. Ziomek, The Foundation of Two Distinct Cell Lineages Within the Mouse Morula, *Cell*, 1981, **24**(1), 71–80.

11. T. Frum and A. Ralston, Cell Signaling and Transcription Factors Regulating Cell Fate During Formation of the Mouse Blastocyst, *Trends Genet.*, 2015, **31**(7), 402–410.

12. A. K. Tarkowski and J. Wróblewska, Development of Blastomeres of Mouse Eggs Isolated at the 4- and 8-cell Stage, *J. Embryol. Exp. Morphol.*, 1967, **18**(1), 155–180.

13. M. J. Wheelock and K. R. Johnson, Cadherin-mediated Cellular Signaling, *Curr. Opin. Cell Biol.*, 2003, **15**(5), 509–514.

14. R. O. Stephenson, Y. Yamanaka and J. Rossant, Disorganized Epithelial Polarity and Excess Trophectoderm Cell Fate in Preimplantation Embryos Lacking E-Cadherin, *Dev. Camb. Engl.*, 2010, **137**(20), 3383–3391.

15. R. Moore, W. Tao, Y. Meng, E. R. Smith and X.-X. Xu, Cell Adhesion and Sorting in Embryoid Bodies Derived from N- or E-Cadherin Deficient Murine Embryonic Stem Cells, *Biol. Open*, 2014, **3**(2), 121–128.

16. C. Lorthongpanich, D. M. Messerschmidt, S. W. Chan, W. Hong, B. B. Knowles and D. Solter, Temporal Reduction of LATS Kinases in the Early Preimplantation Embryo

Prevents ICM Lineage Differentiation, *Genes Dev.*, 2013, **27**(13), 1441–1446.

17. H. Sasaki, Position- and Polarity-dependent Hippo Signaling Regulates Cell Fates in Preimplantation Mouse Embryos, *Semin. Cell Dev. Biol.*, 2015, **47**, 80–87.
18. Y. Hirate, S. Hirahara, K. Inoue, A. Suzuki, V. B. Alarcon, K. Akimoto, T. Hirai, T. Hara, M. Adachi, K. Chida, S. Ohno, Y. Marikawa, K. Nakao, A. Shimono and H. Sasaki, Polarity-dependent Distribution of Angiomotin Localizes Hippo Signaling in Preimplantation Embryos, *Curr. Biol.*, 2013, **23**(13), 1181–1194.
19. P. Huypens, S. Sass, M. Wu, D. Dyckhoff, M. Tschöp, F. Theis, S. Marschall, M. H. de Angelis and J. Beckers, Epigenetic Germline Inheritance of Diet-induced Obesity and Insulin Resistance, *Nat. Genet.*, 2016, **48**(5), 497–499.
20. E. J. Nestler, Transgenerational Epigenetic Contributions to Stress Responses: Fact or Fiction?, *PLOS Biol.*, 2016, **14**(3), e1002426.
21. E. Heard and R. A. Martienssen, Transgenerational Epigenetic Inheritance: Myths and Mechanisms, *Cell*, 2014, **157**(1), 95–109.

Section 2: Concluding Remarks

Conventional wisdom dictates that the love of your life should also be your best friend and the overlap is mirrored in the biochemistry of romance. Lust and friendship is an incautious cocktail and the romantic and platonic varieties of love are polar opposites, but when the bewitching thrill of romance mellows, the companionate love we aim to foster in its wake thrives on the same blend of biochemicals that binds us to friends and family members.

All of these drives identified by Fisher—lust, attraction and attachment—seem to tap into the brain's reward network. Our brains mete out pleasure not only for actions that promote childbirth, but also social bonding. We need not fall in love with someone to derive pleasure from interacting with them. We are a social species and our ability to work together buffers us from the hardships of the solitary existence eked out by sea turtles.

This has interesting ramifications for the position advanced by Johann Hari in Chapter 3. His conviction that love is the true antidote to addiction is completely compatible with the affliction's biochemical model. Indeed, Liebowitz directly compares romantic attraction to an amphetamine high and rejection to the come down that follows. If romantic love and social bonding in general tap into our reward networks, this provides biochemical support for Hari's theory: people without love in their lives may turn to drugs to cadge a compensatory pleasure fix from a hoodwinked reward network.

The Chemistry of Human Nature
By Tom Husband
© Tom Husband 2017
Published by the Royal Society of Chemistry, www.rsc.org

These findings show us another way we can cheat the self-replicators. We have evolved to feel pleasure as a reward for actions that have previously furthered their interests. In spite of the questionable claims of the odd fakir, eating is not something we can live without doing. But we can love without rearing children and why should we not? We have evolved to feel pleasure in association with lust and romantic love because those emotional states were previously very likely to culminate in reproduction. But the advent of contraception has empowered us to choose when and even if to have children. We are not obliged to reproduce simply because we are adapted to gain pleasure from its pursuit. On the other hand, many people derive deep pleasure from raising children, which is also absolutely consistent with the biochemical model here advanced.

Section 3:
The Chemistry of Character

Introduction

If the replicators make us feed and breed, what makes us human? All animal species have to eat to stay alive and all reproduce, but so far we have not considered much that is distinctly human. Romantic love probably qualifies, although prairie voles use the same neurotransmitters as us to stoke the flame of monogamous pair bonding.

This section will consider more human-specific aspects of our behaviour, starting with arguably the most defining human trait: creativity. After that, we will consider violence and dominance. Readers might take exception to the idea that violence is part of our nature but, if anything, Chapter 11 better supports the hypothesis that we are not naturally violent. Meanwhile, dominance is common to a huge swathe of species extending far beyond vertebrates. But, as a triumvirate, the trio of traits is uniquely human. Other species use violence to gain dominance and some primate species have tool-making capabilities, but in no other species can the creative drive of an organism propel it to dominance status over its troop. Silicon Valley is a testament to the power of creativity to propel hard-working innovators into the stratosphere of human society. Moreover, the two methods of elevating status are not mutually incompatible. In *The Art of War* the ancient military strategist Sun Tzu counselled generals to be creative in their approach to battle, to create misinformation and to achieve the element of surprise by acting in unpredictable

The Chemistry of Human Nature
By Tom Husband
© Tom Husband 2017
Published by the Royal Society of Chemistry, www.rsc.org

ways. One of the legendary military battles in Chinese history was won by a general named Cao Cao, who had the idea to burn down his opponent's grain store, depriving the enemy troops of the sustenance they would need to mount a long-term military campaign. This section will consider how our chemistry enables us to demonstrate these traits.

CHAPTER 10

The Chemistry of Creative Intelligence

In the 1980s, life imitated the art of a very unlikely pair of criminals. Identical twins June and Jennifer Gibbons were incarcerated in Broadmoor secure mental-health facility after being convicted of arson in a Welsh college. An unusually tame addition to the facility, the sisters had long felt imprisoned in an inescapable relationship with each other. June was finally liberated following an eerie revelation by her sister. Born to an RAF pilot of West Indian heritage and raised in a small Welsh town, the twins quickly came to the attention of various teachers and educational psychologists. They refused to speak to anyone but each other and yet they fostered a love/hate relationship of murderous proportions.

Their descent into crime began with artistic disappointment. In addition to chronicling every detail of their lives in journals, each of the twins wrote a novel, as well as various short stories. Expecting to hit the big time, they were disappointed that all the publishers they approached rejected their novels. Eventually, the two sisters pooled their resources to pay a vanity publisher to release June's *Pepsi Cola Addict,* a cautionary tale in which the protagonist is jailed for stealing to feed his habit. Their failure triggered wild behaviour. Each of them vied for the attentions of the same boy, a love rivalry which ended up with the twins

The Chemistry of Human Nature
By Tom Husband
© Tom Husband 2017
Published by the Royal Society of Chemistry, www.rsc.org

plunged into a river with June's hands clamped around Jennifer's throat. Having made up, they turned to arson. Their first crime was in a tractor store, which the authorities might have been able to overlook, but then they struck at Pembroke Technical College.

Biographer Marjorie Wallace told the world their story[1] after succeeding where so many had failed. She may have got to know the twins better than even their own family. She read everything they had ever written and became a regular visitor at Broadmoor, where she managed to draw them into conversation by talking about their writing. In time, they opened up to her, so much so that Jennifer shared an astonishing insight. A week before they were to be released, Jennifer revealed that she had chosen to sacrifice herself so that June could be released from their self-destructive relationship. On the bus ride home after her release from Broadmoor, June found Jennifer slumped against her shoulder, eyes glazed. She was rushed to hospital but died shortly afterwards. The cause of death was myocarditis but the real mystery is the cause of the cause. There were no signs that she had been poisoned, no recent changes had been made to the drugs she was prescribed and she had not had the sort of serious illness that could trigger myocarditis.[1] Her death remains a mystery.

June and Jennifer's story shows an interesting facet of creativity that represents a much deeper truth about human nature. By the prevailing definition of current research, the twin's creativity was in a sort of superposition, like Schrödinger's famous cat, because they simultaneously were and were not creative. Producing fiction would generally be considered an example of creative writing, but academics might find this a careless use of the term. Creativity is defined as the production of something both novel and useful.[2] Certainly, the twin's fiction was novel, but since publishers were rejecting it, it is reasonable to assume that it was not useful. But that situation was to change.

Creativity is a two-way street. Chopsticks equipped with noodle-cooling fans were only a hit in Japan. We can only judge their inventor to be creative based on the fact that Japanese society was receptive. If creativity combines novelty with usefulness, we cannot consider an individual creative in isolation from their milieu.[3] Furthermore, each is subject to influence by the other.

June and Jennifer manipulated the receptivity of their milieu in such a way as to bequeath their work with usefulness. Their crime spree brought them notoriety that gave June's *Pepsi Cola Addict* a cult following. Suddenly, their milieu had become more receptive to their fiction, which shaped their characters in return, because now, by virtue of this acceptance, their fiction can be deemed creative after all.

The Gibbons twins' story reflects the situation with the nature–nurture dichotomy. For more than a century, scientists have struggled to parse the effects of a person's genes from their environment in an attempt to fathom the causes of complex traits, including intelligence and creativity. Emerging evidence is increasingly showing that the two cannot be parsed. Just as classical physicists were astonished to realise that energy and matter overlap, so to do nature and nurture. Our genes can affect our receptivity to our environment, while our environment influences the way that our genes are expressed.

So far, this book has rationalised traits shared by all animals. This chapter will explore arguably the most defining feature of human nature: creativity. Starting with the birth of the nature–nurture paradigm, it will investigate how the epigenetic mechanisms introduced in Chapter 9 may blur the boundaries between nature and nurture. Finally, it will consider the role of the pleasure centre, which eternally motivates us in service of the self-replicators.

10.1 THE BIRTH OF A PARADIGM

A convenient starting point for our story is the Victorian poly-math and intrepid explorer Francis Galton. A cousin of the great Charles Darwin, Galton left a remarkable legacy. A youthful jaunt in search of Lake Ngami in Africa paid for itself, partly because he produced and sold maps of the area but also when he immortalised his derring-do in a series of travel-writing articles. Amongst other achievements, he devised the statistical concept of regression to the mean, to be explained in Chapter 12, he laid the groundwork for fingerprinting to become a major tool of forensic analysis and he coined the term "nature–nurture". The one considerable blot on his record was that he was the father of eugenics.[4]

His eugenic proposals were not idly made. He invented the idea, he christened it and he established a fellowship in eugenics with an associated laboratory and journal. His overwhelming conviction that intelligence was almost entirely inherited gave him to think that a super-race could be developed by adopting the techniques with which breeders produce champion race-horses. He reasoned in a letter to the magazine *Macmillan's* that poor people would surely understand what a public service they would do by not breeding, while smarter citizens could be encouraged to spread their superior genes more liberally by issuing tax cuts for each child they had.[4]

Galton hatched this idea after completing some research of questionable value. Galton was the first person to use twins as a means to dissect the ancient riddle of nature *versus* nurture. The scientific historian Nathaniel Comfort has dubbed Galton a poster boy of the dichotomy[4] because he benefitted from several years of tuition from an invalid elder sister. Apparently heedless of this, he ultimately concluded that intelligence was almost entirely dictated by an individual's genetic profile. His evidence was sometimes tenuous and based on methodology since discovered to be flawed.

Galton's views were based on two sets of twin studies, each of which was conducted in different contexts.[5] In one study he considered how the character of twins altered after leaving home. Now they were spending less time together, their environments would be sufficiently different to measure the impact of nurture. He concluded that their similarity remained constant in spite of their separate lives. Next, he compared the characters of twin children who had always behaved very differently since birth. In spite of their shared environment, he found such twins to remain very different in character. Taken in tandem, Galton concluded from his studies that nature had a much larger impact than nurture on the development of intelligence.

A retrospective problem with Galton's argument is that no one understood the genetic distinction between identical and non-identical twins at that time.[5] Monozygotic twins are so-named because they grow from a single zygote, the name given to a fertilised ovum, which subsequently splits in two. Conversely, dizygotic twins grow from two separate ova, each of which happened to be fertilised at roughly the same time. It is monozygotic twins who have since come to be considered genetically identical (although even that now seems in doubt, as we shall see), while dizygotic twins are no more genetically similar than normal, singly born siblings. No one knew this in Galton's day. Nowadays, it is commonly said that monozygotic twins share 100% of their genes, while dizygotic twins share 50%. This refers not to the entire genome but to the small proportion of genes that can vary. Chapter 5 explained how roughly 99% of any two human genomes are the same. The 100% and 50% figures refer to the all-important variation possible in that remaining 1%.

Two components of Galton's legacy proved distressingly enduring: eugenics and the supremacy of nature over nurture in dictating an individual's intelligence. The development of intelligence tests was strongly motivated by eugenic aims, basically as a means to decide which individuals should be encouraged to have children and which should be discouraged. Lewis Terman, developer of the Stanford–Binet test, was one of the first researchers to propose that the "dullness" of certain ethnic groups must be racial, which constituted "a grave problem because of their unusually prolific breeding".[6] Such ideas were then taken in justification of the vilest practices of the Nazis in one of the darkest chapters in human history. Even after the atrocities of the World War II, progressive luminaries such as Julian[7] and Aldous Huxley[8] still supported eugenic practices. In fact, even in the 21st century, compulsory sterilisation has been practiced on convicted criminals in California in the US[9] and other countries too.

Intelligence testing was frequently controversial even after eugenic principles were abandoned. Often, the controversy centred around accusations that researchers sought to revive such principles. Edward Wilson's publication of *Sociobiology* provoked a furore spanning decades. Wilson suggested a genetic basis for behavioural trends, implying, to name a particularly contentious example, that women were evolutionarily adapted to be housewives.[10] Controversy hit again in the 1990s when Charles Murray and Richard Herrnstein published a diatribe entitled *The Bell Curve*. Among its more odious proposals was to close US borders for fear that immigrants would pollute the country with their inferior genetic pedigree.[11] No matter how methodology evolved, there were always researchers willing to perpetuate Galton's dogma that nature dominates nurture in the allocation of smarts.

Such claims centred around the findings of heritability estimates. Heritability is the proportion of variance in a particular trait, such as intelligence, which results purely from genetic differences.[12] In any society, intelligence varies between individuals. The classic bell curve shows that most people are of average intelligence and that the further anyone deviates from this average, the fewer of them exist, making people with profound learning difficulties approximately as rare as geniuses. Statistical manipulation reduces the data contained in a bell curve to a single figure called variance, which indicates the total "differentness" in the trait between all the members of the group. Variance indicates the extremes as well as the tally. For example, concerning intelligence, it would incorporate not only the highest and lowest IQ values in the cohort but also the comparative number of participants exhibiting each IQ score. Next, heritability is the fraction of that variance attributed to differences in genes. So, what heritability does *not* mean is that, say, 20% of *your* intelligence results from your genes, while the other 80% is due to your upbringing. Rather, it means that the spread of IQ scores between you and your friends is attributable by a measurable amount to the differences in your genetic profiles.

Heritability has classically been measured using an updated version of the twin studies pioneered by Galton. Science is all about controlled tests, the idea being to change just one variable and keep all the others the same. And much of the field of behavioural genetics has been built around the belief that different twin types offer the perfect control measures by which to parse nature from nurture. The rationale is as follows: if we consider only those children who are raised by their biological parents, both twin types experience the same nurturing. Where they differ is in their genetic profile. As mentioned, monozygotic twins are thought to share 100% of their genes, compared to dizygotic twins, who share 50% of the genes that can vary. In this sense, dizygotic twins share the same proportion of genes in common as do normal siblings. To measure the heritability of intelligence, researchers start by measuring the intelligence of all the twins in a particular region. Next, they measure the correlation in intelligence between the identical, monozygotic twins and the correlation between the non-identical, dizygotic twins. For example, if all the identical twins also had identical IQ scores, the intelligence correlation for the monozygotic twins would be +1. After that, subtract the dizygotic correlation from the monozygotic correlation, double the difference and the result is heritability.[12]

Heritability estimates have been calculated for a variety of complex traits other than intelligence, including how we vote,[13] how we worship,[14] the sexuality we develop,[15] our criminality[16] and even how much television we choose to watch.[17] Most of these studies have used vast databases of twins, in particular the Minnesota Center for Twin and Family Research. Basically, the details of twins are collected, after which they are sent questionnaires to ascertain how much television they watch, how much money they earn, their religion, their political affiliation and so on. Then, the numbers are crunched as per the explanation above and out pops a heritability figure.

Most of these traits were found to be heritable at least to some extent. This of course suggests—and frequently formed the next phase in research—that there were specific genes that might promote such traits. This has evolutionary implications. If genes underpin certain behaviours, it suggests they must have been advantageous at some point in our evolutionary history. This has been particularly contentious on traits such as violence, a point to be considered in the next chapter.

This genetic basis for certain traits can be baffling at face appearance. Does the fact that our television-viewing preferences have a heritability of 45%[17] arise from some mysterious television-watching gene? In one of *The Simpson's* Halloween specials, the family was depicted in fairy tale times. Instead of watching television, Bart whiled away his leisure time staring at a fire. It's a well-observed joke because fire is oddly beguiling. Does the flickering screen of a television tap into our promethean instincts? And what about homosexuality? If genes proliferate by enhancing reproductive capacity, how could a gene be adaptive that romantically inclines its carrier *against* the gender with whom they can procreate? Tim Spector has shared some interesting insights into a genetic basis for homosexuality, suggesting that while such genes decrease reproductive fitness in members of one gender, they may increase fertility when they appear in members of the opposite sex.[18]

Television viewing has been associated with broader themes in human character. In fact, the researchers never imagined that an individual's television-watching habits reflected analogous, adaptive behaviour from our ancestors' days in the savannah. Rather, they linked it with the heritability of neurosis. This is one of the so-called "big five" traits, between which all of our characters are said to be divided by differing proportions. In addition to neurosis, the big traits include openness to experience, conscientiousness, happiness and likeability. Each of these has also been shown to be heritable by different amounts, all of which exceeded 40% in one study.[19] These softer gauges of personality are more readily conceived in genetic terms. For example, neurosis is a sort of hangover of the vigilance that would have protected us before we conquered the food chain. In Chapter 7 we explored the biochemical mechanism of the fight-or-flight response, several stages of which could be modified by genetic variation. Such variants could include a greater density of receptors in the amygdala, augmenting the individual's sensitivity to potential threats, the production of greater amounts of the stress hormones, such as adrenaline, and so on. Subsequent research has found that neurotic people can be more inclined to watch television.[20]

10.2 HOW HERITABLE IS CREATIVITY?

As both intelligence and creativity are hard to measure, it is perhaps unsurprising that heritability estimates vary for both. Twin studies for the heritability of creativity have turned out a wide range of figures between 20%[21] and 92%, but these figures have adopted a variety of instruments by which to measure creativity. The 92% heritability figure was calculated in a twin study in which the participants had to rank their own musical ability against that of the general population.[22] Unfortunately, self-assessment is notoriously unreliable. Ben Goldacre notes in *Bad Science* that, in studies of self-reported intelligence, more than half of people consider themselves to be above average, demonstrating the very human tendency to overrate our own abilities.[23] Sensible of this weakness, researchers have developed the Creative Achievement Questionnaire, in which respondents rate how true are objective prompts such as: *"My work has been cited by other scientists in national publications."* Researchers Davide Piffer and Yoon-Mi Hur used the criteria to show that creative achievement was 43% to 67% heritable.[21]

Another research team suggested that the emergenic nature of creativity arises because the combination of the relevant genes is greater than the sum of its parts. Previous researchers have already identified sub-traits, all of which are supposedly required in order for true creative genius to emerge, including a hard-working temperament, self-confidence bordering on ego-maniacal and a greater regard for ideas than people.[24] Niels Waller, Thomas Bouchard, David Lykken, Auke Tellegen and Dawn Blacker suggested the genes underlying such traits combine in a multiplicative way.[25] The genes for height are additive, meaning the more of the height-promoting genes you have, the taller you grow. Conversely, multiplicative genes can be useless on their own but combine to produce desirable traits. For example, certain species of bee have an advantageous trait of jettisoning diseased grubs from the hive. This requires the bees to eat away the wax plug that seals the grub into its gestation chamber, as well as then to detect its malady and toss it out. Research discovered that different genes underpinned these separate behaviours, but that each was useless without the other. Only when bees had the genes for both behaviours did they display this useful trait.[26]

This would explain why genius-level creativity is not so easily passed from parent to child. In support of their hypothesis Waller's team notes that intellectual giants, such as Einstein, Newton and Darwin, did not have famous grandfathers. (However, Darwin did have a fairly famous son, who originated the theory that the moon was once a part of Earth. Also, Darwin's bulldog Thomas Huxley was grandfather to the considerable talents of Aldous, Andrew and Julian Huxley.) When parents reproduce, 50% of each of their genes is selected at random and passed on to the child. This means that complex combinations of genes are unlikely to be replicated in the offspring. Meanwhile, dizygotic twins also share 50% of their genes and monozygotic twins have long been said to share 100%. Waller's team argued that if creativity was a genetically emergenic trait, it should correlate much better between identical twins. Yet another creativity-gauging instrument was deployed, the Creative Personality Scale, which rates respondents on sub-traits, such as *egotistical, confident, unconventional, inventive* and *original*. They found that heritability of creativity in reared-apart twins was 54% for monozygotic twins and −6% for dizygotic twins, supporting their hypothesis that creativity is an emergenic trait.

10.3   THE VALIDITY OF HERITABILITY STUDIES

Frequent attacks have been launched on twin studies, questioning the validity of their findings.[27] Heritability estimates have been enormously useful for tracking down the causes of complex diseases,[28] but as so often happens in science, the numbers do not quite add up. High heritability estimates for any characteristic—whether a personality trait or proneness to a disease—suggest that particular genes must be responsible. But attempts to track down the genes have been difficult. Even where studies have successfully linked gene variants with diseases, they have failed to account for the prevalence of those diseases predicted by the heritability measurements. This disparity has become known as the missing heritability.[28]

Evan Charney mounted a particularly fervent attack on this type of inheritance in 2012 when he published a wide-ranging critique of twin studies. He began by listing the assumptions on which twin studies are based before systematically dismantling each one in turn. The three key assumptions are as follows. First, monozygotic twins share 100% of their genes, whereas dizygotic twins share just 50%. Second, these figures hold true throughout the life of each twin, where monozygotic twins have exactly the same genome from birth until death. Third, complex traits originate entirely from either genetic or environmental influences.

A fascinating point to arise from the critique is the fact that "jumping genes" have the power to alter our genomes as we live and breathe. These unconventional genes were yet another discovery of Barbara McClintock introduced in Chapter 5. Since her discovery, we have learned that certain genes, called retrotransposons, code for a peculiar brand of protein that returns inside the nucleus, runs off another copy of its genetic blueprint

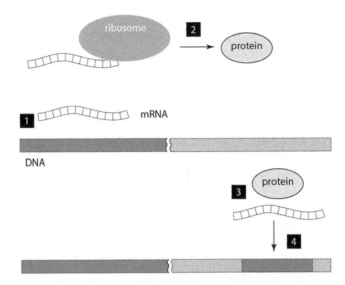

Figure 10.1 Jumping genes. (1) The gene is transcribed into mRNA, which is then (2) translated into the corresponding protein. (3) The protein reverse transcribes the original strip of mRNA back into DNA, which is then (4) reinserted into a different section of the chromosome.

and then reverse engineers the mRNA back into DNA, which it promptly splices into a different region of the genome, as shown in Figure 10.1. Basically, these genes make copies of themselves that insert themselves elsewhere into our existing stock of genes. Originally, researchers thought this only happened in the germ line, the tissue which produces sex cells, but over the last decade leaders in the field, including Haig Kazazian Jr, have been astonished to learn that it also happens in the somatic DNA.[29] What that means is that the complement of genes you have in any of your cells may be slightly different from the blend you inherited when you were first conceived. Also, the set of genes you have in one cell may be slightly different to the set you have in another, such as in neurons compared to liver cells. Haig Kazazian is among the researchers trying to establish how often this happens and the consequences when it does. Mercifully, the vast majority of alterations seem to be benign, but as Charney notes, there are ramifications for twin studies.

Charney poses a variety of uncomfortable questions for heritability estimates.[30] The jumping genes that all of us carry in our

DNA mean monozygotic twins will not share 100% of their genes. Also, the degree of similarity between their genomes will alter through their lifetimes because of continued jumping gene activity. Another criticism concerns placentas. It was traditionally assumed that monozygotic twins always share a single placenta, while dizygotic twins develop one each. In fact, either variety of twins can develop either arrangement. Sharing a placenta exposes both foetuses to all the same material from the mother's bloodstream. Consequently, they are much more likely to develop the same viral infections, which can significantly influence development.

A key strand in Charney's argument concerns the epigenetic mechanisms introduced in the last chapter. Researchers have realised that our interaction with the environment has the power to influence the genes we express. Charney is not the only commentator to point out the potential impact of this fact on heritability estimates.[31] Twin studies have linked the stock of genes twins share to the traits they exhibit, but now there is a complicating factor: for each twin, how do we know which genes are switched on and which are switched off? The next section will consider potential ways that our environment could shape creativity *via* these epigenetic mechanisms.

10.4   THE PRICE OF CREATIVITY

A small and diverse group of mourners attended the funeral of George Price. Among them were homeless people and the great evolutionary biologists William Hamilton and John Maynard Smith.[33] Price made a fantastic contribution to the discovery of inclusive fitness, but ultimately his brilliance cost him his life.

Legend has it that the story began in a Cambridge pub, where Hamilton was talking with another key player in the discovery. John Haldane was asked if he would sacrifice his life to save another human being. The apocryphal tale tells it that he answered he would, but only if he could save two brothers, four half-brothers or eight first cousins. Hamilton went away and gave Haldane's jape a mathematical treatment and the result was the theory of inclusive fitness.[32]

The concept of fitness, introduced in Chapter 5, can be separated into personal and inclusive fitness. Suppose we consider a male squirrel. Its personal fitness is the number of offspring it produces, but its inclusive fitness includes all of his children, as well as any other young squirrels that share his genome, such as nieces and nephews, as fitness can also be applied at the genetic level. For any given allele the squirrel has, the fitness can be expressed as the number of times that allele is replicated in future offspring. Since its children will receive 50% of their genes from each parent, there is a 50% chance of any of the squirrel's alleles being replicated in each of its children. Inclusive fitness accounts for the fact that the squirrel need not have any children in order to promote the replication of its genes. Since the squirrel inherited 50% of its genes from each of its parents, for any allele there is a 50% chance that any of its siblings carries the same allele. Accordingly, there is a 50% chance that the same allele will be passed on to any of that sibling's children. So, there is an overall 25% chance that any of the squirrel's nieces or nephews carries that allele. Inclusive fitness takes count of all the copies of genes expected in future generations, whether they be in the squirrel's offspring or those of family members.

This has been used to explain extreme examples of altruistic behaviour. For example, red squirrels, *Sciurus vulgaris* sometimes shriek an alarm call when they see a predator approach, which is obviously beneficial to the clan but extremely dangerous for the shrieker. Assuming the behaviour has a genetic basis, it would likely involve several genes, but suppose that a single gene suffices. How could such a gene replicate itself in future generations when it threatens to kill carriers before they have produced any children? If a shrieking squirrel is killed by the predator, its personal fitness will certainly suffer, but its inclusive fitness may increase. If the act saves two nieces, there is a 25% chance that each of those nieces carries the same alarm-call gene as their self-sacrificing kinsman. Thus, the theory of inclusive fitness demonstrates how an individual's genes can prosper through even suicidally altruistic behaviour.

George Price had left his two daughters and their divorced mother in New York to make a highly speculative bid to achieve greatness in London. Troubled but inspired by the notion that altruism could be selected by evolution, he spent a fortnight researching Hamilton's work, the result of which was his landmark equation. The story goes that he was employed at University College London within 90 minutes of showing this mathematical breakthrough to Cedric Smith, a professor at the university.[33]

The Price equation summarises the relationship between traits, such as altruism, and the fitness of the individuals that exhibit them. It shows how the prevalence of a trait changes according to the changing fitness of successive generations. Where fitness increases with the prevalence of the trait, obviously that trait becomes more common and *vice versa*. His stroke of genius was to uncouple the concept of relatedness from the closeness of the blood relationship between group members and instead define it in terms of the resemblance between their genomes.[34]

These discoveries were a huge step forward for evolutionary biology. Haldane, Hamilton and Price had explained a long running mystery of evolutionary biology. Evolution rewards traits that enhance the individual's capacity to produce offspring, in which the underlying genes will be replicated. How, then, could traits that actively reduce an individual's fitness—such as the squirrel's warning shriek—be considered evolutionarily advantageous? Inclusive fitness provided the answer and laid the groundwork for the emergence of sociobiology and evolutionary psychology. Here was evidence that human altruism originates not in philosophy or religion but at the deeper level of our genetic code. Richard Dawkins' also drew heavily on these findings as he espoused his theory that the selfish gene, and not the individual, is the unit of natural selection.[26] Genes thrive when they equip their parent vehicles to bid more efficiently for the finite resources of fuel and building materials, thus perpetuating the ancient and life-originating process of self-replication.

Although he may not have realised it, Price had absolutely succeeded in his ambition to make an outstanding contribution to science. Sadly, his life started to unravel. He became obsessed with coincidence, which led him to abandon his atheism and embrace Christianity. While turning his mathematical acumen to the search for a bible code, he also tried to foster an altruistic spirit by steadily donating the entirety of his worldly wealth to North London's homeless population. Consequently, he ended up destitute himself in a squat, where the tragic arc of his life culminated in suicide.[33]

The precise root of Price's mental health problems is a matter of speculation. Days before he killed himself he saw psychiatrist Dr Christopher Lucas, who told the inquest that his erratic intake of thyroxine tablets might have fostered depression, a common complaint when people forego the medication. There was a suggestion that Price may have had schizophrenia, but Lucas could not be sure. Price's biographer, Oren Harman, suggests he was on the autistic spectrum based on characteristics, including his obsession with numbers, his social awkwardness and his insensitivity towards other people.[33] Meanwhile, Dr Christopher Badcock, author of *The Imprinted Brain: How Genes Set the Balance of the Mind Between Autism and Psychosis*, builds on these ideas[35] with the observation that calendar-counting savantism, which he links to Price's search for bible ciphers, and religious delusion are symptomatic of autism[36] and psychosis,[37] respectively.

To what extent did Price's mental health problems overlap with his genius? This question was first posed by the Greek philosophers, who described genius as the *divine madness*. Attempts to firm up the now commonplace association erupted into the so-called mad genius controversy,[38] one difficulty is parsing the mob jeers from the clinical diagnosis. Unconventional behaviour is often mockingly called mad, but it is precisely the defiance of convention that enables a genius to interpret old truths with new eyes. But mere eccentricity should hardly send psychiatrists reaching for their prescription pads.

Genius has variously been linked with bipolar disorder, schizophrenia and autism. Many well-known performers have been diagnosed with bipolar condition, such as Stephen Fry and Vivien Leigh, the iconic actress who played Scarlett O'Hara in *Gone with the Wind*. Such performers have sometimes shown themselves reluctant to take medication because they value the inspiration they find during manic episodes.[38] The film *A Beautiful Mind* depicted Nobel laureate John Nash's battle with schizophrenia. Following from an illustrious career, the Japanese artist Yayoi Kusama currently resides as an outpatient of a mental health facility, and Van Gogh is one of many intellectual giants to be retrospectively diagnosed with since-discovered conditions. In a study of 291 eminent individuals, severe psychopathology was diagnosed for 18% of scientists, 31% of composers, 38% of the artists and 46% of the writers.[24]

George Price is not the first scientific genius to be retrospectively diagnosed with Asperger's syndrome. Psychiatrist Michael Fitzgerald and his co-author Brendan O'Brien suggest in *Genius Genes* that great visionaries, including Charles Darwin, Isaac Newton and Albert Einstein, all had Asperger's syndrome,[39] which falls on the spectrum of autism disorders. Some people have questioned his findings[40] but leading autism researcher Simon Baron-Cohen is "fairly certain" that at least Newton and Einstein had Asperger's.[41] A growing body of evidence also shows that people working in science, technology, engineering or maths (STEM) are more likely to have autistic traits,[42,43] while students on the autistic spectrum are more likely to pick STEM subjects at university.[44]

But does this alleged overlap between STEM and autism extend to creativity? Whether or not they produce paintings or write great works of literature, the scientifically minded are great innovators, making original observations about reality, like Newton with his apple, or pioneering new inventions. Being on the autistic spectrum is not a prerequisite for scientific originality but the two sometimes coincide. However, a study of more than a million people found that those with autism were not more likely to work in creative scientific or artistic professions, although their siblings were.[45] Moreover, another study found that autistic people demonstrated less, rather than more, imagination.[46] However, one research group worried that typical test conditions might constrain the imagination of autistic people, especially when creative output was limited to what they could do with a pencil and paper. So, they ran a study in which 27 participants created three-dimensional designs, which a panel of experts from Google judged to be unusually original,[47] which fits with the now firmly established idea that spatial reasoning ability is vital in STEM careers.[48]

But is there a relationship between insanity and genius? The reality is that you can have mental health problems without being a genius and you can be a genius without having mental health problems. But as Stephen Durrenberger notes in the *Encyclopedia of Creativity*, mental health problems are more common among eminently creative people than average.[38]

The latest theory frames creativity and neurosis as co-dependents. A research team led by Adam Perkins and Dean Mobbs argues that self-generated thoughts are the engine of neurosis.[49] Self-generated thoughts are ideas that arise not because of what we are currently seeing, hearing or otherwise sensing but in response to memories of previous experience. They are not inherently negative, but the team believes that a certain brain circuit, when overactive, can colour them to create what they call a "camel's hump of misery", such that they fret even though no immediate threat exists. The default mode network (DMN) combines two separate regions of the brain: the medial prefrontal cortex, introduced in Chapter 7, and the posterior cingulate cortex. Studies including functional magnetic resonance imaging (fMRI) scans have shown not only that these regions activate during the origination of self-generated thoughts but also when those thoughts take a negative turn. A subcomponent of the DMN, called the medial prefrontal network, connects with various mood-regulating regions, including the amygdala. Finally, they argue that the ability of neurotic individuals to imagine different realities to their present reality underpins the creative flair commonly associated with the neurotic personality type.

I cannot keep writing about this as if from an entirely objective viewpoint. Many of my closer friendships were forged in the crucible of a shared understanding of the nightmare of neurosis. I am entirely aware of how inventive a neurotic brain can be because, left unattended, mine will invent any manner of threats, none of which actually warrant attention. I think the moral psychologist Jonathan Haidt might belong to this club after his comments in *The Happiness Hypothesis*.[50] (Who else but a neurotic would write a book about happiness?) Haidt describes what he calls "the Imp", that inner voice that constantly suggests inappropriate, dangerous or terrifying ideas. He cites the example that during dinner parties, the Imp suggests terrible things for him to say to his fellow diners. On top of cliffs it tells him to jump. That one doesn't work on me anymore, I've progressed to thinking: *why don't you throw someone* else *off the cliff?*

One of my fellow neurotics rejoiced upon learning of "the Imp", but what was really interesting was when he tried to explain it to other friends. They had no idea what he was talking about. The imp employs a devious imagination in the constant invention of unrealistic threats, so the best thing to do is set it to work hatching ideas for creative pursuits.

Proving the link between neurosis and creativity is difficult. Examples come readily to mind of people who demonstrate both traits; Woody Allen's whole career is an exploration of his neurotic tendencies. But as with the mad genius controversy, just because a person is neurotic does not mean they will also be creative. There is no clear relationship between neurosis and creativity.[51–53]

10.5 NATURE–NURTURE DUALITY

Another thing creativity and neurosis have in common is the difficulty of pinning down their heritability. Earlier, we saw how the elusive heritability of creativity may result from the multiplicative combination of underlying genes. Meanwhile, studies have recorded a range of values of the heritability of neurosis.[54] Behavioural genetics has striven to parse the effects of genes (nature) from those of the rearing environment (nurture). But evidence is accumulating that the two influences overlap. An individual's environment has the power to alter the genes they express, just as their genetic makeup can change their susceptibility to their environment.

A highly relevant example is the discovery that childhood cuddles mitigate adult stress. Not only might this explain the questionable heritability of neurosis and creativity, it may also show that heritability estimates are questionable full stop. Another strand of Evan Charney's aforementioned assault on the heritability paradigm is the fact that cuddling seems to influence the genes we express.[30] Multiple studies on rats have shown that the more they are groomed in infancy, the more oxytocin receptors they express as adults. Not only does this incline them to demonstrate the same grooming habits with their own pups, it also reduces their stress levels, for, as we saw in Chapter 8, oxytocin signalling is yet another trigger for our pleasure centre in the brain's reward network.

Several studies of rats confirm that mothers emulate the parenting they received as pups. Rats are categorised by how much licking and grooming they lavish on their children. When pups are often licked and groomed, they go on to exhibit the same grooming tendencies when they produce their own offspring. That might not sound that surprising. After all, we all grow into our parents. But the relationship holds even when the mother and pup are completely unrelated. When rats groom foster litters, the adopted pups grow up to demonstrate the same grooming habits. So, this important behaviour is differentiated not by the different genes they inherit but how their genes are switched on or off by the way they are nurtured.[30]

Further studies have confirmed that the maternal grooming habits change the expression of genes. Pups that receive high levels of licking and grooming in infancy grow up to express more oxytocin receptors, as well as more glucocorticoid receptors. Last chapter we saw how methylation of CpG islands and histone tails has the effect of placing genes into storage. Methylated genes are packed away beyond the reach of the transcriptional machinery, while un-methylated genes are readily unspooled from their histones, facilitating their expression. When pups receive a lot of grooming, as adults they exhibit low levels of methylation on the promoter regions for both the oxytocin receptor genes and the glucocorticoid receptor gene.[55] In other words, the genes are out of storage and available for transcription.

Both of these receptors are associated with the body's stress response. Chapter 8 showed how oxytocin can trigger the pleasurable response of the brain's reward centre, but what about the glucocorticoid receptor? Glucocorticoids are a class of hormone, best known amongst which is cortisol. The steroid is vital for proper functioning of the body, exerting influence over processes such as blood pressure and heart rate. It also plays a role in the fight-or-flight response, previously described. Brief cortisol spikes are healthy, but problems occur when levels stay high for long periods. In the short term, high levels help to prepare the body for fight or flight, by dampening the immune response while jacking up energy levels. But over longer periods, problems such as depression and sleeplessness may kick in. One reason for this is that cortisol normally suppresses the secretion of two other stress hormones: corticotrophin-releasing hormone (CRH) and adrenocorticotropic hormone (ACTH). When levels remain high, the hypothalamic–pituitary–adrenal axis gets stuck in stress response mode.[56] The number of glucocorticoid receptors an individual has will influence their stress response. If they express fewer of the receptors, they will be less sensitive to cortisol, which will heighten the activity of the hormones it normally supresses: CRH and ACTH. Consequently, the pups who receive less grooming in infancy will grow up to have a heightened stress response.

How well do these findings translate to human subjects? More evidence exists for the density of glucocorticoid receptors than for oxytocin receptors. A grim study led by Patrick McGowan and Michael Meaney investigated receptor density in childhood abuse victims who had committed suicide.[57] Post-mortem analyses found these unfortunate souls had fewer glucocorticoid receptors than individuals in the control group. Subsequent studies have extended these findings to other transcripts of the associated gene.[55] Childhood abuse is in a very different category from the extent to which rodent pups receive grooming, but the findings support the idea that a person's childhood environment exerts influence lasting right through adulthood *via* epigenetic modification.

This may or may not have anything to do with creativity. The link between neurosis and creativity is only hypothetical and both traits will originate from the combination of a huge array of factors. One study could be interpreted utterly to refute the idea that reduced oxytocin receptor density predicts greater creativity. A team led by Carsten De Dreu and Simone Shamay-Tsoory used a variety of methods to show how oxytonergic circuitry boosts creativity.[58] Firstly, higher plasma levels of oxytocin predict a greater tendency for participants to seek novelty. Secondly, issuing oxytocin nasal spray reduced analytical thinking, while enhancing divergent thinking. Finally, carriers of a different allele of the oxytocin receptor gene were found to be more original.

A potential link with autism is supported by stronger evidence. A team led by Simon Gregory and Margaret Pericak-Vance used the polymerase chain reaction (PCR; explained in Chapter 12) to show that the oxytocin receptor gene was transcribed less frequently in people on the autistic spectrum,[59] which suggests they have fewer oxytocin receptors. Meanwhile, another study used the same technique in post-mortem analysis to show that autistic people had fewer glucocorticoid receptors in the middle frontal gyrus region of the brain.[60] These results should be interpreted with caution, particularly for the latter study, which had just 13 members in the treatment and control groups, and which was so recent that the results are unlikely yet to have been replicated. What this particular pathway can tell us about the development of creativity sits within a broader framework. There may or may not be a link between neurosis and creativity, which may or may not be linked with autism, along with various other mental health conditions. Whether or not this specific pathway influences the development of creativity, the point is made that environmental factors have the power to shape character. In his assault on heritability studies, Evan Charney argues that this is just one of many ways in which our interaction with the world can switch genes on or off.[30] When this happens in our brains, of course it has the power to affect the way we think and act.

But the oxytocin receptor gene demonstrates the nature–nurture duality in the opposite direction too. Not only may the environment in which we are raised influence the number of oxytocin receptors we express, but the specific allele we carry may influence our susceptibility to that environment. The study mentioned above investigated two alleles of the oxytocin receptor gene, the commonly carried A allele and its rarer variant, the G allele. A study of 288 university students led by Robyn McQuaid and Hymie Anisman found that oxytocin-related behaviours, such as empathy, optimism and trust, were enhanced in carriers of the G polymorphism, but the flipside was that such individuals were more likely to develop symptoms of depression in response to childhood adversity.[61] In the simplest terms the carriers of the G allele were more sensitive. This is a perfect example of how a person's nature influences their susceptibility to their nurture.

Just as the character of the Gibbon sisters made society more receptive to their writing, so our genes and environment exert mutual influence over each other. It is far from clear as to how this interaction might affect the development of an individual's creativity, but it looks to be a vital component in the equation.

10.6   A ROLE FOR THE REWARD NETWORK?

A final interesting point is the role of the reward network in creativity. Various studies have shown that creative pursuits activate those parts of the brain introduced in Sections 1 and 2, such as the ventral tegmental area, known to mediate reward.

Back in 2012 a group of rappers turned up to a research facility to freestyle in the name of science. A research team led by Siyuan Liu and Allen Braun was keen to pinpoint the part of the brain associated with creativity, so they asked the artists to alternate between memorised and improvised lyrics.[62]

Scans of the freestyling rappers registered alternating activity between two different brain regions. As they improvised, the scans recorded activity in either the dorsolateral or the medial prefrontal cortex but never both at the same time. Earlier, we saw how the medial prefrontal cortex forms part of the DMN associated with self-generated thoughts, those reflections that are unconnected with what is going on immediately around us. Conversely, the dorsolateral prefrontal cortex is associated with the cognitive control network (CCN), which processes incoming sensory data. This supports the proposed role of the DMN in creativity.

But the scans also implicated the reward network. When the rappers were freestyling, activity in the DMN was accompanied by activity in the ventral tegmental area and substantia nigra, implicating our motivational reward network in yet another human trait.

Building on similar findings, Alice Flaherty has proposed a dopamine-fuelled model of creativity as a successor to the now outdated left-brain/right-brain model.[63] Adding to the evidence in support of a role for the reward network, the neurologist notes that dopamine antagonists not only alleviate symptoms of psychosis, but they also reduce creative output. She suggests that the motivation provided by the dopamine system is a better predictor of quality creative output than cognitive ability, reasoning that an unquenchable compulsion to produce will ultimately elevate skill as practice makes perfect.

We are a long way off understanding how creativity develops. The so-called "mad genius controversy" might be desperately in need of a more enlightened name, but the hypothesised overlap between psychopathology and creativity continues to generate interesting lines of inquiry. The DMN looks like a promising candidate for the origin of ideation, but it does not seem to work in isolation. A huge list of factors could influence creativity. Different alleles may change the number of receptors produced or alter the shape or receptors, making them more or less sensitive to their associated neurotransmitters. Finally, the involvement of the brain's reward network fits with the developing thesis of this book. Creativity has hugely advanced the average fitness of *Homo sapiens*, enabling us to manipulate our environment to eliminate many of the dangers that preyed on our ancestors, so it is no surprise that we find creativity a pleasurable pursuit. It is just another reward paid out for services to the self-replicators.

REFERENCES

1. M. Wallace, *The Silent Twins*, Vintage, London, 1996.
2. R. E. Jung, B. S. Mead, J. Carrasco and R. A. Flores, The Structure of Creative Cognition in the Human Brain, *Front. Hum. Neurosci.*, 2013, 7, 330.
3. B. Barbot, M. Tan, and E. L. Grigorenko, "Chapter 4: The Genetics of Creativity: The Generative and Receptive Sides of the Creativity Equation", in *Neuroscience of Creativity* ed. O. Vartanian, A. S. Bristol and J. C. Kaufman, The MIT Press, Cambridge, Massachusetts, 2013.
4. N. C. Comfort, Zelig: Francis Galton's Reputation in Biography, *Bull. Hist. Med.*, 2006, **80**(2), 348–363.
5. M. Bulmer, *Francis Galton: Pioneer of Heredity and Biometry*, Johns Hopkins University Press, Baltimore, 2003.
6. L. M. Terman, *The Measurement of Intelligence: An Explanation of and a Complete Guide for the Use of the Stanford Revision and Extension of the Binet–Simon Intelligence Scale*, Houghton Mifflin, 1916.
7. J. Huxley, *Man in the Modern World*, Chatto and Windus, London, 1947.
8. F. Dikotter, Race Culture: Recent Perspectives on the History of Eugenics, *Am. Hist. Rev.*, 1998, **103**(2), 467.
9. C. G. Johnson, "Female Inmates Sterilized in California Prisons without Approval | The Center for Investigative

Reporting". Available at: http://cironline.org/reports/female-inmates-sterilized-california-prisons-without-approval-4917. [Accessed: 24-Apr-2016.]

10. E. O. Wilson, *Sociobiology: The New Synthesis*, The Belknap Press of Harvard University Press, Cambridge, Mass, 1975.
11. R. J. Herrnstein and C. Murray, *The Bell Curve: Intelligence and Class Structure in American Life*.
12. R. Plomin, J. C. DeFries, G. E. McClearn and P. McGuffin, *Behavioral Genetics*, Worth Publishers, 5th edn, 2008.
13. J. R. Alford, C. L. Funk and J. R. Hibbing, Are Political Orientations Genetically Transmitted?, *Am. Polit. Sci. Rev.*, 2005, **null**(2), 153–167.
14. B. M. D'Onofrio, L. J. Eaves, L. Murrelle, H. H. Maes and B. Spilka, Understanding Biological and Social Influences on Religious Affiliation, Attitudes, and Behaviors: A Behavior Genetic Perspective, *J. Pers.*, 1999, **67**(6), 953–984.
15. R. C. Pillard and J. M. Bailey, Human Sexual Orientation has a Heritable Component, *Hum. Biol.*, 1998, **70**(2), 347–365.
16. T. Frisell, Y. Pawitan, N. Långström and P. Lichtenstein, Heritability, Assortative Mating and Gender Differences in Violent Crime: Results from a Total Population Sample Using Twin, Adoption, and Sibling Models, *Behav. Genet.*, 2011, **42**(1), 3–18.
17. R. Plomin, R. Corley, J. C. DeFries and D. W. Fulker, Individual Differences in Television Viewing in Early Childhood: Nature as well as Nurture, *Psychol. Sci.*, 1990, **1**(6), 371–377.
18. T. D. Spector, *Identically Different: Why You Can Change Your Genes,* Weidenfeld & Nicolson, London, 2012.
19. K. L. Jang, W. J. Livesley and P. A. Vemon, Heritability of the Big Five Personality Dimensions and Their Facets: A Twin Study, *J. Pers.*, 1996, **64**(3), 577–592.
20. C. Persegani, P. Russo, C. Carucci, M. Nicolini, L. L. Papeschi and M. Trimarchi, Television Viewing and Personality Structure in Children, *Pers. Individ. Dif.*, 2002, **32**(6), 977–990.
21. D. Piffer and Y.-M. Hur, Heritability of Creative Achievement, *Creat. Res. J.*, 2014, **26**(2), 151–157.
22. A. A. E. Vinkhuyzen, S. van der Sluis, D. Posthuma and D. I. Boomsma, The Heritability of Aptitude and Exceptional Talent Across Different Domains in Adolescents and Young Adults, *Behav. Genet.*, 2009, **39**(4), 380–392.

23. B. Goldacre, *Bad Science*, Fourth Estate, London, 2008.
24. C. Martindale, *"Genetics"*, *Encyclopedia of Creativity*, Academic Press, vol. 1, 2 vols., 1999.
25. T. J. Bouchard Jr, D. T. Lykken, A. Tellegen, D. M. Blacker and N. G. Waller, Creativity, Heritability, Familiarity: Which Word Does Not Belong?, *Psychol. Inq.*, 1993, **4**(3), 235–237.
26. R. Dawkins, *The Selfish Gene: 30th Anniversary Edition*, Oxford University Press, Oxford, 2006.
27. D. M. Evans and N. G. Martin, The Validity of Twin Studies, *GeneScreen*, 2000, **1**(2), 77–79.
28. E. E. Eichler, J. Flint, G. Gibson, A. Kong, S. M. Leal, J. H. Moore and J. H. Nadeau, Missing Heritability and Strategies for Finding the Underlying Causes of Complex Disease, *Nat. Rev. Genet.*, 2010, **11**(6), 446–450.
29. H. Kazazian, Interview, 21-Dec-2015.
30. E. Charney, Behavior Genetics and Postgenomics, *Behav. Brain Sci.*, 2012, **35**(5), 331–358.
31. M. Trerotola, V. Relli, P. Simeone and S. Alberti, Epigenetic Inheritance and the Missing Heritability, *Hum. Genomics*, 2015, **9**(1), 1.
32. P. Kitcher, *Vaulting Ambition: Sociobiology and the Quest for Human Nature*, MIT Press, Cambridge, Mass, 1987.
33. P. S. Harper, Oren Harman: The Price of Altruism. George Price and the Search for the Origins of Kindness, *Hum. Genet.*, 2010, **128**(6), 649–650.
34. S. A. Frank, George Price's Contributions to Evolutionary Genetics, *J. Theor. Biol.*, 1995, **175**(3), 373–388.
35. C. Badcock, "Who WΔz George Price?", *Psychology Today*. Available at: http://www.psychologytoday.com/blog/the-imprinted-brain/201005/who-w-z-george-price. [Accessed: 21-Apr-2016.]
36. L. Heavey, L. Pring and B. Hermelin, A Date to Remember: The Nature of Memory in Savant Calendrical Calculators, *Psychol. Med.*, 1999, **29**(1), 145–160.
37. G. E. Getz, D. E. Fleck and S. M. Strakowski, Frequency and Severity of Religious Delusions in Christian Patients with Psychosis, *Psychiatry Res.*, 2001, **103**(1), 87–91.
38. S. D. Durrenberger, Mad Genius Controversy, in *Encyclopedia of Creativity*, Academic Press, vol. 2, 2 vols., 1999.

39. M. Fitzgerald and B. O'Brien, *Genius Genes: How Asperger Talents Changed the World*, AAPC Publishing, 2007.
40. S. Dosani, Autism and Creativity: Is There a Link between Autism in Men and Exceptional Ability?, *Br. J. Psychiatry*, 2005, **186**(3), 267.
41. I. James, Singular Scientists, *J. R. Soc. Med.*, 2003, **96**(1), 36–39.
42. S. Baron-Cohen, S. Wheelwright, R. Skinner, J. Martin and E. Clubley, The Autism-spectrum Quotient (AQ): Evidence from Asperger Syndrome/High-functioning Autism, Males and Females, Scientists and Mathematicians, *J. Autism Dev. Disord.*, 2001, **31**(1), 5–17.
43. E. Ruzich, C. Allison, B. Chakrabarti, P. Smith, H. Musto, H. Ring and S. Baron-Cohen, Sex and STEM Occupation Predict Autism-spectrum Quotient (AQ) Scores in Half a Million People, *PLOS ONE*, 2015, **10**(10), e0141229.
44. X. Wei, J. W. Yu, P. Shattuck, M. McCracken and J. Blackorby, Science, Technology, Engineering, and Mathematics (STEM) Participation Among College Students with an Autism Spectrum Disorder, *J. Autism Dev. Disord.*, 2013, **43**(7), 1539–1546.
45. S. Kyaga, M. Landén, M. Boman, C. M. Hultman, N. Långström and P. Lichtenstein, Mental Illness, Suicide and Creativity: 40-Year Prospective Total Population Study, *J. Psychiatr. Res.*, 2013, **47**(1), 83–90.
46. B. Crespi, E. Leach, N. Dinsdale, M. Mokkonen and P. Hurd, Imagination in Human Social Cognition, Autism, and Psychotic-affective Conditions, *Cognition*, 2016, **150**, 181–199.
47. M. L. Diener, C. A. Wright, K. N. Smith and S. D. Wright, Assessing Visual–Spatial Creativity in Youth on the Autism Spectrum, *Creat. Res. J.*, 2014, **26**(3), 328–337.
48. J. Wai, D. Lubinski and C. P. Benbow, Spatial Ability for STEM Domains: Aligning Over 50 Years of Cumulative Psychological Knowledge Solidifies its Importance, *J. Educ. Psychol.*, 2009, **101**(4), 817–835.
49. A. M. Perkins, D. Arnone, J. Smallwood and D. Mobbs, Thinking Too Much: Self-generated Thought as the Engine of Neuroticism, *Trends Cogn. Sci.*, 2015, **19**(9), 492–498.
50. J. Haidt, *The Happiness Hypothesis: Putting Ancient Wisdom to the Test of Modern Science*, Arrow, London, 1st edn, 2007.

51. M. Giampietro and G. M. Cavallera, Morning and Evening Types and Creative Thinking, *Pers. Individ. Dif.*, 2007, **42**(3), 453–463.

52. J. Hoseinifar, M. M. Siedkalan, S. R. Zirak, M. Nowrozi, A. Shaker, E. Meamar and E. Ghaderi, An Investigation of the Relation Between Creativity and Five Factors of Personality In Students, *Procedia Soc. Behav. Sci.*, 2011, **30**, 2037–2041.

53. L. A. King, L. M. Walker and S. J. Broyles, Creativity and the Five-factor Model, *J. Res. Personal.*, 1996, **30**(2), 189–203.

54. G. Andrews, G. Stewart, R. Allen and A. S. Henderson, The Genetics of Six Neurotic Disorders: A Twin Study, *J. Affect. Disord.*, 1990, **19**(1), 23–29.

55. P.-E. Lutz and G. Turecki, DNA Methylation and childhood Maltreatment: From Animal Models to Human Studies, *Neuroscience*, 2014, **264**, 142–156.

56. E. R. Kandel, J. H. Schwartz, T. M. Jessell, S. A. Siegelbaum and A. J. Hudspeth, in *Principles of Neural Science*, McGraw Hill, New York, 5th edn, 2013.

57. P. O. McGowan, A. Sasaki, A. C. D'Alessio, S. Dymov, B. Labonté, M. Szyf, G. Turecki and M. J. Meaney, Epigenetic Regulation of the Glucocorticoid Receptor in Human Brain Associates with Childhood Abuse, *Nat. Neurosci.*, 2009, **12**(3), 342–348.

58. C. K. W. De Dreu, M. Baas, M. Roskes, D. J. Sligte, R. P. Ebstein, S. H. Chew, T. Tong, Y. Jiang, N. Mayseless and S. G. Shamay-Tsoory, Oxytonergic Circuitry Sustains and Enables Creative Cognition in Humans, *Soc. Cogn. Affect. Neurosci.*, 2014, **9**(8), 1159–1165.

59. S. G. Gregory, J. J. Connelly, A. J. Towers, J. Johnson, D. Biscocho, C. A. Markunas, C. Lintas, R. K. Abramson, H. H. Wright, P. Ellis, C. F. Langford, G. Worley, G. R. Delong, S. K. Murphy, M. L. Cuccaro, A. Persico and M. A. Pericak-Vance, Genomic and Epigenetic Evidence for Oxytocin Receptor Deficiency in Autism, *BMC Med.*, 2009, **7**(1), 62.

60. N. Patel, A. Crider, C. D. Pandya, A. O. Ahmed and A. Pillai, Altered mRNA Levels of Glucocorticoid Receptor, Mineralocorticoid Receptor, and Co-Chaperones (FKBP5 and PTGES3) in the Middle Frontal Gyrus of Autism Spectrum Disorder Subjects, *Mol. Neurobiol.*, 2016, **53**(4), 2090–2099.

61. R. J. McQuaid, O. A. McInnis, J. D. Stead, K. Matheson and H. Anisman, A Paradoxical Association of an Oxytocin Receptor Gene Polymorphism: Early-life Adversity and Vulnerability to Depression, *Front. Neurosci.*, 2013, 7, 128.
62. S. Liu, H. M. Chow, Y. Xu, M. G. Erkkinen, K. E. Swett, M. W. Eagle, D. A. Rizik-Baer and A. R. Braun, Neural Correlates of Lyrical Improvisation: An fMRI Study of Freestyle Rap, *Sci. Rep.*, 2012, 2, 834.
63. A. W. Flaherty, Frontotemporal and Dopaminergic Control of Idea Generation and Creative Drive, *J. Comp. Neurol.*, 2005, **493**(1), 147–153.

CHAPTER 11

The Chemistry of Violence

In June, 2010, Mark Soliz was arrested after a high-speed police chase ended with the collision of his vehicle into an 18-wheel truck. Shortly afterwards, he was charged with the murder of 61-year-old grandmother Nancy Weatherly. Jurors heard how Soliz also killed a beer delivery man named Ruben Martinez during an eight-day, methamphetamine-fuelled crime spree, involving robberies and a drive-by shooting. Cornered after the chase, his girlfriend at the time subsequently testified that he planned either to shoot one of the officers or himself but dropped his gun and so ran for it instead.[1] She also testified that Soliz laughed at the grandmother's accent as she pleaded for her life and also later when he recounted killing her.

The evidence was much too damning for the defendant to plead innocent to any crime, rather his defence team argued that he had diminished responsibility because he had partial foetal alcohol syndrome, a neurological condition associated with greatly impaired learning, impulsivity and aggressive behaviour.[2] The jury was asked to consider whether any mitigating circumstances would justify the reduction of the sentence from death to life in prison. They agreed within an hour that he should be executed.[3]

A fatal dispute in Italy ended with a more lenient outcome. Algerian Abdelmalek Bayout admitted to stabbing and killing

The Chemistry of Human Nature
By Tom Husband
© Tom Husband 2017
Published by the Royal Society of Chemistry, www.rsc.org

Walter Perez after the Colombian visitor insulted his kohl eye make-up. Bayout, who claimed he wore the cosmetic for religious reasons, was schizophrenic,[4] so the nine-year jail sentence was already much lower than might otherwise have been expected. But, in a subsequent appeal, another year was knocked off the punishment when expert witnesses testified that Bayout carried several genes that predisposed him to violent outbursts in stressful situations.[5] This is a curious twist because activity of one of those genes was associated in Chapter 7 with obsessive romantics but, in this case, the protein monoamine oxidase A (MAO A) was linked with violence.

Many scientists questioned the decision in the Bayout case. Critics noted that studies found that the effects of the allele varied depending on the person's ethnicity. Renowned geneticist Steve Jones argued in even simpler terms: "90% of all murders are committed by people with a Y chromosome—males. Should we always give males a shorter sentence?"[5]

Both of these cases centre on biochemical bases of violent behaviour. This chapter will consider the science behind the expert testimonials to explore the chemistry of violence. This is an important research area with the potential for unsettling findings. What if there was a gene, or more likely a suite of genes, that coded for a violent phenotype? Evolutionary psychologists have suggested that human violence could be a gene-coded adaptation.[6] Sharks and other vicious predators demonstrate that violence can be adaptive but what about humans? Is it a basic instinct that must find an outlet as surely as our other drives? Alternatively, is it something we learn? And, in either case, can the instinct be overcome or the behaviour unlearned?

11.1   FOETAL ALCOHOL SPECTRUM DISORDER

Mark Soliz was diagnosed with *partial* foetal alcohol syndrome because, although he met a variety of behavioural indicators, his facial features were not affected in certain ways emblematic of the condition. Jurors in the case heard that, while he had the close-set eyes that characterise the syndrome, he did not have a flat upper lip or flattened area between the nose and lips.[7] But another set of mitigating circumstances concerned the horrific nature of his upbringing.

Foetal alcohol syndrome seems to inflict its effects *via* epigenetic mechanisms. One problem is that the developing foetus cannot metabolise alcohol, as the genes for the relevant proteins remain switched off at this early developmental stage. Alcohol dehydrogenase and aldehyde dehydrogenase are the critical enzymes required for metabolism. Members of many cultures lack the genes for the latter, a phenomenon linked with their greater tendency to get red-faced when drinking.

The second problem is how the alcohol acts on the developing brain. The wiring of the brain is a fantastically complicated process, during which nascent neurons stretch out and migrate towards chemical signals,[8] demonstrating the chemotaxis described in Chapter 5. This stretching out is achieved by the actin–myosin complex, which pulls the edges of the cell along tracks of actin filament. As with the earlier stages of foetal development, the precise concoction of proteins necessary to achieve this feat is generated thanks to programming of the epigenome. Research has discovered that alcohol poisoning strips the methyl markers from the CpG islands,[9] which suggests that developing neurons would be flooded with unnecessary proteins. Methylation has the effect of placing genes in storage, so removal of those markings ought to draw various genes out of storage. The tightly regulated ecosystem of neural development is thus disturbed.

An important question is whether the syndrome causes violence. The condition is linked with various negative outcomes, such as inappropriate sexual behaviour, problems at school and with learning in general, difficulties in understanding the actions of others, criminal behaviour and problems with drugs or alcohol. Violent behaviour is a far from universal symptom, but it is much more likely when the individuals have been victims of violent abuse in childhood.[10] This was a point on which Soliz's defence team drew heavily.

Soliz was himself a victim of the most appalling neglect as a child. His mother not only drank during her pregnancy but also took drugs, including regularly sniffing paint. After his birth, she continued to fund her habit by working as a prostitute, sometimes even pushing her son out of the only available bed to do so. He was roaming the streets alone as young as four years old. By age 10, he was telling a psychiatrist that he heard voices telling him to kill people.[7]

This confluence of factors obscures what the story shows about the chemistry of violence. His pregnant mother's drug abuse would have inflicted lasting damage on Soliz, but this cannot be considered in isolation as a potential cause of his aggressive character. The continued abuse he endured in childhood may have left him with a dearth of glucocorticoid receptors like the individuals described in the previous chapter. Foetal alcohol syndrome itself can afflict individuals without them developing violent character, provided they are properly cared for.[11] The abuse is the stronger causal agent because many conditions are linked with violence only on the condition of coincidental abuse.

MAO A has been dubbed the warrior gene because of its apparent implication in violent behaviour. As ever, the moniker is misleading because research has centred on abnormal functioning of an important gene on which we all rely. Abdelmalek Bayout carried a variant called the L-allele,[4] which has long been linked with violent character, but contingent on coincidental abuse, as with the Soliz incident.

11.2   WHAT IS A GENE ANYWAY?

Genes have proven a slippery concept to define. We have known about them for a long time but working definitions required constant revisions in light of emerging knowledge. Once they were thought of like beads on a necklace, but new findings suggest that any piece of jewellery made to gene specs would create too grotesque a trinket to gift even a most hated mother-in-law.

If anything, the tortuous cookbook analogy presented in Chapter 9 was inadequately convoluted. Imagine making dinner for friends like this: you grab five cookery books and consult several different pages of each, combining many sections of different recipes to make a single dish. One of the books also contains a set of instructions advising you how to season the pan but even that you cherry pick and splice with alternative instructions from a rival tome. Likewise, you amalgamate snippets from several sets of instructions on how best to remove the remnants of the dish burned to the bottom of the oven tray. Now you have an idea of the complexity of genomic function. Proteins are just one variety of many biochemicals coded by DNA, some of which facilitate protein production, others of which obstruct it.

A group of researchers led by Mark B. Gerstein and Michael Snyder has suggested the following updated definition for a gene: "A gene is a union of genomic sequences encoding a coherent set of potentially overlapping functional products."[12] No longer can a gene be considered a stretch of consecutive nucleobases, nor does it code a single protein, nor does it only code proteins. In addition to the mRNA transcripts translated to make proteins, a wide array of non-coding RNA transcripts exist, which do other things besides dictating the sequence in which for amino acids to be connected. Some of these directly facilitate the production of biomolecules, such as transfer RNA (tRNA), which helps to line up the amino acids. Others actually obstruct protein production, such as micro-RNA, which binds to complementary stretches of mRNA such that they cannot be translated into proteins. As if anyone needed convincing that biochemistry was complicated, our attempts to fathom it endlessly yield additional bafflement.

In short, a gene is not simply a code translating to a sequence of amino acids. It also contains auxiliary regions that regulate when and how the nucleobase sequence is transcribed with supporting information about whether to translate the resulting transcription into an amino acid sequence or do something else with it.

One consequence of this is that mutations can have very different effects, as the ongoing research into the violence phenotype shows. Figure 11.1 shows a range of possible outcomes. Some mutations cause proteins to be assembled such that they differ by a single amino acid from the most common variants. This can have the effect of making that protein very slightly more of less efficient. As we saw last chapter, carriers of the GG oxytocin receptor allele seem to be slightly more sensitive and empathic than carriers of the more common AA allele. Another more drastic thing that can happen is that a single nucleotide mutation absolutely wrecks the corresponding protein. This is like the difference between adding lemon juice when the recipe called for lime and downing tools midway through a recipe and bunging everything in the oven. Rather than coding for the wrong amino acid, the mutation inadvertently creates a stop

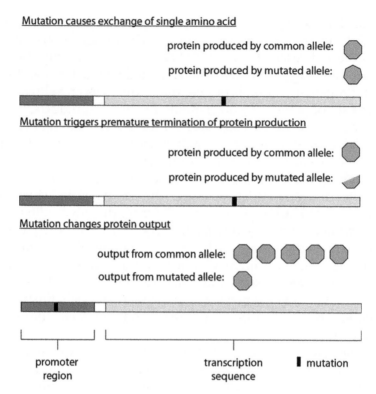

Mutation causes exchange of single amino acid

protein produced by common allele:

protein produced by mutated allele:

Mutation triggers premature termination of protein production

protein produced by common allele:

protein produced by mutated allele:

Mutation changes protein output

output from common allele:

output from mutated allele:

promoter   transcription   ∎ mutation
region   sequence

Figure 11.1   Different consequences of genetic mutations.

signal that prematurely terminates production of the protein. This is what happens in Brunner Syndrome, symptoms of which include violently aggressive outbursts.

But mutations can also strike non-coding sections of genes. In such cases, pristine specimens of the associated protein might trundle off the production line, the problem is that too many or too few of them get made. Problems like this occur when mutations alter the promoter region, one of the gene's "on–off" switches described in Chapters 6 and 9. Hence, the name of the *MAO A* allele, which helped to reduce the sentence of Abdelmalek Bayout: the "L" allele is named for low activity.

11.3   BRUNNER SYNDROME

Last century, genetic mutation caused turmoil in a Dutch family. Exclusively affecting male descendants, carriers of the mutation demonstrated serious problems with aggression, violence and sexually inappropriate behaviour. One family member was institutionalised and subsequently stabbed his supervising warden with a pitchfork in the chest. Another would threaten his sisters with a knife so as to force them to undress, while a third tried to run over his boss in response to criticism of his work. Cases of arson, attempted rape and exhibitionism were also documented.[13] Their condition was named Brunner Syndrome after one of its discoverers.

Brunner Syndrome is an extreme opposite of the hysteroid dysphoria described in Chapter 7. Where Liebowitz's lovesick patients suffered from overactive MAO, victims of Brunner Syndrome suffer from a paucity of the enzyme. Consequently, they are unable to metabolise serotonin, noradrenaline, dopamine and β-phenylethylamine, suggesting that excessive levels of these neurotransmitters may conspire to trigger violent outbursts, as well as learning difficulties, with which the syndrome is also associated.

The mutation was uncovered following analysis of the affected males' urine. As we saw in Chapter 7, MAO metabolises the target neurotransmitters into various products, which are then further metabolised into end products, including one called vanillyl mandelic acid (VMA). (Resemblance of the name to the fragrant essence of vanilla is no coincidence as the two share a hallmark molecular feature with other members of the vanillyl group, as shown in Figure 11.2.) Conspicuously absent from the urine samples were signs of VMA or any other of the expected metabolites of the monoamine neurotransmitters.[13]

Although the mutation constituted a very minor alteration to the gene, the impact was profound. The researchers discovered that the exchange of a single nucleobase was causing the problems.[14] Where the gene usually featured a cytosine nucleobase,

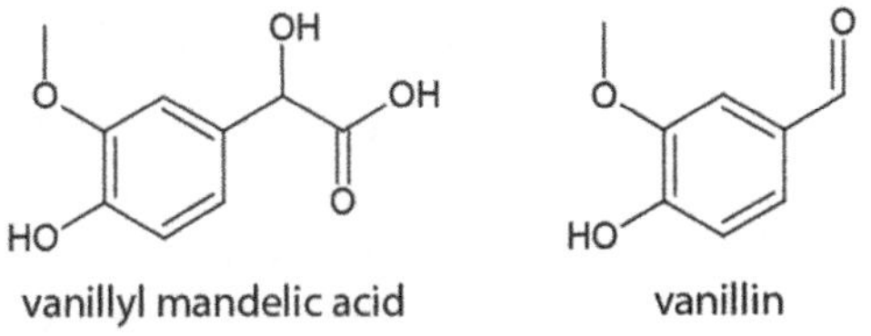

vanillyl mandelic acid   vanillin

Figure 11.2 Vanillyl mandelic acid and vanillin, the main ingredient of synthetic vanilla flavouring.

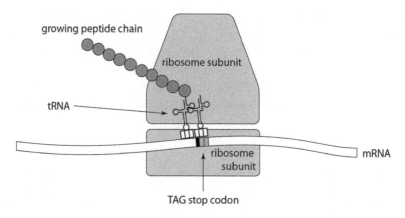

growing peptide chain

ribosome subunit

tRNA

ribosome subunit

mRNA

TAG stop codon

Figure 11.3   The single nucleobase mutation in Brunner Syndrome, switching cytosine for thymine, creating a TAG stop codon. Production of the protein is terminated even though untranslated mRNA still remains.

in the affected males it had been replaced with thymine. A blip like this might normally lead to the inclusion of one wrong amino acid during construction of the protein, but this case was more serious. Instead of creating a codon that selected for a different amino acid, the new combination (TAG) was identical to a stop codon, which is the sequence of nucleobases that tells the mRNA translation apparatus when protein assembly has finished. So, the gene produced an mRNA transcript that differed by just a single nucleobase, but when the blueprint was whisked off to the ribosomes for translation into a protein, the production process was terminated long before the protein was actually finished, as shown in Figure 11.3. This stunted version of MAO A was unable to metabolise its target neurotransmitters, which the researchers blamed for the aggressive behaviour.

11.4   OTHER PROBLEMS WITH THE *MAO A* GENE

Brunner Syndrome is thankfully very rare, but the discovery promoted research into other possible malfunctions of the *MAO A* gene. Not long afterwards, researchers reported the discovery of the low activity allele (*MAO A-L*) invoked in the case of Abdelmalek Bayout.

Carriers of this allele had exactly the right recipe for the MAO A protein, but researchers reasoned their bodies did not use it often enough. In Brunner Syndrome, a mutation in the coding sequence of DNA truncates the protein, whereas the mutation in the low activity allele affected the promoter region, discouraging transcription of the coding region. Consequently, people with Brunner Syndrome produced the right amount of the wrong protein, while carriers of *MAO A-L* produced the wrong amount of the right protein.

The site of the mutation was comparatively far from the region of DNA that codes the protein. In Chapter 9 we saw that promoter regions can be located more than a thousand nucleobases upstream from associated genes and this is a perfect example. Researchers Sue Sabol, Stella Hu and Dean Hamer found discrepancies located 1200 nucleobases upstream from the start site of the coding sequence for MAO A.[15] The finding centred on the following oblique sequence of 30 nucleobases:

ACCGGCACCGGCACCAGTACCCGCACCAGT

It was not the presence or absence of the sequence that distinguished the alleles. Every genome investigated contained it. In fact, they all contained more than one copy of it. This entire sequence was copied up to five times over and that was what set the participants apart. While some people had just two consecutive copies of the sequence, others had three, four, five or even three *and a half* copies, as shown in Figure 11.4. Researchers concluded that the more copies a person's genome contained, the more

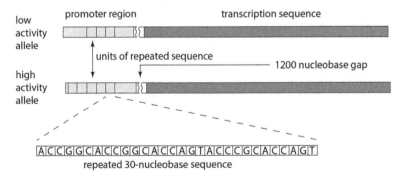

Figure 11.4   The 30-nucleobase sequence is repeated different numbers of times in the low- and high-activity alleles.

MAO A their cells would express. Accordingly, the alleles were categorised as high- or low-activity alleles.[15]

Obviously, the next thing to do was investigate whether low-allele carriers were more inclined to violent behaviour. Researchers, including Avshalom Caspi, conducted a long-running study tracking the development of more than a thousand New Zealand boys up to the age of 26.[16] No correlation was found between carrying the allele and developing a violent character, but the researchers did spot a different correlation.

Carriers of the low activity allele seem to be more inclined to develop violent characteristics if they suffer abuse during their childhood. In the New Zealand study roughly a tenth of the boys were maltreated in their youth and also possessed the low-activity *MAO A* allele, but they were responsible for nearly half the convictions for violent crime. These findings were replicated in several follow-up studies, one meta-analysis of which noted a "robust" relationship;[17] however, other studies have failed to confirm these findings.[4] But the evidence remains compelling that, as with foetal alcohol syndrome, gene level anomalies are not sufficient to trigger violent character, rather both conditions make it harder for victims of childhood maltreatment to escape the destructive cycle of abuse.

11.5   A ROLE FOR THE EPIGENOME

Recent research may have uncovered a role for epigenetic phenomena in the link between the low-activity allele and violence, which may further explain some of the difficulties of replicating the findings of studies like Caspi's.

Comparative activity of the various alleles may have more to do with epigenetic markings than the number of copies of the upstream promoter sequence. Cytosine and guanine appear side by side several times in the 30-nucleobase sequence, making it a prime region for methylation of CpG islands. This has been used to explain a discovery that poses difficult questions for the idea of high- and low-activity alleles. If the alleles live up to their names, there ought to be a corresponding high or low activity of MAO A in the brains of people who carry them. But a recent experiment found no such evidence.

The study found that epigenetic markings were a far superior predictor for MAO A activity. Using positron emission

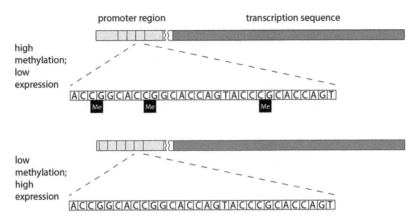

Figure 11.5   When the CpG islands in the promoter region are methylated, the gene is expressed less frequently than when the CpG islands are not methylated.

tomography (PET) scanners, a research team led by Elena Shumay and Joanna Fowler gauged the number of molecules of the metabolising enzyme residing in different people's brains.[18] What they found was that the supposed low-activity and high-activity alleles had no effect on the actual levels of MAO A in the brain. The degree of methylation on the CpG islands of the promoter region proved a far more reliable indicator of activity, as shown in Figure 11.5. They also found that there was no correlation between the number of copies of the promoter sequence and the degree of methylation. This finding has also been replicated in a study of prisoners, which found that hyper-methylation of the promoter region was a strong predictor of antisocial personality disorder in a cohort of 86 incarcerated male offenders.[19]

Different mechanisms distinguish these three strands of re-search but they all lead to the same outcome. Victims of Brunner Syndrome are unable to make MAO A. Competing hypotheses also demonstrate how carriers of a perfectly functioning recipe for the protein just make less use of it. But, in all cases, violent behaviour is linked with lower levels of this vital enzyme.

11.6   A ROLE FOR THE REWARD NETWORK

The role of MAO A in violent behaviour hints at the involvement of the brain's reward network. The enzyme is used to maintain appropriate levels of the monoamine neurotransmitters, including dopamine and serotonin. Dopamine especially is a key player in the brain's reward network and serotonin also interacts with it. This raises the possibility that mesolimbic circuitry rewards us for violent behaviour and, if that is true, does that make it a natural human instinct? Are violent outbursts an act for which we are rewarded by virtue of the services they have previously rendered to the self-replicators?

Research indeed implicates the reward centre in violence but generally as the result of malfunction. A study on rats detected unusual activity in the nucleus accumbens in association with violent behaviour but the rodents were deliberately bred to exhibit extremely low anxiety. In the study led by Daniela Beiderbeck and Inga Neumann the rats were highly aggressive when strange intruders were placed in their cages, but they were much less aggressive after treatment with dopamine antagonists, including haloperidol.[20] Not only do these findings support the case for a role of the reward network in violent behaviour, they also suggest that the trait thrives on fearlessness.

Another line of research is that psychopaths suffer from hypersensitivity of the dopamine reward system. In fact, the research has been led by the clear evidence that psychopaths are more prone to drug addiction, in which we have already considered the role of dopamine in detail. Not all psychopaths are violent, of course, but studies have shown they are more inclined towards the behaviour.[21,22]

A study led by Joshua Buckholtz and David Zald found evidence of hypersensitivity but only for a specific strain of psychopathy.[23] Questionnaires were used to rank participants in terms of their tendency for impulsive antisocial behaviour as opposed to another brand of psychopathy: fearless dominance. Next the individuals underwent brain scans while playing a game called a monetary incentive delay task. Contestants play to win or avoid losing money by clicking a button to hit a target, but they need quick reactions because the crosshair used to take aim appears and disappears suddenly.[24] Altering the settings enables researchers to zero in on different aspects of the reward process, such as the anticipation compared to actually getting the reward.[25] In this case, the researchers recorded an unusually strong signal in the nucleus accumbens for the impulsively antisocial psychopaths, which occurred during anticipation of the reward but not when it was actually paid out. These data suggest an unusual sensitivity towards the possibility of reward among certain kinds of psychopath.

Studies seem to be converging on the finding that psychopaths might misread cues when it comes to predicting reward. Researchers Andrea Glenn and Yaling Yang have proposed the hypothesis based on a review of a variety of studies.[26] These experiments have consistently shown psychopaths have an unusually large striatum, which houses various reward-mediating brain modules. But it is the *getting* rather than the *having* about which these individuals seem hypersensitive. They have a hair trigger when it comes to the anticipation of reward but not to the reward itself. As such, psychopaths often continue responding to certain stimuli, even when it should be clear that no reward will be forthcoming.

These findings collectively suggest that the brain's reward network could be implicated in motivating violence, but the relationship is far from well-established. Furthermore, these data apply exclusively to psychopaths and this likely reflects the fact that they are more likely to be studied because they are more likely to be guilty of violent crimes. While evolutionary psychologists have suggested that psychopathy itself may be an evolutionarily adaptive behaviour,[6] these findings do not represent broad trends in human behaviour.

11.7   CONCLUSION

Violence is an evolutionarily adaptive behaviour in many species, but is it a natural instinct for humans? We certainly use violence in pursuit of evolutionarily adaptive gains. One could interpret the findings of Napoleon Chagnon's study in Chapter 5 as showing that violence has become almost a peacock's feather for members of the Yanomomö tribes. Surviving the frequent battles with rival clans demands a certain fitness, which could be taken as indicative of a generally stronger genetic stock. Throughout history, violence has been used to seize power, which, as we will see next chapter, could also be considered an evolutionarily adaptive outcome.

But this chapter suggests we are not hard-wired for violence. Certain genes may or may not be implicated in violent behaviour, but far from inclining their carriers towards it, they enhance sensitivity to the environment. When pain is inflicted on these individuals, they become more likely to inflict pain on others. However, the chapter has taken a narrow focus, ignoring strands of research into other genes linked with violence. Even if such genes are uncovered, we are gaining ever more understanding of the competing influences of nature and nurture. Research into violence gives another example of how the world around us has the power to shape our character by manipulating the very genes we express. Tragedy begets tragedy when our needs go unmet.

REFERENCES

1. D. Hunt, "Woman Begged for Her Life before Defendant Shot Her, Witness Says", *Star-telegram*. Available at: http://www.star-telegram.com/living/family/moms/article3830628.html. [Accessed: 29-Apr-2016.]
2. T. Leibson, G. Neuman, A. E. Chudley and G. Koren, The Differential Diagnosis of Fetal Alcohol Spectrum Disorder, *J. Popul. Ther. Clin. Pharmacol.*, 2014, **21**(1), e1–e30.
3. M. Smith, "Soliz Receives Death Penalty", *Cleburne Times-Review*. Available at: http://www.cleburnetimesreview.com/news/local_news/soliz-receives-death-penalty/article_adf7a825-213b-5f8a-aa49-dbdda955de17.html. [Accessed: 29-Apr-2016.]
4. M. L. Baum, The Monoamine Oxidase A (MAOA) Genetic Predisposition to Impulsive Violence: Is It Relevant to Criminal Trials?, *Neuroethics*, 2013, **6**(2), 287–306.
5. E. Feresin, Lighter Sentence for Murderer with "Bad Genes", *Nat. News*, 2009, DOI:10.1038/news.2009.1050.
6. S. Pinker, *The Blank Slate: The Modern Denial of Human Nature*, Penguin, London, New edn, 2003.
7. D. Hunt, "Fetal Alcohol Syndrome May Spare a Murderer", *The Seattle Times*, 14-Jul-2012. Available at: http://www.seattletimes.com/nation-world/fetal-alcohol-syndrome-may-spare-a-murderer/. [Accessed: 29-Apr-2016.]

8. S. F. Gilbert, *Developmental Biology*, Sinauer Associates, Sunderland, Mass, 9th edn, 2010.
9. P. C. Haycock, Fetal Alcohol Spectrum Disorders: The Epigenetic Perspective, *Biol. Reprod.*, 2009, **81**(4), 607–617.
10. A. P. Streissguth, F. L. Bookstein, H. M. Barr, P. D. Sampson, K. O'Malley and J. K. Young, Risk Factors for Adverse Life Outcomes in Fetal Alcohol Syndrome and Fetal Alcohol Effects, *J. Dev. Behav. Pediatr.*, 2004, **25**(4), 228–238.
11. A. Raine, Biosocial Studies of Antisocial and Violent Behavior in Children and Adults: A Review, *J. Abnorm. Child Psychol.*, 2002, **30**(4), 311–326.
12. M. B. Gerstein, C. Bruce, J. S. Rozowsky, D. Zheng, J. Du, J. O. Korbel, O. Emanuelsson, Z. D. Zhang, S. Weissman and M. Snyder, What is a Gene, Post-ENCODE? History and Updated Definition, *Genome Res.*, 2007, **17**(6), 669–681.
13. H. G. Brunner, M. R. Nelen, P. Van Zandvoort, N. G. Abeling, A. H. Van Gennip, E. C. Wolters, M. A. Kuiper, H. H. Ropers and B. A. Van Oost, X-linked Borderline Mental Retardation with Prominent Behavioral Disturbance: Phenotype, Genetic Localization, and Evidence for Disturbed Monoamine Metabolism, *Am. J. Hum. Genet.*, 1993, **52**(6), 1032.
14. H. G. Brunner, M. Nelen, X. O. Breakefield, H. H. Ropers and B. A. van Oost, Abnormal Behavior Associated with a Point Mutation in the Structural Gene for Monoamine Oxidase A, *Science*, 1993, **262**(5133), 578–580.
15. S. Z. Sabol, S. Hu and D. Hamer, A Functional Polymorphism in the Monoamine Oxidase A Gene Promoter, *Hum. Genet.*, 1998, **103**(3), 273–279.
16. A. Caspi, J. McClay, T. E. Moffitt, J. Mill, J. Martin, I. W. Craig, A. Taylor and R. Poulton, Role of Genotype in the Cycle of Violence in Maltreated Children, *Science*, 2002, **297**(5582), 851–854.
17. A. Taylor and J. Kim-Cohen, Meta-analysis of Gene–Environment Interactions in Developmental Psychopathology, *Dev. Psychopathol.*, 2007, **19**(4), 1029–1037.
18. E. Shumay, J. Logan, N. D. Volkow and J. S. Fowler, Evidence that the Methylation State of the Monoamine Oxidase A (MAOA) Gene Predicts Brain Activity of MAO A Enzyme in Healthy Men, *Epigenetics*, 2012, **7**(10), 1151–1160.

19. D. Checknita, G. Maussion, B. Labonté, S. Comai, R. E. Tremblay, F. Vitaro, N. Turecki, A. Bertazzo, G. Gobbi, G. Côté and G. Turecki, Monoamine Oxidase A Gene Promoter Methylation and Transcriptional Downregulation in an Offender Population with Antisocial Personality Disorder, *Br. J. Psychiatry*, 2015, **206**(3), 216–222.
20. D. I. Beiderbeck, S. O. Reber, A. Havasi, R. Bredewold, A. H. Veenema and I. D. Neumann, High and Abnormal Forms of Aggression in Rats with Extremes in Trait Anxiety – Involvement of the Dopamine System in the Nucleus Accumbens, *Psychoneuroendocrinology*, 2012, **37**(12), 1969–1980.
21. G. T. Harris, M. E. Rice and C. A. Cormier, Psychopathy and Violent Recidivism, *Law Hum. Behav.*, 1991, **15**(6), 625–637.
22. J. Blais, E. Solodukhin and A. E. Forth, A Meta-analysis Exploring the Relationship Between Psychopathy and Instrumental Versus Reactive Violence, *Crim. Justice Behav.*, 2014, 797–821.
23. J. W. Buckholtz, M. T. Treadway, R. L. Cowan, N. D. Woodward, S. D. Benning, R. Li, M. S. Ansari, R. M. Baldwin, A. N. Schwartzman, E. S. Shelby, C. E. Smith, D. Cole, R. M. Kessler and D. H. Zald, Mesolimbic Dopamine Reward System Hypersensitivity in Individuals with Psychopathic Traits, *Nat. Neurosci.*, 2010, **13**(4), 419–421.
24. J. M. Bjork, S. J. Grant, G. Chen and D. W. Hommer, Dietary Tyrosine/Phenylalanine Depletion Effects on Behavioral and Brain Signatures of Human Motivational Processing, *Neuropsychopharmacology*, 2014, **39**(3), 595–604.
25. K. Lutz and M. Widmer, What Can the Monetary Incentive Delay Task Tell Us About the Neural Processing of Reward and Punishment?, *Neurosci. Neuroecon.*, 2014, 33.
26. A. L. Glenn and Y. Yang, The Potential Role of the Striatum in Antisocial Behavior and Psychopathy, *Biol. Psychiatry*, 2012, **72**(10), 817–822.

CHAPTER 12

The Chemistry of Dominance

George Orwell was shot in the neck volunteering to fight fascism, but his ideals were the casualties of his personal battle. He loathed power, but he sought to elevate his status. He waged war on money, but he amassed tons of the stuff. He despised inherited wealth for the eternity to which it condemned the poor, but he left his fortune to his family. In evolutionary terms, he thwarted the self-replicators, for while he greatly enhanced his reproductive fitness, he never sired descendants. But any accusations of hypocrisy would be vastly outweighed by his uncompromising integrity. He never pandered to commercial viability and, most importantly, he won success with creativity in an age where violence was the preferred path to power.

The celebrated writer was much concerned with power. He blasted dictatorship in his books *Animal Farm* and *1984*. The latter portrayed a dystopic vision of immutable totalitarianism, his major foe in the turbulent era through which he lived. His hypothesis was that emerging technology would enable a ruling elite to permanently consolidate a ruthlessly repressive power-base. But he candidly admits that part of his motivation to write was to exert influence over the direction taken by human affairs.[1] His family had endured declining fortunes for three generations, from a great-grandfather who married the daughter of the Earl of

The Chemistry of Human Nature
By Tom Husband
© Tom Husband 2017
Published by the Royal Society of Chemistry, www.rsc.org

Westmorland to young Eric Arthur Blair,[2] Orwell's real name, who suffered the indignity of attending Eton on a scholarship. Again, he admits his ego partly motivated him to claim his place in the upper echelons of humanity,[1] which he achieved with spectacular aplomb.

An interesting question is whether the abolition of inherited wealth would actually increase social mobility. Orwell felt the practice was the major mechanism by which the rich stayed rich and the poor stayed poor. His own dynasty is a dead end for speculation on this matter because the son who ultimately inherited his wealth was adopted. As such, if any talent was passed down from father to son, none of it could have been transmitted genetically. But there is no shortage of commentators who believe that success is passed down *via* genes rather than wills. The theory has some biological merit because anything that elevates status should also increase reproductive fitness. Richer parents can select superior mates, so as to maximise replication of their genes in future descendants.

This chapter will consider whether the brain's reward network motivates the human thirst for power. We have seen how the brain's nucleus accumbens metes out pleasure as a reward, essentially, for behaviours that promote reproductive fitness. We enjoy eating because it provides us with the fuel and building materials with which to perpetuate the self-replication of our genes. We enjoy reproduction because it produces new vessels in which for self-replication to continue after our own have expired. The next question is whether we are rewarded for elevating our status because it increases our access to the primary needs of food and mates.

The claim is uncontroversial, but the case is far from closed. It makes intuitive sense that our brains should reward behaviour that increases our access to desired, finite resources. Indeed, many researchers are pursuing the hypothesis.[3,4] However, proving the hypothesis is exceptionally difficult. Status is yet another complex trait that is difficult to measure accurately. As with the difficulties of creativity and violence considered in earlier chapters, it is also inextricably bound to other traits, such as dominance and aggression, which also defy straight-forward definition. Furthermore, creativity and violence themselves are both means by which to elevate status.

We will first consider the difficulty of defining these traits, as well as the relationship between them. Next, we will look at the evolutionary case for reward-mediated status elevation in the context of heritability. Finally, we will consider the findings from contemporary research, which has been uncovering the neurological secrets of the will to power.

12.1 WHAT IS DOMINANCE?

What is dominance, how is it done and what does it achieve? These are central questions for this chapter. Dominance has traditionally been linked with aggression, but research has increasingly shown that the two may not be so closely linked. Another question is precisely which spoils go to the dominant victor.

Dominance is a nebulous concept and subject to fluctuations. In an early episode of *The Simpsons*, Homer is a snivelling sycophant towards his boss Mr Burns until the miserly tyrant sells his power plant. During a subsequent chance encounter, jobless Homer no longer needs to suck up, so instead he blasts Mr Burns, calling him a greedy, lonely, loveless old man, to the defeated man's utter devastation. Without the clout of an employment infrastructure, Burns becomes the subordinate. Similarly, in Tom Wolfe's *Bonfire of the Vanities*, protagonist Sherman McCoy is a master of the universe in his financial job but gets much less cocky when he accidentally drives into the wrong side of town.

In 1993 researcher Carlos Drews waded into a field of conflicting views and offered the following definition:

> *"Dominance is an attribute of the pattern of repeated, agonistic interactions between two individuals, characterized by a consistent outcome in favour of the same dyad member and a default yielding response of its opponent rather than escalation. The status of the consistent winner is dominant and that of the loser subordinate."*[5]

Two points suggest themselves. First of all, what is meant by a "consistent outcome"? In order for a group member to be identified as dominant is it necessary that they emerge as victor from *all* of their agonistic interactions? And what value is dominance so defined? It makes no mention of what the dominant member gains. Do they enjoy increased access to mates or food?

Philip Kitcher raised such points in his masterful critique of what he labelled "pop sociobiology". His book *Vaulting Ambition*[6] meticulously dismantled many of the claims abounding in the wake of Edward Wilson's *Sociobiology*. One of many assertions to which Kitcher took exception was the idea that human males are more dominating than women and that it results from a genetically transmitted evolutionary adaptation. The reasoning went that primates have dominant males, and therefore it makes sense that humans should too. Kitcher objected in virtually every way imaginable. He listed various primate species in which females, rather than males, dominate. He also attacked the assumption that a trait that is evolutionarily adaptive for one species should automatically be considered adaptive to another species. Finally, he noted that even male-dominated primate groups often demonstrated a fluid hierarchy, in which many factors influenced which male should be considered dominant at any given moment.

Kitcher references an interesting study in support of this last point. Primate ethologist Hans Kummer recounted the details of an agonistic interaction between two male hamadryas baboons,[7] a species noted for its violent aggression. The younger and the older male disagreed over which direction to head in search of food and the older one got its way. But, while the older baboon dominated in decision-making, it was the younger baboon who "owned" more females. An isolated incident, perhaps, but it challenges the idea that dominance necessarily increases reproductive fitness. Conversely, reproductive fitness is not measured solely by the number of offspring produced. What is the point of producing a huge litter of children if they are born in the firing line of a pride of ravenous lions? For optimum reproductive fitness it is necessary that successive generations continue to thrive reproductively. Moreover, it may be that the young baboon was so unthreatened by his elder that he felt secure enough to take a rest from being dominant. The phenomenon that some high-powered businessmen like to pay dominatrices to give them a taste of submission[8] demonstrates a craving to cast off the yolk of authority.

Another interesting point is whether dominance is always achieved with aggression. Traditionally linked, research has turned up various exceptions, including the following.[3] In one study there was no correlation between a chicken's aggressiveness and its place in the pecking order. Similarly, differently ranked paradise fish were all equally aggressive. As with the hamadryas baboons above, a ranking of rhesus monkeys by their aggressive dominance was completely out of kilter with how well they competed for female mates, whilst other primate studies have noted the importance of forging social alliances to becoming the alpha male. This latter point bears a striking resemblance to findings in human children. Patricia Hawley has argued that toddlers may use coercive strategies to get what they want from peers, by being bossy, interrupting, showing off or fighting. But growing children soon tire of such tactics and, by the end of school, leaderly behaviour is associated with being supportive, sociable and independent.[9]

Researchers investigating animal models use a variety of indicators to gauge dominance.[3] In rodent studies they compare the areas about which mice feel confident to spread their urine, the share of a finite supply of food each mouse can obtain and how persistently they vie for the attention of females. They also record how often one mouse plucks another's whiskers. Another interesting test is the tube paradigm, where mice are fed into either end of a tube. A snug fit, the tube is too narrow for one mouse to pass the other. Instead, when they meet in the middle, one forces the other backwards, establishing himself as the dominant mouse. This can be used to rank several rodents based on the results of a tube tournament.

This last instrument probably resonates the most strongly with a human audience. I have often pondered the laws dictating who gets out of whose way when two pedestrians suddenly find themselves on a collision course. It can be frustrating when you alter your path but the other person just keeps moving forward, fostering the concern that had the one party not moved, the two would have collided. But this very rarely happens. One person generally gets out of the other's way, but people differ in how long they hold on before stepping aside, just as if they were playing a game of chicken—pedestrian chicken. The tube paradigm might tempt us to suggest that the sidestepper has been dominated, which certainly rings true, but the pavement and the tube are quite different. There is a greater cost to being pushed backwards than being forced to one side, so it is quite plausible that someone ambivalent about stepping to one side would fight much harder to avoid being pushed backwards. Thus, a pavement dominator might go submissive in the tube.

12.2   DOMINANCE AND STATUS

An important question for this chapter is the extent to which dominance influences status in humans. The desire to elevate status is a pretty universal characteristic. Is it that greater dominance leads to greater success? Probably to some extent but not entirely.

Marrying dominance and status gets very messy in the diverse sphere of human activities. There is an argument that the diversity of activity stems from our universal thirst for power, exploiting whichever talents avail themselves. If you cannot succeed with brute force, try brains and *vice versa*. Nowadays, we can seek success in the hierarchy that best matches our talents. If you cannot be a political leader you can be a religious leader or, if that does not suit, you can become a football leader or a wealthy entrepreneur or a famous musician or a respected academic. There exist many ladders up that, to chase status and one's ability to dominate, must surely influence how quickly one climbs. Not only that but one ladder might have different peaks. Even in one field, success might mean prestige, wealth, popularity or critical acclaim, or all of them at once.

The hit TV series *The Wire* demonstrates how fractious our human hierarchy has become. Crime-ridden Baltimore is carved between the competing hierarchies of government and crime, each of which is splintered into rival factions. Power is highly fluid. In theory, the city mayor is in charge of everything beneath him, but wily police chiefs can gain leverage against him, as can political rivals. Scaling the ladders in the government camp involves courting the media, while the criminals fight to avoid such attention. Meanwhile, the media itself is partial to any interpretation of world events that progresses its chances of winning a status-enhancing Pulitzer Prize.

All of these worlds collide to forge an unruly terrain—a blizzard of ladders resembling an MC Escher drawing. Your situation in life influences the ladders you are likely to climb, as well as the rate at which you ascend. Young people born in "the projects" struggle to avoid a life of crime, in which the skill set for success is the ability to thrive in lethally violent territorial disputes. Politicos and businessmen need to be articulate, charismatic and well-educated. Police officers need to be tough and willing to bear the mantle of law enforcement. Some people start off on a higher rung of the ladder, while other times Escher's crooked perspective just makes it seem that way.

One interesting character is Marlo. An utterly ruthless drug dealer, he refuses to bow to anyone and kills anyone who gets in his way. Friends are sparse for an individual who understands the danger of forming attachments. Ultimately, the police destroy his operation but not with sufficient evidence to convict him. Instead, the district attorney strikes the deal that Marlo walks free provided he commits no further crimes. The slightest future infraction will entail a crushing sentence. Taken under the wing of the crooked politician Clay Davis, he attempts to join the world of legitimate business. But one evening of schmoozing prompts him to flee back to the street, where he can resume his criminal activities. Is it the thrill of violent conflict that pulls him away, or the indignity of going from top cat to small fry that pushes him away? Does he lack confidence in his ability to cultivate a new skill set? Specifically, is it that his method of dominating does not apply in this new world? Alternatively, there are many higher echelons to access in the criminal world but perhaps their occupants are too wary of this solitary killer and the *de facto* suspended sentence he violates. His considerable ability to dominate has clunked into a very personal glass ceiling.

We should be careful of inferring too many universal truths from a work of fiction, however feted for its gritty realism. But it illustrates several relevant points. It is very difficult to measure traits as complex as dominance. A person who dominates in one scenario may be subjugated in another. There are different ways of being dominant and one's ability to dominate does not guarantee that their status will increase. As we saw from the definition, dominance merely requires that one consistently triumphs from dyadic encounters, but who cares about that if there is no corresponding acquisition of resources? If you force someone out of your way on the street, is it dominance? Will you earn anything from it? Does it predict your ability to earn in other contexts? And finally, does any of that necessarily entail better representation of your genes in future generations?

12.3   HERITABILITY OF STATUS

George Orwell's feelings about inherited wealth link to a fascinating study recently conducted by researcher Gregory Clark. In his book *The Son Also Rises*[10] he investigates the tendency of offspring to inherit their parents' success. No one should be too surprised that people of high status tend to produce children of high status; the interesting question is the mechanism of transmission. Orwell believed that inherited wealth gave the children of wealthy parents an unfair advantage, while Clark suggests that their genes play a more important role. If we wish to prove the hypothesis that dominance is an evolutionarily advantageous adaptation, it makes sense to investigate whether the trait has a genetic component.

An obvious starting point would be to consider the heritability of dominance. Carlos Drews noted the importance of definition in such inquiry.[5] Simply put, dominance is heritable defined in some ways but not when defined in others. For example, he notes several breeding studies that found dominance to be heritable on the condition that dominance was defined as aggressiveness. But, as we have seen, aggression and dominance are distinct. Other studies found that dominance ranking was heritable, but, as Drews points out, dominance ranking is dependent on the group of individuals against which each subject is ranked. Like Marlo, individuals may be dominant in one group but submissive in another. Coupled with the concerns raised in Chapter 10, traditional heritability studies of this particular trait may not yield much fruit.

Enter Clark's study. He has used an entirely different method to investigate how the fortunes of families run between the generations, although it is neither a direct study of genetic heritability, nor of the transmission of dominance.

Clark uses an ingenious method to track the fortunes of families across time and around the globe. Using registries of birth, death and taxes, and the historical rolls of prestigious institutions, he gauges the representation of rare surnames in high-status professions. Imagine 90% of a population has the surname Smith. In this society there would be nothing remarkable about 90% of doctors bearing the name Smith. Now suppose 10% of the population is named Smith, but it is still true that 90% of doctors bear the name. This would be noteworthy. Smiths would be nine times better represented in the medical establishment than in society at large. Conversely, if 90% of the population was called Jones but only 10% of doctors had that surname, Joneses would be much worse represented.

When these data are expressed as ratios, they start to tell a story. If 90% of all doctors are called Smith, while 10% of all citizens are called Smith, the ratio of these figures is 90 divided by 10, which is nine. The corresponding ratio of medical Joneses to all Joneses is 0.11. Such a snapshot suggests that Smiths have an advantage over Joneses when it comes to entering the profession. Clark's inference is that good genes underlie the above average intelligence required to enter the profession but why?

The important information is not the snapshot but the change over time. Clark's data show that the next generation of doctors will contain a lower proportion of Smiths but a higher proportion of Joneses. That's not all. Over the following 10 to 16 generations the trend will continue until the ratio for both Smiths and Joneses is one, which is to say that neither Smiths nor Joneses are better or worse represented in the profession than they are in society at large. At this point, we say that each has regressed to the mean.

Regression to the mean was discovered by Francis Galton, the eugenicist polymath we met in Chapter 10. He knew that tall parents tend to have tall children, while short parents usually have short children, and we are pretty confident that genetic factors explain this. But within his data sets he noted an important twist: very tall people tend to have children slightly shorter than themselves, while very short people tend to have taller children. This is applicable in many fields. Regression to the mean dictates that if you get an extreme data point the first time you measure something, the second measurement is likely to be closer to the average.

Random chance plays an important part here. A person's height is decided by competing factors, including the genes they inherit and the availability of nutrition while they are growing up. Within this there are many random components. Do extremely tall people insist on having children only with other extremely tall people? No. If an outlier on the data set has children with someone much closer to average height, there is nothing remarkable about the resulting child being shorter than the very tall parent. Next, even if two very tall people do have children, there is no guarantee that their height-promoting genes will perfectly synchronise for optimal height. As we have seen, when babies are made, half of each parent's genes are selected at random and spliced together. The "tall genes" may not even make the cut and, even if they do, any that are recessive will need to be inherited from both parents, just as children will only be blue-eyed if they inherit the blue-eyed gene suite from both parents. These factors of chance underlie the regression to the mean.

The stand-out finding of Clark's analysis is the chilling immutability of the rate at which family fortunes regress to the mean. Our hypothetical situation of the Smiths and Joneses converging on mediocrity is representative of real data from a variety of contexts. Clark has tracked data from Communist China, from caste-segregated India, from progressive Sweden and even the rise of artisanal communities in medieval England. He has considered not only doctors but also the judicial establishment, the roll of Mandarin bureaucrats in pre-revolutionary China and prestigious lists of noble families in Sweden. Using a range of rare surnames unique to each culture, he finds regression to the mean always takes 10 to 16 generations.

Clark finds it noteworthy that the rate has endured many changes in social attitudes and institutions. The fortunes of the privileged families who thrived as Chinese mandarins did not decline any faster when Mao Zedong established the communist republic. Modern Sweden is hailed as a masterpiece of progression, but nothing changes each family's progress towards mediocrity. Modern and medieval England are alike in their rates of regression to the mean.

This leads Clark to claim that "nature dominates nurture", in other words, that success is largely in the genes. If nurture played any significant role, then the interventions of social institutions, such as the welfare state and free education, ought to have accelerated the rate at which poorer families have found fortune. Meanwhile, the fortunes of wealthy families endure no matter how many children they have. It used to be that wealthier people had much larger families, but that trend has been in decline since the Victorian era. Even rich families have finite means by which to bestow a superior education on their progeny. Therefore, the children of larger families ought to come out with less money each. If nurture plays a significant role, this ought also to accelerate the rate of regression to the mean, but no.

There is a potential problem with this argument. Let's suppose that making the transition from large to small families does confer an advantage. The family can afford to send their few children to a better school than they could have afforded for a larger brood. The problem is that if all wealthy parents have stopped having big families they will all feel the same advantage. In the race for success they can run faster but so can all their competitors. An economic analysis is also possible. If decreasing family size increases the resources to spend on children, the sensible option is to look for better, more expensive schools. But if all families are doing the same thing, then this increased demand will drive up the price of schooling. Mysteriously, the cost of privately educating three children is suddenly equivalent to the previous bill for 10.

A theory gains merit when it makes accurate predictions, which Clark's theory achieves. The only thing that can change the rate of regression to the mean is what he calls endogamous marriage. This effectively means to breed exclusively with people on the same tier of society. He makes the prediction that India ought to have a slower regression to the mean because of the historical influence of the caste system, which strongly encourages people to marry within their caste. Accordingly, the persistence rate is a higher 0.9 in India.

All of this leads Clark to suggest that no measures have ever appreciably improved social mobility. This stands in contrast to a variety of studies with data showing otherwise, which points to a familiar problem: how to measure social mobility. Various indices are used, such as comparing the average educational attainment of one generation to the next. Such measures have shown that educational attainment typically increases, but success at school is no guarantee of career success. Furthermore, as university drop-outs, technocrats Bill Gates and Mark Zuckerberg would register as "low status" in such measures, when they are both astronomically successful. Clark offers the theoretical alternative of an underlying competence. No matter our start in life, the availability of inherited wealth or the ideology of our society, he believes that status depends chiefly on general competence, which comes mostly from our genes.

This gloomy conclusion is mitigated by his prognosis, which is that we should embrace the Nordic style of government practiced in countries like Sweden. He believes status is decided by genetic lottery and, thus, that the winners ought to compensate the losers *via* high taxes to fund public health, education and other important services.

The finding is interesting, but it rests on some questionable assumptions. Can we be certain that a family has regressed to the mean simply because they cease to flourish in a certain profession? In the earlier introduced snapshot, the ratio of medical Smiths to all Smiths declined from a starting point of nine. The implication is that doctors called Smith start off above average in status but become less so over the course of the following generations. Just because fewer doctors are Smiths does not require that Smiths as a whole have regressed to average status. Quite the contrary could be true. Doctors may enjoy above average status, but there are many echelons above them. What if the Smiths, at the apex of their stronghold on the medical establishment, decided to chase even loftier achievements, as famous writers, entrepreneurs or footballers? Medical Smiths would dwindle while the respective households prospered. It is true that Clark considers other professions besides medicine, such as law, but there are so many other ways to elevate status. Indeed, for high-achieving school students, medicine and law are quite safe choices. Does Clark's index gauge underlying competence or an underlying aversion to risk-taking? Whatever he has measured, its ominous constancy is compelling.

12.4 FINDINGS FROM NEUROSCIENCE

The argument for genetic heritability of dominance and/or status looks pretty wobbly so far, which may have implications for any complicity of the brain's reward network. An important part of the question is whether the brain actually does reward the elevation of status.

Dominance has previously been linked with sex hormones in general and with testosterone in particular. We have seen how the androgens prime us for sex. Meanwhile, animals often use dominance to decide who gets the highest quality mate. The idea that the same hormone could fuel both drives makes sense.

Two researchers, David Terburg and Jack van Honk, have investigated the phenomenon of staring matches as a means to gauge the role of testosterone in the formation of social hierarchies.[11] Primates often establish or maintain dominance by outstaring more submissive rivals. Having observed the same behaviour in humans, Terburg and van Honk surmise that testosterone mediates the phenomenon. They argue the hormone reduces fear in social situations whilst also tapping into the brain's reward network. Two studies into "angry eye contact" have been particularly illuminating.

One study found a way to measure how quickly people avert their eyes from a menacing gaze.[12] Participants faced a screen on which was shown a grey dot in the centre and a row of three coloured dots beneath. After some time, the grey dot changed colour to match one of the three below, at which point the participants had been directed to shift their gaze to the matching dot. What they were not told is that they were subliminally flashed images of a facial expression, which would be angry, neutral or happy. What the researchers found was that the participants shifted their gaze to the second dot more rapidly when presented with an angry expression.

The next study established that a dose of testosterone would affect how quickly the participants looked away.[13] In the placebo-controlled experiment, 20 female volunteers, aged 20 to 25, repeated the eye-tracking experiment. The data showed that testosterone helped women to steel their gaze against angry expressions for longer.

The use of subliminal presentation of the facial expressions helped the researchers to reach another conclusion. They hypothesise that the formation of social hierarchies, not only in humans but in all vertebrates, is an unconscious primordial brain mechanism.[13] This runs contrary to the prevailing view that testosterone fuels dominance in humans by producing conscious feelings of superiority, strength and anger in the brain's neocortex.[13] By using subliminal images they were able to time the aversion of gaze without the women even realising what was happening. Furthermore, they asked the participants in the second study to complete questionnaires about their emotional state. What they found was that the responses were the same whether the participants had been issued with testosterone or the placebo. Their data suggest that feelings of superiority in the neocortex are not needed for individuals to dominate with eye contact.

They have further supported their hypothesis by proposing neural mechanisms featuring two familiar neurotransmitters—vasopressin and dopamine. We have already seen how testosterone modulates dopamine currents by promoting the expression of nitric oxide synthase (NOS). This activity mediates sexual behaviour in the brain's medial prefrontal cortex, but Terburg and van Honk suggest a similar mechanism alters dopamine activity in the orbitofrontal cortex, a known part of the reward network introduced in Chapter 7 in conjunction with romantic love and obsessive-compulsive disorder (OCD). The region also influences activity in the brain's fear centre—the amygdala. The researchers suggest the classic reward-seeking behaviour associated with dopamine could stem from the orbitofrontal cortex.[11] But testosterone also promotes the expression of vasopressin and, in the amygdala, it has the effect of repressing fear. As such, it seems that testosterone mediates dominance by switching off fear and by linking status to the brain's reward mechanism.

12.4.1 Processing Information

Another part of the brain with a role in social affairs is the medial prefrontal cortex (mPFC). The prefrontal cortex is the brain's centre for executive behavioural control and its medial section is involved with various functions, such as decision making, pursuing goals, paying attention and working memory.[14]

One relevant study investigated the effect of manipulating neural circuitry in the mPFC. Synaptic plasticity refers to the tendency of synapses to transmit signals more or less readily between neurons, which can be understood by an analogy with road traffic. When highways become so popular that they start to jam with traffic, civic leaders might decide to expand them, which is analogous to what happens to neural circuits in our brains. The concept is strongly associated with learning *via* the formation of long-term memories. It used to be that everyone could remember a multitude of telephone numbers but then autodial scotched that. Personally speaking, if I dialled a telephone number four or five times within a 24-hour period, I would remember it indefinitely. Neuroscientists believe that repeatedly thinking about the same information encourages its promotion from working memory to long-term memory, a process that has been connected with synaptic plasticity. In the simplest sense, the more a neural circuit gets used, the more readily it is activated. At the cellular level, repeated depolarisation of neurons activates certain receptor-modifying proteins, while also triggering the production of additional receptors. All of this has the effect of sensitising the neuron's "trigger" so that it can fire more easily, the process of which is often described as a strengthening of the synapse.

Manipulating the strength of synapses in the mPFC of rats caused the rodents to switch ranks. The study led by Fei Wang and Hailan Hu[15] used gene therapy, a novel method that hijacks viruses and uses them to deliver genes directly into cells. First of all, the team produced two sets of genes coding for proteins that would either strengthen or weaken the synapses. The genes were inserted into viral cells, both varieties of which were injected directly into the mPFC. The experimenters investigated whether the intervention had affected the rats' social ranking and found that strengthening the synapses had made rats more dominant, while weakening them promoted submissive behaviour. Rats that had previously yielded in the tube test now drove their rivals backwards and spent more time serenading unfamiliar females with ultrasonic vocalisations, another recognised method of gauging dominance. Taken together, these results convinced the team that the mPFC is involved with dominance.

Results such as this are interesting when we consider which other parts of the brain communicate with the mPFC. A multitude of studies have confirmed that two-way traffic exists between the brain's reward network and the prefrontal cortex.

The ventral tegmental area, which lit up in scans of brains in love in Chapter 7, is a key input to the mPFC.[14] Meanwhile, other studies, using scanning techniques such as positron emission tomography (PET) and functional magnetic resonance imaging (fMRI), have found the *ventral* medial prefrontal cortex, a structure that incorporates the mPFC, directly innervates the brain's pleasure centre, the nucleus accumbens.[16]

Based on these findings, it is no surprise that monetary rewards have also been strongly associated with the mPFC. Several studies have used a game called a monetary incentive delay task. Contestants are given prior warning of what they can expect from each coming round, where they may be told they will have the opportunity to win £5 or warned that they risking losing £5. Then a target is presented and the participants need to "hit" it by pressing a button within a varying time limit, meaning that the test requires quick reactions. Afterwards, the players are informed whether they were successful. The task is designed to compare the sensations of anticipating and then actually gaining rewards. An fMRI study led by Brian Knutson and Daniel Hommer[17] found that, when participants expected a $5 win and got it, activity in the mPFC increased, while the converse was true when their expectations were dashed. Several studies have recorded a similar mPFC response to monetary rewards.[16]

It seems the reward network is closely concerned with our status and why should it not be? If the mPFC is concerned with goal-orientated behaviour, it makes sense that the reward network should provide the motivation to pursue such goals. Meanwhile, one definition of social dominance is the tendency to pursue desirable outcomes from social encounters.[18] What would make outcomes desirable other than the salience that primes us to pursue food and mating opportunities? The idea that social dominance is mediated by the dopamine-fuelled reward network is intuitive and backed by an increasing body of evidence.

12.5   COCAINE STUDIES

Cocaine has been used in several studies to gauge the role of dopamine in social dominance. Chapter 3 explained how cocaine shuts down the dopamine transporter, causing molecules of the neurotransmitter to stay outside the neuron, where it can go on activating the receptors of other neurons. This influence over dopamine channels in the brain makes cocaine an attractive drug by which to investigate reward mechanisms.

One study investigated how social housing affected the appetite for cocaine of female macaque monkeys. In the study led by Drake Morgan and Michael A. Nader[19] the monkeys spent the first stage of the study living alone but were then rehoused in groups of four, leading to the establishment of a pecking order. Dominant monkeys were identified by their more frequent bouts of aggression, their less frequent submission to cell mates and by the fact that they were groomed more often by the others. During this time, the monkeys had access to cocaine and could self-administer whenever they liked. The subordinate monkeys took more cocaine than the dominant monkeys. Furthermore, the study produced interesting findings with PET scans. What the scans showed was that dominant monkeys had a greater number of dopamine receptors than the submissive monkeys, so their neurons were more sensitive to activation by dopamine. By using before and after PET scans, the researchers found that the extra dopamine receptors had been created during the establishment of the social hierarchy.

Contrary findings arose from another study, in which rats who secured a larger share of strawberry milkshake also took more cocaine. The study carried out by Bianca Jupp and Jeffrey W. Dalley[18] housed rats in groups of four. The rodents were issued with a single source of Yazoo strawberry milkshake and those that managed to spend the most time drinking it were judged to be the dominant rodents. But, unlike the macaque study, this time it was the dominant rats that took the most cocaine. The studies both recorded elevated activity of dopamine receptors in brain regions associated with reward. Post-mortem dissections confirmed that the dominant rats demonstrated reduced dopamine levels, but greater binding of the dopamine transporter in the brain's pleasure centre, the nucleus accumbens.

The apparent contradiction underlines the problem of studying dominance. Dominant monkeys were identified according to the levels of aggression they displayed, along with how much grooming they received. But the rats were considered dominant based on the share of milkshake they got. The authors of the rat study cite previous findings demonstrating this disparity. Just because a group member is more aggressive, does not mean they get the bigger share of whatever they are competing for.[20] Aggression surely contributes to the ability of organisms to establish themselves as the dominant group member but aggression is not dominance. Quite apart from the fact that the two studies investigated different species, they also measured different versions of dominance, which may explain the contrary preferences of dominant rats and monkeys for cocaine usage.

A human study threw out a curveball. The experiment led by David Matuske and Robert T. Malison found that fewer dopamine receptors were available as the status of participants rose.[21] The volunteers were ranked using an instrument called the Barratt Simplified Measure of Social Status (BSMSS), which evaluates respondents based on the level of education they have achieved and their occupation. Only half of the 32 participants were cocaine users. After PET scans, the data showed that the higher their social status, the lower the availability of the D_2 and D_3 varieties of dopamine receptors. This inverse correlation was specific to particular brain regions. For the healthy volunteers, the lower receptor availability was observed in the ventral tegmental area and the substantia nigra, both of which are associated with the reward network, whereas the cocaine users had fewer vacant receptors in the amygdala region.

The same problem arises of seeking like traits in unlike species. Matuske and Malison rightly cite Morgan and Nader's relevant study on macaques, but there are limitations to what the comparison can achieve because the human study investigated the effect on receptor availability of social status, rather than the ability of respondents to dominate each other. On this point, the BSMSS questionnaire ranks accountants higher than military officers. Presumably, accountants boast a higher average salary than officers, but there is no age distinction to account for the fact that recently graduated accountants must surely earn less than army generals awaiting retirement. If they were dominance ranked on their ability to compete for strawberry milkshake, my money would be on the military officers (provided they all shared the same taste for the drink).

The cocaine studies broadly suggest that dopamine and reward may underpin the elevation of status. We know that cocaine interferes with dopamine currents in the reward network and each of these studies found a distinction between the brain chemistry of dominant and submissive individuals, which also correlated with differences in their cocaine usage. Nevertheless, their seemingly contradictory findings pose new questions. This is, at best, a hint rather than a confirmation that the brain's reward network motivates our eternal lust for one or another brand of power.

12.6   SEROTONIN

The case is not yet made for dopamine, but other evidence makes the hypothesis worth pursuing. Long before the cocaine studies, a role in dominance had already been reasonably well established for another familiar neurotransmitter—serotonin. Several studies have managed to modify hierarchies in animal groups by manipulating serotonergic neurons using drugs that humans use as anti-depressants. Chapter 7 explained that selective serotonin reuptake inhibitors (SSRIs) do for serotonin what cocaine does for dopamine—they jam the transporters so that molecules of the neurotransmitter remain in the synaptic cleft. As serotonin is known to mediate the reward system, this provides indirect evidence of a likely role for the reward kingpin, dopamine.

Researchers Earl Larson and Cliff Summers used SSRIs to manipulate dominance rankings in reptiles.[22] *Anolis carolinensis* is a tree-dwelling lizard that can change colour from green to brown. They also have throat fans, inflatable flaps of skin that hang beneath their throats. Males puff themselves up when trying to attract females or during territorial disputes. The researchers paired up 60 lizards, and established the dominant and the subordinate male in each dyad. They carried out a controlled test in which lizards received either a placebo or sertraline, an SSRI marketed as Zoloft. Various different combinations were used: in one trial both the dominant and the subordinate lizard in each pair received sertraline, while in each of two more trials either the dominant or subordinate lizard was given the SSRI. In both of the trials where subordinate lizards received sertraline there was no effect on hierarchy. In the trial where only the dominant lizard took the SSRI they lost their dominant status, becoming either equal with or subordinate to their rival. The researchers reasoned that elevating synaptic serotonin levels reduced status by dampening aggression. In order to parse the effects on the distinct traits of aggression and dominance it would be necessary to show that reducing serotonin activity elevated the lizards' dominance ranking.

An experiment designed to do exactly that had previously discovered the opposite effect on vervet monkeys. The experiment led by Michael Raleigh and Arthur Yuwiler investigated how previously subordinate vervets responded to the creation of a power vacuum.[23] The primates, characterised by a black face encircled with white hair, lived in groups of three males, at least three females and their children. Experimenters investigated what happened when the dominant monkey was removed. The team used two kinds of drugs, one that increased serotonin activity, including the SSRI Prozac, and another that decreased activity. When the dominant male was removed, one of each of the two remaining males received one or other of the drugs. The researchers found that increasing serotonin activity caused the monkey to become dominant, while decreasing activity caused them to become subordinate. It is interesting that SSRIs had contrary effects on the lizards and vervets, decreasing and increasing dominance status, respectively, but it is not too surprising for such distantly related species. Of greater interest is the finding that manipulating serotonin activity has the power not only to decrease but also to elevate status, suggesting a clear role in dominance rather than mere aggression.

Studies such as these strengthen the case for the role of the reward network in motivating dominant traits. Other research has shown that humans respond much like vervets to changes in serotonin traffic.[24] While the role of serotonin in dominance behaviour is better evidenced than dopamine's,[25] all findings point to a potential role for the reward network. Both of the neurotransmitters modulate the reward network, as shown by the recreational drug, ecstasy. The main ingredient, 3,4-methylenedioxy-methamphetamine (MDMA), originally patented as an appetite suppressant, primarily affects serotonin activity. Just as amphetamine affects the dopamine transporter, MDMA does not just stop the serotonin transporter but actually flips it into reverse. Rather than vacuuming up serotonin from the synapse, the MDMA-hijacked transporter spews out extra serotonin from inside the neuron. So, while MDMA also interacts with the dopamine and noradrenaline transporters, it has the biggest impact on serotonin levels.[26] But since ecstasy only has its name by virtue of the euphoria it triggers, it is clear that the reward network must be activated. Studies on rats have shown that MDMA usage does indeed elevate dopamine levels in the nucleus accumbens,[24] and both the nucleus accumbens and the ventral tegmental area receive inputs from serotonergic axons.[27] So, while we may better understand the effect of serotonin over dopamine on dominance, there is no reason to ditch the hypothesis that the trait is mediated by the reward network.

12.7 THE WINNER EFFECT

In 1997 the hopes of a nation were mounted on the shoulders of a rising tennis star who seemed like Britain's best hope to win Wimbledon since 1936. In his drawn out "Super Sunday" clash with Paul Haarhuis, Tim Henman eventually won a 93-minute deciding set. Sadly, he was knocked out in the quarter-finals and, despite making the semi-finals four times subsequently, never satisfied the British public's hunger for a Wimbledon triumph. He was the number one ranked UK tennis player in the world and secured winnings across his career, amassing some £11m, but Wimbledon remained out of reach. Could it be that the fabled loser effect was at work?

Victory and defeat each carry a certain momentum. The winner effect is the phenomenon that recent experience of winning makes you more likely to win, while the loser effect is the corresponding likelihood that one defeat predicts another. This is not an idly forged theory, nor a hangover from hopelessly biased folklore. Countless experiments have controlled for the obvious factors that might interfere with accurate measurement of the phenomenon. Could there have been psychological or even physiological reasons for Henman's failure?

In the animal kingdom the winner effect is usually investigated with physical fights. Experiments have to be carefully conducted to control for obvious factors that might prejudice the enquiry. For example, it is not simply that superior fighters win fights. Typical tests work as follows: one pair of evenly matched conspecifics (members of the same species) fights, and the winner and loser are established. For their next bout, each organism fights a conspecific that has never fought before. If it was simply that the previous winner beats the naïve fighter, we could chalk it up to greater experience. But many experiments have also shown that the previous loser fails to beat the naïve fighter. This is well-established across a vast range of species, including vertebrates—such as Syrian hamsters, blue-footed booby birds, the copper snake and Siamese fighting fish—and invertebrates, including crickets, crab spiders and cray fish.[28] Across these disparate species, the common finding emerges that winning one bout improves an organism's chances of winning the next one, while losing predicts future defeat.

One of the many studies into the winner effect in humans focused on tennis matches. A researcher called Lionel Page interpreted the results of 600 000 tennis matches. Within this impressive data set, he identified 72 294 tennis sets that were won by what he called a "close tie break". A tennis set is won either by the first player to reach seven points or, if both players reach six points, by the first player to win two points in a row. In either case this is a tie break situation. What Page defines as a close tie break is, in effect, one that goes on for a long time. The rationale for this is that the more drawn out the tie break, the more evenly matched the players and, therefore, the greater the role of random events in the ultimate victory. According to his analysis, players who won a close tie break in the first set would win, on average, one game more in the second set.[29]

Despite these findings, the winner effect is unlikely to explain Henman's Wimbledon losses. For one thing, Wimbledon is an annual event. Many other matches would take place between the tournaments and a loss directly preceding Wimbledon would be necessary to even consider the winner effect. Furthermore, the above study investigated pairs of consecutive sets, while winning the entire tournament would involve vast numbers of sets, of which even the champion would be expected to lose at least a few. On the other hand, it could be that Henman experienced the winner effect more weakly than competitors, such that he was never able to sustain a winning streak.

Our understanding of the winner effect has now extended to physiological effects. It now seems that when an individual wins a competitive encounter their brain demonstrates increased activity of androgens, including testosterone. Meanwhile, losers experience more activity of stress hormones, such as cortisol.

One recent study that supports this hypothesis fostered the winner effect with a rigged game of Tetris. Participants in the study, run by Samuele Zilioli and Neil V. Watson,[30] played the game on two consecutive days, but skill was irrelevant on day one, with the rigged game assigning winners and losers at random. The 88 participants were invited to come back the next day to compete for real. By measuring the testosterone levels in the players' saliva before and after each game the researchers reproduced the findings of past studies. The winners registered a spike in testosterone, while the losers' testosterone count plunged. By the start of the next day's battles their testosterone levels were back to normal. It was not the case that every previous winner went on to recreate their victory. But the researchers did find an interesting correlation between testosterone levels and final score: the higher a player's testosterone on the first day, the higher their score was on the second, even if it was not a winning score. In other words, winning on the first day seemed to make them play better on the second.

These findings strengthen the case for the role of testosterone and dopamine in dominance status. We have already seen how testosterone is thought to reduce fear *via* its action in the amygdala, while it may potentially modulate reward-eliciting dopamine currents in the orbitofrontal cortex. Now there is a growing body of evidence suggesting that the hormone, along with other androgens, promotes a winning streak, which may also be mediated by the reward circuit. Having said that, research has not reached a consensus about the mechanism of the winner effect. Some scientists have argued that the testosterone spike enhances competitive aggression, but another promising theory is that the hormone may enhance learning of effective strategies.[28] The reward system could mediate either of these mechanisms.

12.7.1 Nobel Legacies Boosting the Hypothesis for Reward Mediation of the Winner Effect

The hypothesis that the reward network could mediate the winner effect has benefitted from novel analytical tools pioneered by two

Nobel Prize winners. Two recent studies on rodents have detected hormonal activity in brain regions incorporated in the reward network, including the nucleus accumbens and the ventral tegmental area. Both studies used immunochemistry techniques, the discovery of which was led by Rosalyn Sussman Yalow, who won the Nobel Prize in 1977. Meanwhile, one of the studies used a variant of a process called the polymerase chain reaction (PCR), for the discovery of which Kary Mullis shared the prestigious award in 1993. Both techniques are frequently used in conjunction with another phenomenon, chemiluminescence.

Chemiluminescence is similar to fluorescence. Both involve the emission of visible light as a result of electrons transferring between energy levels. When an electron absorbs energy it can be excited to a higher energy level. When it inevitably falls back to its ground state the same amount of energy is released in the form of electromagnetic radiation, and if this happens to be in the visible region of the spectrum the substance in question will glow. A fluorescent material will only glow having been bathed in light, but chemiluminescent materials glow as a result of chemical reactions.

Chemiluminescence can support immunochemistry techniques to help researchers filter out the background noise of biological systems. If you want to confirm that a particular chemical process is taking place in a cell, the judgment is hampered by the vast array of other chemical goings-on around it. So, measuring the concentration of a biomolecule of interest, such as testosterone, is like a colossal game of *Where's Wally*, except Wally is too small to see, like everyone around him, and he has thousands of identical twins. To combat this problem, researchers hijack the foot soldiers used by the body's immune system—antibodies—and reprogram them to act as "Wally finders". First of all, they identify an antibody that binds exclusively with the biomolecule of interest and customise the antibody with a chemiluminescent side group. When the antibody binds with the target biomolecule, the side group glows, as shown in Figure 12.1. When the modified antibody is mixed with a sample, the concentration of the target biomolecule can be determined by how brightly the solution glows.

This technique was used to investigate a potential role for dopamine in the nucleus accumbens of fighting Syrian hamsters. The rodents were trained to win fights and their brains

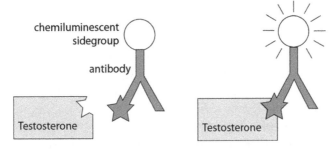

Figure 12.1 Chemiluminescence. A suitable antibody is modified with a chemiluminescent side group. When the antibody binds the biomolecule of interest, in this case testosterone, the side group emits light, the intensity of which can be measured to determine the concentration.

subsequently compared to hamsters that had never fought, nor even socialised. Using immunohistochemistry, the researchers Jared J. Schwartzer, Lesley A. Ricci and Richard H. Melloni Jr. discovered unusually high concentrations of the enzyme tyrosine hydroxylase in the brain's nucleus accumbens.[31] As we saw in Chapter 7, the amino acid tyrosine is converted into dopamine and tyrosine hydroxylase is the critical enzyme in its production. We also saw that dopamine may then be converted into its sister monoamine, noradrenaline. Was tyrosine hydroxylase working overtime to produce dopamine or noradrenaline? Either may play a role in reward, especially since the activity was detected in the reward-dispensing nucleus accumbens. The other issue is whether it was the fighting that elevated the activity or some other aspect of the winners' experience that the naïve hamsters had not shared, such as simply being around other hamsters.

The second study used both of these newly introduced techniques to find further evidence of a reward-mediated winner effect. Matthew J. Fuxjager and Catherine A. Marler used quantitative PCR (qPCR) technology, as well as immunohistochemistry, to investigate California mice.[32] They found that the winner effect modifies the expression of genes in a pair of brain regions including the nucleus accumbens.

The PCR process was invented by Kary Mullis. He is a larger than life character who used the platform granted by the accolade to share a plethora of contentious views. In his autobiography, *Dancing Naked in the Mind Field*,[33] Mullis relates how the invention itself devastated a minibreak that he was taking with his then sweetheart. The idea hit him on the road and left him too excited to sleep for some time afterwards. When the prize was announced, paparazzi snapped him returning from a surfing session, immortalising him as potentially the coolest ever winner. His autobiography is also remarkable for his willingness to oppose broadly held views, evident variously from his denial of climate change, ozone depletion and AIDS, and his support for the ancient art of astrology. The cases he makes are not devoid of substance. He notes, for example, that the international ban on ozone-depleting chlorofluorocarbons (CFCs) coincided uncannily with the expiry of their patents. An unashamed proponent of psychoactive drugs, including LSD and nitrous oxide, he further claims that the astrally projected soul of a female stranger saved him from suffocating himself on laughing gas, by plucking the feed tube from his lips as he lay unconscious. He would definitely make the cut for my fantasy dinner party list.

The PCR technique is an invaluable method by which to amplify nucleobase sequences, such as genes. One of its many uses has been to improve crime-fighting by making it easier to convict felons with DNA evidence. In 2015, 22-year-old Patryk Filipski was successfully linked to a burglary,[34] during which he had helped himself to his victim's beer, but left the empty bottle at the scene of the crime. The police made a successful case using DNA evidence from saliva left on the bottle. The difficulty is that, while the saliva may contain the suspect's DNA, there is very little of it with which to make an iron-cast case. Forensic scientists use PCR effectively to run off a few photocopies of a particular sequence of DNA. Now they have more to play with as they seek a match with DNA samples taken directly from the suspect following arrest.

The PCR works as follows: first of all, researchers take a sample of some DNA, probably an entire genome, which they may have extracted from a cell. They identify a target gene they would like to amplify. Stage two requires a primer. This is a miniature sequence of DNA nucleotides that are complementary to a small portion of the target gene. When added to the sample, these primers will clamp to the corresponding section of the target gene, which has the effect of seeding polymerisation. The process gets its name from DNA polymerase, the enzyme used to construct complementary strands of DNA. The primers guide DNA polymerase so that only the target gene is copied. Copies of the target gene increase exponentially in number (each number in the sequence is twice the last), allowing researchers to determine how many strands of the target gene were present in the sample to start with (Figure 12.2).

As mentioned, the study of the winner effect in California mice used qPCR, which is a variant of traditional PCR. This method can be used to measure the rate at which genes are expressed. As we saw in Chapter 5, production of a protein begins when the corresponding gene is transcribed into a strip of mRNA, which then dictates the sequence in which for amino acids to be connected in the protein. Therefore, it is possible to determine how frequently a cell is expressing a particular gene by counting the number of molecules of its mRNA transcript, which can be achieved with qPCR. First of all, the mRNA is reverse transcribed back into DNA, just like the jumping gene activity described in

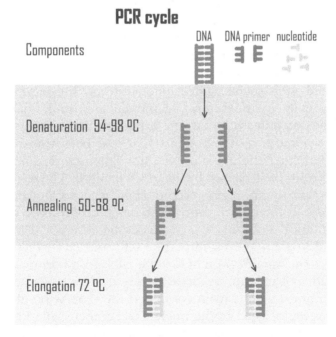

Figure 12.2   The polymerase chain reaction. The reaction runs off copies of a specified sequence of DNA. Primers bind to the DNA and seed replication of the gene sequence under investigation.
(Image from Shutterstock, artwork © WhiteDragon).

Chapter 10. These new strands are called complementary DNA (cDNA). The quantity of cDNA molecules is measured using a chemiluminescent dye, engineered to emit more light when it is bound to DNA, and thus the more light emitted then the more cDNA has been produced. This gives the comparative quantity of mRNA present to start with.

A combination of these techniques threw up an interesting twist in the study of California mice.[32] What Fuxjager and Marler's team discovered was that the rodents' brains were differently affected depending on where they fought. Each mouse won a trio of rigged battles to promote the winner effect before undergoing a single bout with a larger, sexually experienced male. What was unique about the experiment was where the rodents fought. Between fights, each of the males was housed with a female, but when it came time to fight, some of the mice stayed in their usual living quarters, whilst others were transferred to unfamiliar surroundings. But the main point of enquiry was how their brains were affected. Using immunochemical techniques the researchers recorded greater concentrations of testosterone receptors in the reward-implicated nucleus accumbens and ventral tegmental area, and also in the medial anterior bed nucleus of the stria terminalis (BNST), which is associated with social aggression. However, the elevated hormone activity in the ventral tegmental area and nucleus accumbens was only evident in the brains of the mice who fought on their home turf. The rodents that had been transferred to unfamiliar surroundings did not show this increased activity.

This finding was backed by the qPCR analysis. The team found that the gene for androgen receptors had been working overtime. As we saw in Chapter 6, testosterone cannot influence gene expression until it binds to a testosterone receptor. The newly bound duo then moves into the cell's nucleus, where it coordinates the modification of histone tails. So, a hormone is only as useful as the number of receptors available. Using qPCR, Fuxjager and Marler's team compared the expression of the androgen receptor gene in the nucleus accumbens, the BNST and another region called the lateral septum. Once again, they found elevated activity in the nucleus accumbens and the BNST but only in the brains of the mice that fought at home. Their data suggest that the reward network may well mediate the winner effect but that its input may be sufficient, but not necessary, for the phenomenon to occur.

12.8 CONCLUSION

Evidence is accumulating that the reward network mediates the trait of dominance and may, therefore, motivate us to elevate our status. But the researchers are rightly cautious in the conclusions they draw. One possibility is that the activity in the reward network may not relate to dominance displays. For example, the hamsters might demonstrate altered dopamine levels because of the winner effect or simply because they have had the experience of socially bonding with other hamsters. Alternatively, the testosterone and dopamine currents could be linked with learning. Perhaps we are not hard-wired to elevate status but rather develop dominance, having associated status with increased access to our primary needs of food and reproductive opportunities. These are important questions for future research.

There is also the issue of the precise relationship between dominance and reproductive fitness. In many species dominance is associated with increased access to mates. More mating means more self-replication of genetic material, but field work has shown that the dominant partner in a dyad is not necessarily the one with the most mates. Carlos Drews' definition does not link the trait with the acquisition of resources; it is simply the ability to come out on top in the majority of conflicts.

There are limitations to what the reward network can achieve. We get a big reward when we have sex, because that was the only way in which to have children when the adaptation evolved. It used to be that sex was necessary but not sufficient for conception, now it is neither necessary, nor sufficient. In humans the reward network can only promote the chances of increased replication of genetic material. This same exception seems to apply both to dominance and mating behaviour.

But does the status of successive generations follow the immutable regression charted by Clark? The findings of the neurological studies suggest that epigenetic factors may be at play, especially the finding that the winner effect elevated expression of the androgen receptor gene. Clark argues that genes outweigh environment in predicting our underlying general competence, but our environment influences the genes we express. Furthermore, any epigenetic modifications may be inherited by offspring. We have seen that the loser effect releases stress hormones. Environment also influences how dominance correlates with status. People from different backgrounds may be comparable in their ability to dominate but follow very different career paths.

The reward network probably plays an important role in dominance. As ever, a complex interplay exists between the nature of our genetic inheritance and the nurture of our environment. For our ancestors, dominance was probably a better predictor of reproductive fitness but none of this is certain.

REFERENCES

1. G. Orwell, Why I Write, *Gangrel*, 1946.
2. M. Shelden, in *Orwell: The Authorised Biography*, ed. Michael Shelden, Politico's, London, 2006.
3. F. Wang, H. W. Kessels and H. Hu, The Mouse That Roared: Neural Mechanisms of Social Hierarchy, *Trends Neurosci.*, 2014, **37**(11), 674–682.
4. R. Kirby, Testosterone and the Struggle for Higher Social Status, *Trends Urol. Mens Health*, 2014, **5**(1), 11–14.
5. C. Drews, The Concept and Definition of Dominance in Animal Behaviour, *Behaviour*, 1993, **125**(3), 283–313.
6. P. Kitcher, *Vaulting Ambition: Sociobiology and the Quest for Human Nature*, MIT Press, Cambridge, Mass, 1987.
7. H. Kummer, *Primate Societies; Group Techniques of Ecological Adaptation*, Aldine · Atherton, Chicago, 1971.
8. A. McClintock, Maid to Order: Commercial Fetishism and Gender Power, *Soc. Text*, 1993, (37), 87.
9. P. H. Hawley, The Ontogenesis of Social Dominance: A Strategy-Based Evolutionary Perspective, *Dev. Rev.*, 1999, **19**(1), 97–132.
10. G. Clark, in *The Son Also Rises: Surnames and the History of Social Mobility*, ed. Gregory Clark [and 11 others], Princeton University Press, Princeton, New Jersey, 2014.
11. D. Terburg and J. van Honk, Approach–Avoidance Versus Dominance–Submissiveness: A Multilevel Neural Framework

on How Testosterone Promotes Social Status, *Emot. Rev.*, 2013, **5**(3), 296–302.

12. D. Terburg, N. Hooiveld, H. Aarts, J. L. Kenemans and J. van Honk, Eye Tracking Unconscious Face-to-Face Confrontations: Dominance Motives Prolong Gaze to Masked Angry Faces, *Psychol. Sci.*, 2011, **22**(3), 314–319.

13. D. Terburg, H. Aarts and J. van Honk, Testosterone Affects Gaze Aversion From Angry Faces Outside of Conscious Awareness, *Psychol. Sci.*, 2012, **23**(5), 459–463.

14. W. B. Hoover and R. P. Vertes, Anatomical Analysis of Afferent Projections to the Medial Prefrontal Cortex in the Rat, *Brain Struct. Funct.*, 2007, **212**(2), 149–179.

15. F. Wang, J. Zhu, H. Zhu, Q. Zhang, Z. Lin and H. Hu, Bidirectional Control of Social Hierarchy by Synaptic Efficacy in Medial Prefrontal Cortex, *Science*, 2011, **334**(6056), 693–697.

16. S. N. Haber and B. Knutson, The Reward Circuit: Linking Primate Anatomy and Human Imaging, *Neuropsychopharmacology*, 2010, **35**(1), 4–26.

17. B. Knutson, G. W. Fong, S. M. Bennett, C. M. Adams and D. Hommer, A Region of Mesial Prefrontal Cortex Tracks Monetarily Rewarding Outcomes: Characterization with Rapid Event-related fMRI, *NeuroImage*, 2003, **18**(2), 263–272.

18. B. Jupp, J. E. Murray, E. R. Jordan, J. Xia, M. Fluharty, S. Shrestha, T. W. Robbins and J. W. Dalley, Social Dominance in Rats: Effects on Cocaine Self-administration, Novelty Reactivity and Dopamine Receptor Binding and Content in the Striatum, *Psychopharmacology*, 2016, **233**, 579–589.

19. D. Morgan, K. A. Grant, H. D. Gage, R. H. Mach, J. R. Kaplan, O. Prioleau, S. H. Nader, N. Buchheimer, R. L. Ehrenkaufer and M. A. Nader, Social Dominance in Monkeys: Dopamine D2 Receptors and Cocaine Self-administration, *Nat. Neurosci.*, 2002, **5**(2), 169–174.

20. G. J. Syme, Competitive Orders as Measures of Social Dominance, *Anim. Behav.*, 1974, **22**(Part 4), 931–940.

21. D. Matuskey, E. C. Gaiser, J.-D. Gallezot, G. A. Angarita, B. Pittman, N. Nabulsi, J. Ropchan, P. MaCleod, K. P. Cosgrove, Y.-S. Ding, M. N. Potenza, R. E. Carson and R. T. Malison, A Preliminary Study of Dopamine D2/3 Receptor Availability and Social Status in Healthy and Cocaine

Dependent Humans Imaged with [11C](+)PHNO, *Drug Alcohol Depend.*, 2015, **154**, 167–173.

22. E. T. Larson and C. H. Summers, Serotonin Reverses Dominant Social Status, *Behav. Brain Res.*, 2001, **121**(1), 95–102.

23. M. J. Raleigh, M. T. McGuire, G. L. Brammer, D. B. Pollack and A. Yuwiler, Serotonergic Mechanisms Promote Dominance Acquisition in Adult Male Vervet Monkeys, *Brain Res.*, 1991, **559**(2), 181–190.

24. S. N. Young and M. Leyton, The Role of Serotonin in Human Mood and Social Interaction, *Pharmacol., Biochem. Behav.*, 2002, **71**(4), 857–865.

25. S. J. Mooney, D. E. Peragine, G. A. Hathaway and M. M. Holmes, A Game of Thrones: Neural Plasticity in Mammalian Social Hierarchies, *Soc. Neurosci.*, 2014, **9**(2), 108–117.

26. C. D. Verrico, G. M. Miller and B. K. Madras, MDMA (Ecstasy) and Human Dopamine, Norepinephrine, and Serotonin Transporters: Implications for MDMA-induced Neurotoxicity and Treatment, *Psychopharmacology*, 2006, **189**(4), 489–503.

27. S. R. Sesack and A. A. Grace, Cortico-basal Ganglia Reward Network: Microcircuitry, *Neuropsychopharmacology*, 2010, **35**(1), 27–47.

28. Y. Hsu, R. L. Earley and L. L. Wolf, Modulation of Aggressive Behaviour by Fighting Experience: Mechanisms and Contest Outcomes, *Biol. Rev.*, 2005, **81**(1), 33.

29. L. Page, The Momentum Effect in Competitions: Field Evidence from Tennis Matches, *working paper, Citeseer*, 2009, 1–30.

30. S. Zilioli and N. V. Watson, Testosterone Across Successive Competitions: Evidence for a "Winner Effect" in Humans?, *Psychoneuroendocrinology*, 2014, **47**, 1–9.

31. J. J. Schwartzer, L. A. Ricci and R. H. Melloni, Prior Fighting Experience Increases Aggression in Syrian Hamsters: Implications for a Role of Dopamine in the Winner Effect: Dopamine Alterations in Trained Fighters, *Aggressive Behav.*, 2013, **39**(4), 290–300.

32. M. J. Fuxjager, R. M. Forbes-Lorman, D. J. Coss, C. J. Auger, A. P. Auger and C. A. Marler, Winning Territorial Disputes Selectively Enhances Androgen Sensitivity in

Neural Pathways Related to Motivation and Social Aggression, *Proc. Natl. Acad. Sci.*, 2010, **107**(27), 12393–12398.

33. K. B. Mullis, in *Dancing Naked in the Mind Field*, ed. Kary Mullis, Bloomsbury, London, 1999.

34. "Burglar who Burgled Blind Pensioner Caught by DNA Left on Beer Bottle", *Mail Online*, 01-May-2015. Available at: http://www.dailymail.co.uk/news/article-3064501/Burglar-ransacked-blind-pensioner-s-home-caught-DNA-left-beer-bottle.html. [Accessed: 03-Mar-2016.]

Section 3: Concluding Remarks

Our understanding of the chemistry underlying complex traits is far from complete. Simply nailing down objectively irrefutable definitions for these constructs is hard enough without then conceiving illuminating experiments that control for all but one of the multifarious factors that influence them.

Nature and nurture can no longer be considered as mutually exclusive influences on our personal development because our nature affects our nurture and our nurture affects our nature. Meanwhile, advances in the field of epigenetics pose questions for the validity of heritability estimates. They have been a useful tool but probably more for tracking down the causes of disease than for deciphering how we develop our unique personalities. No matter how heritable these traits have been measured to be, identifying the underlying genes has proved extremely challenging. That does not mean they do not exist, nor that heritability estimates are completely wrong. As is usual in science, the relevant models will constantly have to be updated or even overhauled as our understanding deepens.

Factors such as the density of receptors in specific neurons almost certainly influence the development of certain traits. But how does this fact mesh with our understanding of neural plasticity? We can reinforce certain neural circuits by repeatedly activating them, one consequence of which is an increase in receptor density. But what are the limits of this process? What

The Chemistry of Human Nature
By Tom Husband
© Tom Husband 2017
Published by the Royal Society of Chemistry, www.rsc.org

proportion of a person's character can be reprogrammed by exploiting neural plasticity?

Finally, it is no surprise that these traits seem to interact with the brain's reward network. Violence and creativity have been used throughout history to elevate status, which in turn advances the individual's capacity to produce viable offspring, allowing for their genes to continue self-replicating. This raises the obvious question to be addressed in Chapter 13: is it possible, truly, to defy the self-replicators?

Section 4:
So What?

CHAPTER 13

The Chemistry of Free Will

At the start of the book we considered the idea that the Mars One volunteers may not have any authentic agency in their decision to take their one-way trip. If the universe is deterministic and the response of their atoms to physical laws is the true reason for their decision, then surely they cannot be said to have freely made that decision. Alternatively, in an indeterminist universe, are their actions not the result of the random movement of electrons? If so, just because these particles demonstrate something like Epicurus' *random swerve*, it does not follow that the astronauts have any greater agency in their decision. Meanwhile, what of the biological drives we have explored impelling them to feed, mate or elevate status? What does this framework reveal about their motivations? These are the questions to be considered in this chapter.

How would the Mars One volunteers feel if, having embarked on their one-way trip, they were shown conclusive proof that their decision had been wholly determined by atomic inter-actions, over which they had absolutely no control?

13.1 UNEASE AROUND EMERGING BIOLOGY

There is a certain unease around the discoveries emerging from science. Fears that we do not make our own decisions and fears that we are ultimately selfish creatures can be unsettling.

The Chemistry of Human Nature
By Tom Husband
© Tom Husband 2017
Published by the Royal Society of Chemistry, www.rsc.org

Richard Dawkins admitted that he might have thought of a better name for his book *The Selfish Gene*[1] because it was broadly interpreted, mainly by people who had not read it, as a description of a particular gene that makes us selfish. What it actually described was how genes could make us altruistic. The self-replicators flourish when they can maintain constant streams of fuel, building materials and a vehicle in which to replicate. We have evolved to serve those needs with altruism, living in social groups to spread the risks that threaten our existence. But understood this way, apparent selflessness is actually selfish. We help others to help ourselves.

In the 30th anniversary edition Dawkins described two letters written to him in response to his book.[1] One was from a teacher describing how they had been forced to counsel a distraught student after she read the book. Another was from a man wishing that he could "unread" the book as its contents had triggered a 10-year malaise.

In fact, this second correspondent was in good company because a similar phenomenon struck one of the very scientists who discovered the evolutionary mechanism of altruism. As we saw in Chapter 10, George Price became increasingly despondent after discovering the equation that showed how altruism could be evolutionarily adaptive.

Similar concerns have been raised around the machinations of romantic love. Helen Fisher argued that investigation of a possible genetic basis for divorce made some people uncomfortable because it threatened their sense of free will.[2] Michael Liebowitz noted the same phenomenon, suggesting that some people do not like to know what makes them tick. A prime example, he felt, was senator for the democrats, William Proxmire, who used to issue a Golden Fleece Award to research studies he felt were a waste of time, including an $84 000 study into why people fall in love.[3]

This thinking clashes with the idea that we want our romantic relationships to be written in the stars, but how much more fatalistic can you get? *Destiny threw us into each other's arms and we have never looked back!* If we want our relationships to be preordained by the cosmic ballet, why should we object to examining the nuts and bolts of romantic love? We love to invoke the actions of the heavens but not those of the atoms that play out their supposed bidding. A similarly romantic ideal glorifies the fabled compulsion of writers and artists to create. Destiny can be poetic, but we are selective about which bits we embrace.

Many studies have been conducted to find out what non-philosophers think about free will and determinism. Eddy Nahmias explains that volunteers are asked to give their opinion on various hypothetical scenarios, including one that recalls the ideas of Laplace from Chapter 1. Participants in the study were told that a super-computer could predict the future with 100% certainty based on the deterministic interactions of atoms. The super-computer stated that an imaginary character, Jeremy, would rob a bank. When asked if Jeremy had acted of his own free will, 76% said yes.[4] This points to the importance of culpability in the free will debate. We saw in Chapter 11 how one convict had his sentence reduced because he had the mutated version of the *monoamine oxidase A (MAO A)* gene, whilst another received no such lenience on the basis of his partial foetal alcohol syndrome. Could we have expected them to act any differently if we accepted that they were biologically predisposed to violent acts? If the alternative is to be pre-emptively incarcerated in a secure mental health facility, philosopher and cognitive scientist Daniel Dennett argues that legal responsibility is the price we pay for our freedom.

13.1.1 Dennett's Position

Dennett communicates beautifully about the relationship between biochemistry and freedom of will. He refers often to a headline written for an article about his work. *"Sì, abbiamo un'anima. Ma è fatta di tanti piccoli robot."*[5] Translation: "Yes, we have a soul. But it's made of many tiny robots". He uses the beautiful analogy that our consciousness can be free, even while it results from a clump of neurons, which have no freedom, just as a red object is composed of atoms that are not red.[5]

One part of the neuron he particularly sees as robot-like is the vesicle transport by myosin. This is the process in which a vesicle full of freshly made biomolecules is transported along actin filaments from the cell's core to its membrane. This is achieved by a pair of myosin molecules that take it in turns to reach along the filament and drag the vesicle forward, as shown in

Figure 13.1. When I show students animated videos of the process—such as the exceptional *Inner Life of a Cell*—they always laugh, presumably because the process looks so life-like.

There was a fascinating documentary about the comedian Rik Mayall called the *Lord of Misrule*.[6] Watching it, I had a strong impression that his phenomenal success came without any real effort. He grew up making his family laugh, then he got to university and made his fellow students laugh. By that point,

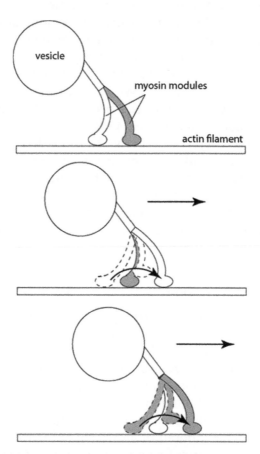

Figure 13.1 Vesicle transport. Molecules of myosin alternately reach along the actin filament, hauling the vesicle around the cell. Dennett believes that opponents of the compatibilist view mistakenly worry about one thing and another. His view is that "determinism doesn't imply that whatever we do, we could not have done otherwise, that every event has a cause, or that our natures are fixed."[5]

it seemed, just being himself was a consistent source of riotous amusement. Word spread, collaborations began, TV execs got wind of it and Rik ended up on television. His shows were hilarious to an appreciable demographic, so new ones were always commissioned. Contrast this with Eddie Izzard, who spent 10 years struggling as a street performer, wondering if he would ever "make it".[7] This is a useful context in which to consider some of the important strands of Dennett's position.

Macro-level analysis yields far more valuable information than micro-level analysis. The point is that, while decisions may ultimately be determined by atomic interactions in the brain, the actual actions themselves are still a more suitable focus for analysis. The manner in which atoms shuttled around Rik Mayall's brain is interesting, but it doesn't tell us why his jokes worked. We may postulate that it was his family's nurturing that fostered his talent, or that a freak mutation produced a gene that overexpressed receptors in the default mode network of his brain, but none of that explains why his jokes are funny. Direct engagement with his humour is how rising talent would have dissected his work, as they sought to forge their own comedic personae in his image. The distinction plays a valued role in Dennett's view of freedom.

Another issue is the difficulty of definitely assigning causality. What caused Mayall's success? Was it that he was born into an era receptive to his anarchic brand of humour or would his talent have turned a success in any period of history? Was it that his family frequently organised plays in which he was able to hone his craft? (In which case, why didn't his brother become a famous comedian?) Was it his chance meeting with comedy collaborators Adrian Edmondson and Ben Elton? But what if we imagine a different, hypothetical world, as Dennett frequently does,[5] with subtle differences? In one such other world, Mayall was not born to parents "in the biz", there were no genetic anomalies, he did not meet Edmondson or Elton, but he was still funny. We would have to conclude that none of those factors was necessary for him to succeed.

We can track this right back to the moment following the Big Bang. If we assume that events are deterministically linked to the birth of the universe, can we say that those events caused Rik Mayall's success? Again, suppose there is a hypothetical world with an almost identical follow-up to the Big Bang but with one or two slight differences, for example, jets of nascent matter were propelled at very slightly different angles into the cosmos compared to following the version of the Big Bang that occurred in our current reality. If Rik Mayall is still a comedian in this alternative world, we cannot claim that Rik Mayall's success as a comedian is linked in an unbroken chain of causally linked events right back to the Big Bang in our current world. The precise angling of the matter jets was not necessary for Mayall's success, casting doubt on their causal role in the outcome. This point is made by Dennett in a more generalised way without reference to Mayall's career.[8]

Another strand in Dennett's argument is what he describes as "evitability". This is predicated on his belief that people have incorrectly concluded that determinism requires the inevitability of future events. If things can be inevitable, can they also be evitable, that is, avoidable? Did Eddie Izzard avoid a life of obscurity? Dennett would argue no.[8] In a deterministic universe, Izzard was always going to succeed, even if he feared that he may not. (Indeed, such fears doubtlessly drove him to succeed.) As such, he cannot be said to have avoided the life of obscurity.

This leads to another point: could Izzard have done otherwise? Was there an event in Izzard's pre-fame years that he fluffed such that, had the meeting gone otherwise, he could have made it earlier? For example, there could have been a meeting with a TV exec in which Izzard might have got what he wanted if he had only acted differently. Such questions are not important to Dennett.[8] He does not see why a person ought to have been able to act in a way other than they did, conditions being as they were. He is much more inclined to wonder how the agent responds to similar events in the future. In this case, if we suppose that Izzard did make a mess of one meeting with a TV exec, did he subsequently evaluate his performance so as to make sure that he did better when another opportunity rolled around?

Both comedians spent a long time honing their craft. Mayall's key advantage may simply have been that he was able to start and finish his 10 000 hours of practice earlier than his contemporary because his family provided so many opportunities for him to perform during his childhood. But Dennett's point, although I am not aware of his having tackled the subject of Mayall and

Izzard's careers directly, is that people can improve themselves, morally, intellectually and, in other ways, by learning from their experience.[8] However easy Mayall made his rise to stardom look, he still had to put in the hours developing his craft.

So, the idea that the prior state of the universe has a causally inviolable effect on our current present does not satisfy Dennett. It is all very easy—especially in the storm's eye of an existential funk—to assume one is living out a destiny preordained by atomic events going back to the Big Bang, but, when you pop the hood and examine the specifics, the situation is more complicated.

13.2   INDETERMINACY

One interesting thing about Dennett's position is the apparent conflict with our indeterminate universe. As we saw in Chapter 1, the random path of the electron confounds our ability to predict exactly how atoms will behave. If we were able to build a super-computer to act as Laplace's demon, it could not predict the future with certainty, as quantum events appear genuinely random.

One philosopher was ideally placed to consider the ramifications for legal responsibility of the freedom of will. David Hodgson was a judge in New South Wales, Australia, but he also took a keen interest in philosophy. He argued that determinism must be false because personal responsibility is the cornerstone of human ethics and the legal system.[9]

Hodgson believed that quantum indeterminacy could play a critical role in human reasoning.[10] While acknowledging the various studies that have concluded that quantum effects are too small to influence, say, the firing of a neuron, he nevertheless thinks it meshes well with another mystery: consciousness.

He firstly asserts that reasoning does not follow rules.[10] "Plausible reasoning cannot be fully explained in terms of rules for good reasoning," a point he notes to have been made by various philosophers, including David Hume. He next argues that classical physics is not adequate to explain the marvel of consciousness, particularly when, in his judgment, "neuro-science itself assumes that the operation of the brain needs no assistance from conscious experiences."[10] One proponent of such a view was the psychologist Daniel Wegner, who argued that human decisions arise from neural processes completely independently from an individual's conscious will. Moreover, he felt that conscious will amounted to nothing more than the illusion that we make our decisions.[11] Hodgson did not agree. He wondered if the mystery of quantum physics could explain the mystery of consciousness.

In Hodgson's view, quantum indeterminacy could provide a means for our will to be free.[10] His view hinges on his belief that our knowledge and experience exist as gestalts, sum-exceeding wholes that exist beyond the rules of neural processes, which nevertheless have the power consciously to affect our decision making. He makes the example that the melody from George Gershwin's *The Man I Love* is greater than the sum of its parts because it does not engage with laws or rules. Although it presumably obeys certain rules of harmony, it does so in a unique fashion that countless other combinations of notes do not. Furthermore, they are understood as gestalts by their listeners, who form their aesthetic judgement based on the gestalt as a whole.

Finally, he invokes the effects of quantum mechanics over a distance.[10] While the causality of classical physics is bound within the locality, quantum effects are observable in separated regions of space. The typically quoted example is the tendency of one subatomic particle to "know" how its spin-paired partner will behave even when they are very far away from each other, a phenomenon famously described by Einstein as "spooky action at distance." Hodgson posits that this quantum entanglement could unite extended regions of the brain in a way that would clear up another revelation to emerge from neuroscience—the lack of a coherent "self" uniting the discrete modules of the human brain.

13.3   PLEASURE AND THE THREE DIRECTIVES

Another way to evaluate the freedom of will is within the confines of our reward system. We are rewarded directly for eating and reproducing, actions that serve the self-replicators. Chapter 12 showed how reward also seems to be meted out for actions that elevate our status. This poses the interesting question of whether it is possible to do anything without increasing or attempting to increase our capacity to replicate our genes in future generations. The novelist Kurt Vonnegut counselled would-be writers that every sentence of a story should do one of two things—advance plot or develop character. Could similar advice be applied to the way we live our lives? Should our every action advance our reproductive fitness?

These questions recall the film *Robocop*. In Paul Verhoeven's futuristic sci-fi film, law enforcer Murphy is brutally murdered by a gang of criminals. The victim's brain is salvaged and placed inside a robot body, creating the cyborg Robocop. Although the cyborg has a human mind, he is obliged to work within guidelines externally imposed by his creators. He has three prime directives: to serve the public trust, protect the innocent and to uphold the law. Thanks to his bionic exoskeleton he has unprecedented freedom in how to meet these objectives. For example, one criminal opens fire with an automatic assault rifle, but his bullets simply ricochet off the titanium-armoured Robocop, who coolly walks towards him and then bends the robber's gun barrel, so that it can no longer fire. He had choice in how to achieve his goal, so long as that goal coincided with the directives to fight crime.

Perhaps we are all like Robocop. Our reward system triggers a pleasurable sensation to encourage behaviours that have previously served the self-replicators, in other words, that have elevated reproductive fitness. In this sense our prime directives would be to eat, mate, elevate status or promote beneficial relations with members of our families and the community. Is it possible to act outside of these "prime directives"?

Dennett labels as trivial, irrelevant or non-functional the following human endeavours: gossip, riddles, poetry and philosophy.[8] I should not write any of these activities off so hastily! Dennett has established himself as a well-respected philosopher, thereby elevating his status. While the precise relationship between status and reproductive fitness is hard to elucidate for modern-day humans, it has a clear biological function in other species. For all the jokes about the unemployability of philosophy graduates, it is nevertheless possible to turn the discipline to material success, as is the case for poetry. Meanwhile, gossip has now been shown to help people avoid selfish members of the community, in favour of those with a good reputation.[12] All of these activities are functional. Richard Dawkins made a bolder claim in *The Selfish Gene*:

> *"We have the power to defy the selfish genes of our birth and, if necessary, the selfish memes of our indoctrination. We can even discuss ways of deliberately cultivating and nurturing pure, disinterested altruism—something that has no place in nature, something that has never existed before in the whole history of the world."*[1]

So, the question is posed: is it possible to act otherwise than to elevate reproductive fitness? The question is almost impossible to answer. Suppose I throw a banana into the air, let it fall to the ground and then leave it there. This pointless act could prove that I can act outside of these boundaries, especially since I would sacrifice an opportunity to eat. But it could also indirectly benefit me. If any friends are present, they may find it entertaining, so the act promotes social bonding. If I use a description of the act as a means to elucidate something on the subject of human nature, it may serve the purpose of elevating status. Even on my own, it may distract me from boredom so as to promote good mental health, upon which foundation I can build further acts to progress my aims.

The consequences for our freedom of will are equally hard to gauge. The more it becomes possible to interpret any act as indirectly promoting reproductive fitness, the less it matters how I choose to behave. Do I get up or stay in bed? If I get up, I will get to work on time, which will please my employers who may promote me. If I stay in bed, I will feel better rested, promoting good mental health so that I can better navigate the many obstacles to a happy and unhindered life.

What Robocop shows us is that prime directives present no obstacle to creative problem solving. In another scene, he finds a man trying to assault a woman. Confronted, the assailant grabs the woman, using her as a shield, while also threatening to stab her. Robocop responds by aiming between the woman's legs and shooting the felon in the genitals. His directives are met but with great creative flair. The question of whether we can break with our directives almost becomes moot when we consider the freedom we have within them.

What if rules are made to be broken? Robocop enjoyed the freedom to act within his prime directives, but he also found ways to break them. The film's climax resolves the mystery of an additional, classified directive. It transpires that Robocop is unable to arrest members of the company that manages his law-enforcement activities. And when the company's vice president Dick Jones comes to be suspected of murder, Robocop has to get him fired before he can arrest him. The sequel depicts an even more drastic departure from observance of his directives. After being reprogrammed, he wipes a new set of directives by elec-trocuting himself. In reality, Robocop only follows his directives when he wants to. Can humans make the same choice?

13.4 BREAKING THE RULES

In fact, there are two ways that we can "break the rules", one is by hoodwinking the self-replicators, while the other consists in the somewhat ironic pursuit of pleasure *via* abstinence from pleasurable activities. They are awkward bedfellows: one is Buddhism, the other is sex and drugs.

13.4.1 Sex and Drugs

One way to break the rules is to hoodwink the self-replicators with sex and drugs. This is a glib reference to the use of psy-choactive substances and to protected sex.

Psychoactive drugs hijack the body's reward system. Chapter 3 showed how amphetamines and cocaine interfere with the brain's dopamine transporters, triggering an ill-gotten dose of pleasure from the reward system. Opiates hoodwink the opioid receptors with similar results. In fact, this is applicable to most recreational drugs, including the more socially acceptable (if not universally legal) drugs, nicotine and alcohol. (Caffeine does appear to interact with receptors in the nucleus accumbens, but doubt re-mains over the extent to which this is felt as reward, particularly at the low doses in which humans typically ingest it.)[13,14]

The point with sex is more complicated. The whole reason we experience pleasure when copulating is because it was the only way to produce children for our ancestors. Now, thanks to con-traception, the activity can be enjoyed with negligible risk of unwanted conception. So, to use Dawkins' term, we "defy our self-replicators" when we enjoy sex with protection, because there is minimal chance of the act leading to the replication of our genes in future generations. Or is there?

Many couples are together for years before they have children. If we assume that a healthy sex life helps to keep couples together, those preceding years of protected sex contribute to the quality of the relationship. While there was no immediate reproductive consequence, it was strengthening the bond between the future parents of the eventual offspring.

Like so many things in science, it ends up being a continuum. It is not that we either do or do not defy the self-replicators because the various sources of reward are inextricably linked. Drinking alcohol is not pleasurable simply because the ethanol activates our reward centre; it is generally enjoyed socially and many people use alcohol precisely to overcome their social inhibitions. The point is that our reward centre is not being entirely duped. Although alcohol does activate our reward system, we enjoy it best when we engage in social bonding, which also activates the reward system.

So, the intentions of our genes are harder to shake off than they might first appear. Although it seems that we can trick our pleasure circuits into paying out unwarranted rewards, actually, those rewards are sweetest in tandem with the warranted variety. Sex and drugs and rock and roll work best when you have company, so it is not really accurate to say that we can hoodwink the self-replicators but rather that we can cadge bigger doses of pleasure than were necessary to keep us feeding and breeding in the distant past.

13.4.2 Buddhism

One group of individuals who would be entirely unfazed by these issues is Buddhists. The illusory nature of the self, agency of the individual in their actions, the cravings that arise in connection with feeding, mating and the pursuit of material wealth are all at the heart of Buddhist doctrine.

Buddhism teaches that the self—or *ego*—is an illusion. Edward Conze, a renowned scholar and practitioner of Buddhism, noted broad agreement on this point between Buddha and certain

Western philosophers, such as David Hume. Bertrand Russell agrees, interpreting the following excerpt from Hume's *A Treatise of Human Nature* as a "repudiation" of the self:[15]

> *"For my part, when I enter most intimately into what I call myself, I always stumble on some particular perception or other, of heat or cold, light or shade, love or hatred, pain or pleasure. I never catch myself at any time without a perception and never can observe anything but the perception."*[16]

Building on the idea that Buddhists agree with Hume about the lack of a self, Conze argues that adherents of the spiritual philosophy differ by their response to the realisation. While the "greed, hate, and attachment" of a Western philosopher remains "practically untouched by his philosophical arguments," a Buddhist reacts by eliminating attachment to the physical world.[17]

To this end, Buddhists seek to overcome the illusion of self. As Conze explains: "The insertion of a fictitious self into the actuality of our experience can be recognized wherever I assume that anything is mine, or that I am anything, or that anything is myself." Thoughts and deeds that foster such illusory beliefs are categorised into five *skandhas*: (1) material, (2) feeling, (3) perception, (4) impulse and (5) acts of consciousness.[17] Consequently, Buddhists eschew material gain, avoid sensual pleasures and eat strictly according to their needs.

This has fascinating consequences for the agency of the individual. If there is no such thing as an individual, which is the logical outcome of the absence of self, how can individuals *do* things? Buddhism teaches that there is action, *karma*, but no agent, *karaka*.[18] This sounds decidedly deterministic, but Buddha is said to have held the same objection to determinism as Democritus. B. Alan Wallace explains that Buddha took a pragmatic opposition to the deterministic viewpoint because such beliefs might erode a person's desire to act with moral fortitude.[19]

For Buddhists, the free will question is not *either/or* but *how much*. The matter does not end with a decision on whether or not free will exists but rather on how to increase the freedom of one's will. Meditation allows practitioners to cultivate the wisdom to make decisions that promote the happiness of themselves and others. Wallace suggests: "When one 'breaks through' ordinary consciousness to pristine awareness, one transcends the realm of the intellect and of causality, and it is here that true, primordial freedom is discovered."[19]

From this practice arises the irony that avoiding pleasure promotes pleasure. Through various practices, notable among which is the abstention from activities that trigger the reward network, such as sexual acts, eating luxury food items or advancing material wealth, Buddhists achieve the blissful state of *Nirvāna*. In this sense, breaking the rules appears not only to be possible but desirable.

The only problem is, as any Buddhist would enthusiastically acknowledge, that we cannot objectively test their claims about enlightenment. The only known way to sample *Nirvāna* is to extinguish the sense of self through diligent practice of the philosophy. Having said that, it is commonly accepted that Buddhist monks look remarkably happy, especially considering that they forego all those little crutches on which so many of us depend, like cups of coffee, biscuits, gossip, booze, social media, cigarettes and so on.

13.4.3 So, Can We Break the Rules or Not?

Perhaps it is not important to definitely answer the question of whether we can or cannot truly break the rules. Certainly Buddhists eschew the triggers to pleasure but not entirely. They still need to eat, even if they subsist on fewer and smaller portions, and that activates the reward centre. Furthermore, it could be argued that where they maintain ties with family members, their wise compassion could be said, on a modest level, to render assistance with the raising of nieces and nephews, in which their own genes may be replicated. Similarly, the methods of hijacking the reward system work best in tandem with social bonding. It is precisely in the absence of social bonding, such as when people drink alone, that the pleasure of hijacking turns into the pain of addiction.

Humans are designed to do a job and that job is replicating our genes in future generations. It is very hard to completely avoid doing the job. Insisting that we only gain pleasure from activities that have zero chance of promoting gene replication becomes an act of spite, to oneself and everyone else. It is better to see it like this: we have the best job going! We have a job like the people who work for Silicon Valley tech companies, where they can wear pyjamas to work and play foosball. Having fun with friends, falling in love and eating delicious food are all pleasures that have evolved in service of the self-replicators. When you are out partying, on some small level you are doing their bidding, forging a strong community in which for your kinsmen, and hence your genes, to flourish.

13.5 A HEALTHY BALANCE?

A less extreme path to happiness than Buddhism might consist of maintaining a balance between our prime directives. In his book *The Conquest of Happiness*, philosopher Bertrand Russell counselled against over-eating, noting that greed inevitably leaves one uncomfortably dyspeptic.[20] Meanwhile, the Action for Happiness organisation recommends maintaining a work–life balance.[21] Such advice can be understood in terms of balancing the different means of activating the brain's reward network. By this rationale, a person could be considered healthy when they succeed in spreading the pleasure they derive from an even spread of eating, loving and elevating status.

This healthy balance model could explain how people develop addictions. Chapter 3 introduced the addiction researcher Johann Hari's view that love is the solution to drug dependency. He argued against the biochemical model of addiction, but the two strands are not incompatible. Section 2 showed how both companionate and romantic love activate the body's reward centre. Could it be that people with an absence of love in their lives compensate by using drugs as an alternative way to activate the brain's reward system?

This is a difficult hypothesis to prove. There is clear evidence that an adverse childhood experience is a strong predictor of unhealthy behaviours in adulthood, including substance addiction and alcoholism.[22] But to equate this category with the absence of love would be a crassly inadequate interpretation.

Alternatively, it provides an interesting lens through which to consider the question of whether money makes you happy.

Once you have enough money to comfortably meet basic needs, additional money correlates indifferently with additional happiness.[23] As the Beatles sang, *Money Can't Buy Me Love*, and so perhaps the pertinent point is not whether money makes us happy, but whether it can make us happy in the absence of love, which seems doubtful.

This may explain why people both rich and poor develop substance addictions. It is often wrongly assumed that poverty drives people to drugs, but recent studies have recorded only weak evidence of the correlation[24] and research has recorded that significant numbers of adolescents from high socio-economic households have been engaging in substance abuse.[25] It is not necessarily the case that substance abuse will lead to addiction, nor that it demonstrates a lack of love, but it does show that status alone is no deterrent to drug use.

The hypothesis that addiction arises from a failure to balance the maintenance of our prime fits well with the existing phenomenon of reciprocity, which is where an addict gives up one addiction, only to switch it for another. If we are all adapted to seek a certain amount of reward, it is clear that a deficit in one area will encourage compensatory behaviour in another. But even if the hypothesis were true, it would not rule out the existence of other causes of addiction.

It is not as simple as saying that someone engages in compulsive behaviour because they have a deficiency of one of their basic needs. Some people have a hair trigger on the circuits concerned with impulsive behaviour. Other times, people may over-indulge because the relevant neurons are less sensitive, for example, because they have fewer receptors and therefore need more stimulus to get the same amount of sensation, as we saw in Chapter 3.

Balancing maintenance of the prime directives will not be useful advice for everybody. First of all, it is all very well urging people to work hard, fall in love and raise children, but those things are not easy. Neither are they universally accessible or desirable. The North Pond Hermit had no desire for human interaction and, by his own judgment, was much happier in isolation from society. When people's neural circuits are wired atypically, the balanced diet of food, love and success may not quite work. This poses a difficulty: all people have functioning reward networks, but not everyone can make use of all of its triggers. In short, balancing maintenance of the prime directives is not a panacea for the difficulties we face as humans.

13.6 MORE ANSWERS FROM BUDDHISM

Buddhism and neuroscience have been enjoying a growing association for some time. Several years ago, landmark studies confirmed that meditation has profound physiological effects on the brain. Since then, His Holiness the Dalai Lama has invited experts from the field to provide updates on the latest findings from neuroscience.[26] Neural plasticity is an important part of the overlap between the philosophy and the science because it is the phenomenon by which meditation actually changes the wiring of the brain.

Novice monks were compared to expert monks who had spent more than 10 000 hours practicing meditation. Brain scans were then performed on both groups, members of which were instructed to use a particular meditational style in which they cultivate compassionate feelings for all living beings. The results showed a striking difference. Compared to novices, experts demonstrated unprecedented levels of gamma waves.[27]

The method used to conduct the scans is called electroencephalography. This technique can be used to diagnose epilepsy and also played an intrinsic role in important discoveries of how the brain changes during sleep. A series of electrodes are placed in a variety of locations on the scalp. Inside the brain, ions flow in and out of neurons as signals shuttle between them. The combined charge of all the ions creates a series of fluctuating electrical fields, which are detected *via* the electrodes. The frequency and voltage of these fields acts as a signature of different states of alertness. Low frequency and high voltage characterise sleep, whereas high frequency and low voltage are detected when subjects are awake.[28]

Quite what the unusual gamma wave activity indicates is not yet understood. Characterised by frequencies between 30 and 80 hertz, it is recorded both during rapid eye movement (REM) sleep and when subjects are awake. Researchers believe gamma waves are generated by the synchronisation of the firing of neurons in separate regions of the brain. Some scientists have even suggested it may provide the key to consciousness itself. Many researchers believe it characterises states of alert vigilance,[29] which fits with the findings of the monks. The huge spikes in gamma activity suggest the expert monks could muster extraordinary levels of vigilance.

Similar studies have yielded a lot of evidence demonstrating the impact of meditation on brain structure. One experiment showed that meditation increased the proportion of grey matter in the brains of monks.[30] A brain scan study recorded a smaller response in the amygdala when expert meditators were shocked by a sudden sound.[31] Another comparison of novices and experts tested the impact of the Vipassana style of meditation on attentiveness.[32] A combination of 19 letters and two numbers were displayed in rapid succession at the same spot of a computer screen. The experienced meditators were much better at catching the second, as well as the first number. Still another scanning study found that meditation increased the cortical thickness of brain regions, including the prefrontal cortex and right anterior insula.[33]

These studies demonstrate the neuroplasticity of the brain. As we saw in Chapter 12, neuroplasticity can refer to the reinforcement of existing connections between neurons, to the formation of new connections or even to the formation of new neurons. Researchers Richard Davidson and Antoine Lutz suggest that the physiological changes triggered by meditation are no different in nature to the ways our brains change in response to other experiences, such as learning and emotion.[31]

In essence, Buddhism and neuroscience agree on the ability we have to change our brains in positive ways. Along with the enhanced attentional focus meditation seems to foster, a range of studies have also confirmed the practice can reduce stress, anxiety and depression.[34] Meditation also shows promise for tackling obesity. Various studies have shown that mindfulness can reduce food cravings and binge eating.[35–38]

An interesting task for neuroscience will be to establish the limits of this plasticity. What sort of transformation might be possible with meditation? Could it boost intelligence or creativity? Can it decrease violent tendencies? Can people with a violent disposition be induced to practice meditation? Could these processes be helped with pharmaceuticals?

13.7 WHAT SHOULD WE CHANGE?

If we can change the physical structure of our brains, it begs the question of how we should seek to change. As a species we find ourselves developing an ever more detailed understanding of how we came to be in the position we are in. This book has advanced the idea that many of our behaviours echo what has previously served the needs of the self-replicators. We should question which of these we might better jettison. Obviously, we cannot stop eating and who would want to? Similarly, social bonding sums up just about the best that life has to offer and, if we wish to continue existing as a species, we need to keep having children.

Many appetites are better managed than eliminated. Research continues a pace with methods to reduce obesity. The condition is strongly linked with a variety of serious diseases, such as diabetes, and, as such, constitutes a drain on public health. Pharmaceuticals have been developed, meditation has been trialled, and rehabilitation clinics have been set up to treat over-eating the same as if it were a drug addiction. This is great. This pesky hangover we have from the days of our ancestors in the Savannah inclines us to eat opportunistically. Never mind if you need it or not, eat it anyway. Most of us no longer need to think this way about eating, so it is good that so many researchers are seeking ways to help more vulnerable individuals overcome it. But why should we stop there? I can think of another characteristic of questionable value.

Diminishing lust for power could be a boon for humanity. Some friends and I were chatting about the super-wealthy once and one of them remarked: "I guess once you'd made a billion pounds you'd stop, because you'd have enough then." When he heard what he had said out loud, he started laughing. What he said was logical, but it does not describe how people behave. No sooner does a person get rich than they want to get richer. Society is stratified into the haves and the have nots, but even though some people have a personal fortune to rival a small country's GDP, the wealthier tend to fight tooth and nail against any political ideology that seeks to redress the balance. Meanwhile, environmental catastrophe looms ever closer, but no government seriously plans to stop acquiring fossil fuels, because energy access effectively equates to power. Is this something we can medicate? Can meditation help a tycoon learn to be satisfied with a fortune of a mere billion?

A heart-warming tale about baboons serves to illustrate the point. A few decades ago, Robert Sapolsky and Lisa Share realised that primate communities had started foraging for food in a rubbish tip.[39] Humans were throwing out enough food on which for all the primates to subsist but without the toil that normally characterises existence. But there was a problem. The alpha males did not want to share. The leaders of these primate packs would shoo away any subordinate males or females or children, hogging all the food to themselves.

In an astonishing twist, all the alpha males were wiped out. A batch of bad meat was thrown onto the tip and they were all poisoned. But what happened next? Was the resulting power vacuum filled by another generation of miserly leaders, stepping up to hoard their ill-gotten gains to themselves? No. After the alphas were killed, the rest of the baboons all started sharing the food and living together in harmony. Thereafter, if any of the primates started developing alpha tendencies, the other baboons would set them straight.

One could immediately counter that their comfortable set up was only sustainable thanks to the daft humans who were throwing away good food. By extension, if a human revolution were able to establish a genuinely equitable society, we could not forego toil because there would be no supply of waste food for us to eat. But, if we suppose we would still need to work, if only to maintain our collective sanity, is it actually necessary for us to have the hierarchical structure that characterises organisations all over the world? If only there were some sort of experiment that were possible to test these ideas.

At the start of this book, I stopped shy of questioning the sanity of the Mars One volunteers, but this is a very un-enterprising spirit. Rather than naysay, it is better to consider the opportunities. How about a study of egalitarian community? Instead of selecting the finest specimens of leadership to dispatch to their Martian eternity, let's send a group of non-alphas. Let's find clever astronauts who do not need endlessly to assert themselves to decide the course of action on behalf of their lessers. Let's send up a group of people who are happy to interact as equals in conditions that make unnatural the emergence of leadership. The Mars One hopefuls might yet seed an extraterrestrial Utopia and the freedom of their will would become a tedious footnote in a great stride forward for mankind.

REFERENCES

1. R. Dawkins, *The Selfish Gene: 30th Anniversary Edition,* Oxford University Press, Oxford, 2006.
2. H. Fisher, The Nature of Romantic Love, *J. NIH Res.*, 1994, **6**, 59–64.
3. M. Liebowitz, *Chemistry Of Love*, New York, Berkley, 1984.
4. E. Nahmias, Intuitions about Free Will, Determinism, and Bypassing, in *The Oxford Handbook of Free Will*, ed. R. Kane, Oxford University Press, Oxford, 2nd ed, 2011.
5. D. C. Dennett, *Freedom Evolves*, ALane, London, 2003.
6. A. Humphries, *Rik Mayall: Lord of Misrule*, 2014.
7. S. Townsend, *Believe: The Eddie Izzard Story*, 2009.
8. D. C. Dennett, *Elbow Room: The Varieties of Free Will Worth Wanting*, New edn, MIT Press, vol. 2015, 2015.
9. A. J. C. Freeman, Responsibility Without Choice. A First-person Approach, *J. Conscious. Stud.*, 2000, 7(10), 61–68.
10. D. Hodgson, Quantum Physics, Consciousness, and Free Will, in *The Oxford Handbook of Free Will*, ed. R. Kane, Oxford University Press, Oxford, 2nd edn, 2011.
11. D. M. Wegner, Precis of the Illusion of Conscious Will, *Behav. Brain Sci.*, 2004, **27**(5), 649–659.
12. M. Feinberg, R. Willer and M. Schultz, Gossip and Ostracism Promote Cooperation in Groups, *Psychol. Sci.*, 2014, **25**(3), 656–664.
13. M. Lazarus, H.-Y. Shen, Y. Cherasse, W.-M. Qu, Z.-L. Huang, C. E. Bass, R. Winsky-Sommerer, K. Semba, B. B. Fredholm, D. Boison, O. Hayaishi, Y. Urade and J.-F. Chen, Arousal Effect of Caffeine Depends on Adenosine A2A Receptors in the Shell of the Nucleus Accumbens, *J. Neurosci.*, 2011, **31**(27), 10067–10075.
14. M. M. Lorist and M. Tops, Caffeine, Fatigue, and Cognition, *Brain Cogn.*, 2003, **53**(1), 82–94.
15. B. Russell, *History of Western Philosophy,* Routledge, London, New York, New edn, 2004.

16. D. Hume, *A Treatise of Human Nature: Being an Attempt to Introduce the Experimental Method of Reasoning into Moral Subjects, etc.*, John Noon, London, vol. I, part iv, section vi, 1739.

17. E. Conze, *Buddhism: Its Essence and Development*, Windhorse, Birmingham, 2001.

18. E. Conze, *Buddhist Thought in India: Three Phases of Buddhist Philosophy*, Munshiram Manoharlal Publishers, New Dehli, 2002.

19. B. A. Wallace, A Buddhist View of Free Will: Beyond Determinism and Indeterminism, *J. Conscious. Stud.*, 2011, **18**(3–4), 217–233.

20. B. Russell, *The Conquest of Happiness*, Routledge, London, 2006.

21. "Action 47 Get a Good Balance Between Work and Life", *Action for Happiness*. Available at: http://www.actionforhappiness.org/take-action/get-a-good-balance-between-work-and-life. [Accessed: 15-May-2016.]

22. V. Felitti, R. F. Anda, D. F. Nordenberg, D. F. Williamson, A. M. Spitz, V. Edwards, M. P. Koss and J. S. Marks, Relationship of Childhood Abuse and Household Dysfunction to Many of the Leading Causes of Death in Adults, *Am. J. Prev. Med.*, 1998, **14**(4), 245–258.

23. E. Diener and R. Biswas-Diener, Will Money Increase Subjective Well-being?, *Soc. Indic. Res.*, 2002, **57**(2), 119–169.

24. J. Z. Daniel, M. Hickman, J. Macleod, N. Wiles, A. Lingford-Hughes, M. Farrell, R. Araya, P. Skapinakis, J. Haynes and G. Lewis, Is Socioeconomic Status in Early Life Associated with Drug Use? A Systematic Review of the Evidence: Socioeconomic Status and Drug Use, *Drug Alcohol Rev.*, 2009, **28**(2), 142–153.

25. J. L. Humensky, Are Adolescents with High Socioeconomic Status More Likely to Engage in Alcohol and Illicit Drug Use in Early Adulthood?, *Subst. Abuse Treat. Prev. Policy*, 2010, **5**(1), 1.

26. C. Koch, Neuroscientists and the Dalai Lama Swap Insights on Meditation, *Scientific American*. Available at: http://www.scientificamerican.com/article/neuroscientists-dalai-lama-swap-insights-meditation/. [Accessed: 07-Apr-2016.]

27. S. Begley, Scans of Monks' Brains Show Meditation Alters Structure, Functioning, *Wall Str. J.*, 2004, http://www.wsj.com/articles/SB109959818932165108. [Accessed: 19-Feb-2016.]
28. Squire, *Fundamental Neuroscience*, Academic Press, Amsterdam, Boston, 4th edn, 2008.
29. A. K. Engel, P. Fries, P. König, M. Brecht and W. Singer, Temporal Binding, Binocular Rivalry, and Consciousness, *Conscious. Cogn.*, 1999, **8**(2), 128–151.
30. P. Vestergaard-Poulsen, M. van Beek, J. Skewes, C. R. Bjarkam, M. Stubberup, J. Bertelsen and A. Roepstorff, Long-term Meditation is Associated with Increased Gray Matter Density in the Brain Stem, *NeuroReport*, 2009, **20**(2), 170.
31. R. J. Davidson and A. Lutz, Buddha's Brain: Neuroplasticity and Meditation, *IEEE Signal Process. Mag.*, 2008, **25**(1), 176–174.
32. H. A. Slagter, A. Lutz, L. L. Greischar, A. D. Francis, S. Nieuwenhuis, J. M. Davis and R. J. Davidson, Mental Training Affects Distribution of Limited Brain Resources, *PLoS Biol*, 2007, **5**(6), e138.
33. S. W. Lazar, C. E. Kerr, R. H. Wasserman, J. R. Gray, D. N. Greve, M. T. Treadway, M. McGarvey, B. T. Quinn, J. A. Dusek, H. Benson and others, Meditation Experience is Associated with Increased Cortical Thickness, *NeuroReport*, 2005, **16**(17), 1893.
34. S. G. Hofmann, A. T. Sawyer, A. A. Witt and D. Oh, The Effect of Mindfulness-based Therapy on Anxiety and Depression: A Meta-analytic Review, *J. Consult. Clin. Psychol.*, 2010, **78**(2), 169–183.
35. H. J. E. M. Alberts, S. Mulkens, M. Smeets and R. Thewissen, Coping with Food Cravings. Investigating the Potential of a Mindfulness-based Intervention, *Appetite*, 2010, **55**(1), 160–163.
36. H. J. E. M. Alberts, R. Thewissen and L. Raes, Dealing with Problematic Eating Behaviour. The Effects of a Mindfulness-based Intervention on Eating Behaviour, Food Cravings, Dichotomous Thinking and Body Image Concern, *Appetite*, 2012, **58**(3), 847–851.
37. J. L. Kristeller and C. B. Hallett, An Exploratory Study of a Meditation-based Intervention for Binge Eating Disorder, *J. Health Psychol.*, 1999, **4**(3), 357–363.

38. J. Kristeller, R. Q. Wolever and V. Sheets, Mindfulness-based Eating Awareness Training (MB-EAT) for Binge Eating: A Randomized Clinical Trial, *Mindfulness*, 2013, 5(3), 282–297.
39. R. M. Sapolsky and L. J. Share, A Pacific Culture among Wild Baboons: Its Emergence and Transmission, *PLoS Biol.*, 2004, 2(4), e106.

Appendix: Thermodynamics, Immiscible Liquids and Heat Receptors

PART 1: WHAT MAKES REACTIONS HAPPEN

The classic analogy for chemical reactions is the boulder rolling down the hill. It needs a little shove to get going and then it carries on down to the bottom of the hill. Having rolled down the slope, the boulder will not roll itself back up. The problem is that sometimes the rock kind of does roll back up the hill, depending on the kind of reaction. A reaction that suits the analogy is the burning of a match. Once the match has been burned, nothing is going to unburn it. Why not? One reason is energy. A match releases energy, which we can tell because we can feel the heat from it and even use it to set other things on fire. This energy comes from the formation of new chemical bonds.

Breaking bonds needs energy, whereas forming bonds releases it. As we considered in Chapter 1, bonds result from the attraction between clusters of positive and negative charge; the positive nuclei of two atoms are simultaneously attracted to the electrons they distribute between them. If we imagine that a sodium atom has donated an electron to a chlorine atom, we now have a positively charged sodium ion and a negatively

The Chemistry of Human Nature
By Tom Husband
© Tom Husband 2017
Published by the Royal Society of Chemistry, www.rsc.org

charged chloride ion. We would expect the two ions to move towards each other unaided, whereas to separate them would require us to use some energy to pull them apart. Their coming together constitutes the forming of a bond, while their separation is the breaking of a bond. As such, we can see that forming bonds produces energy, while breaking them requires it.

In the case of the match more energy was produced when the new products were formed than was required to break the bonds in the reactants. Such reactions are described as exothermic and, overall, they are more likely to happen than their counterparts, endothermic reactions, which suck in heat from the surroundings and make them colder. This way of considering reactions is referred to as the enthalpy change. Exothermic reactions have a negative enthalpy change, whereas endothermic reactions have a positive enthalpy change.

So, is it that a reaction can only happen if it gives out energy? No. There are many reactions and processes that are endothermic, for example, dissolving sugar in water. The difference is incredibly subtle, but tea will decrease ever so slightly in temperature when sugar is added—even correcting for the fact that the sugar will be colder than the tea. As shown in Figure A1, the molecules in a crystal of sugar are held together by hydrogen bonds, the intermolecular force in which partially positive hydrogen atoms of one molecule are attracted to partially negative atoms in neighbouring molecules, which is, in this case, oxygen. In fact, when the sugar dissolves, new hydrogen bonds form between the molecules of sugar and the water molecules in the tea. But more energy is required to break the bonds between the sugar molecules in the crystal than is released by the formation of the new bonds they form with the water molecules.

This being the case, what makes the sugar dissolve? Oddly, the answer is probability. A crystal of sugar is a highly ordered substance. All of the sugar molecules are lined up next to each other, all in the exact same orientation to one another. Once they dissolve, they are mixed up higgledy–piggledy with the water molecules in a completely random arrangement and demonstrating no pattern (Figure A2). When the universe is viewed as a whole, increasing disorder is always the most probable outcome. This is actually very intuitive. Imagine you are handed a deck of cards arranged in numerical and suit order. A moment of

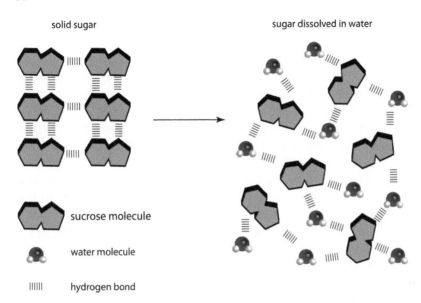

solid sugar

sugar dissolved in water

sucrose molecule

water molecule

IIIIII   hydrogen bond

Figure A1 Table sugar dissolving in water. In the solid state, sucrose molecules are bound together by hydrogen bonds. When it dissolves, hydrogen bonds form between the sucrose and water molecules, but the average strength of the hydrogen bond is weaker than in the solid.

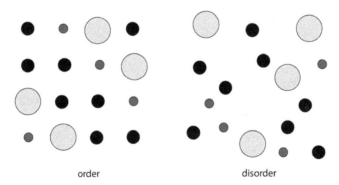

order

disorder

Figure A2 Order *vs* disorder.

shuffling will destroy the pattern, but what length of time spent shuffling would restore them to their ordered origin? In practical terms it would never happen, although there is a probability—no matter how miniscule—that it would eventually occur. This is not a difficult concept for which to find analogies. A bedroom naturally gets messier, dust naturally accumulates and beauty is gradually ravaged by time.

This way of considering reactions is called entropy. Reactions that lead to an increase in disorder are said to have a positive entropy change, whereas an increase in order is a negative entropy change. From the last paragraph, we can see that positive entropy changes are more favourable than negative entropy changes, and hence more likely to happen. Unfavourable does not mean impossible. Obviously, disordered atoms can become more ordered, otherwise beautiful crystals would never form and clouds would endlessly expand as the oceans were irreversibly transformed to water vapour.

How do the thermodynamic laws account for processes accompanied by an increase in order? Returning to the example of sugar, not only can it dissolve, but it can also be recrystallised. This is achieved by increasing the proportion of sugar to water in solution, which can be achieved by boiling off the water. Although the entropy of the sugar decreases, the process is accompanied by a larger increase in the entropy of the water as it boils into steam. This leads nicely to the second law of thermodynamics, which Cambridge researchers James Keeler and Peter Wothers define as follows:

> *"In a spontaneous process, the entropy of the universe increases."*[1]

What this means is that the house always wins. When a casino gambler wins a huge pay-out, the company stays in profit on the back of the accompanying losses. If a casino pays out more in winnings than it makes from losses, the business is no longer viable. The roulette wheel maintains its profitability thanks to additional slots on which *the gambler cannot bet*. Bets are taken on slots 1 to 36, but not on 0 or 00. This transforms the odds in a way that keeps the house in profit. Chemistry is the same; the only way a system can increase in order is if an even larger amount of disorder accompanies it.

Returning to the second law, a spontaneous process is one that happens with minimal provocation. Thomas Engel and Philip Reid define a spontaneous process as one that is likely to occur if any obstacle to the change is transcended.[2] A match will not light unless struck, but after this small input of activation energy, it continues to burn unaided. In this case, the striking of the match overcomes the barrier to its combustion and so we label the process spontaneous.

Entropy and enthalpy are key factors when deciding if a process is spontaneous. Earlier, we saw how processes are more likely to happen if they give off energy, like the burning match, and if the final product is less ordered than the starting material. Different combinations are possible. Explosives and fuels are favourable in both ways. They release energy while producing chaotic, gaseous products (which explains the derivation of the word *gas* from the Greek *khaos*). Transforming coal into diamond is not favourable in either way. It not only absorbs energy but also leads to a more ordered product. This explains why diamond is rare but also poses the question of how the stone can exist at all. Superman provides a scientifically legitimate answer. In the superhero's third outing he crushes a piece of coal so forcefully that it turns into diamond. Pressure is another factor in the thermodynamic mix but can be left out of our deliberations, which exclusively concern processes occurring at normal atmospheric pressure.

Enthalpy and entropy can be favourable in opposite directions. If we review our findings so far with sugar, when it dissolves, the enthalpy is unfavourable, but the entropy is favourable. Conversely, when it is recrystallised, the enthalpy is favourable, but the entropy is unfavourable. This is the hallmark of reversible processes. As we shall soon see, the freezing of water is an exothermic process with a negative entropy change, while melting ice is an endothermic process with a positive entropy change. Chemical reactions can also be reversible. Carbon dioxide is forced into fizzy drinks under pressure. Some of its molecules will react with water molecules to produce carbonic acid. Once the drink is opened, the pressure plummets and much of the carbonic acid dissociates back into carbon dioxide and water. Again, the formation of carbonic acid is enthalpically favourable but entropically unfavourable, while the converse is true for its decomposition.

This poses questions for the rolling boulder. The analogy says that it rolls down the hill and stays put, yet there are some processes in which the boulder seems to roll back uphill. In fact, the rolling boulder is best viewed as an emblem of a spontaneous process, and these can be reversible or irreversible. The burning of a match is spontaneous once the barrier to change is overcome. If a sugar cube is dropped into boiling water, it will spontaneously dissolve. But the water will spontaneously cool, after which it will spontaneously evaporate; so, if it is left long enough, the sugar molecules will spontaneously regroup back into ordered crystals.

Finally, we see why we cannot unburn a match. Not only does the process release energy, but the highly ordered structure of the wood is transformed into disordered molecules of carbon dioxide and water vapour. In order for the reaction to go backwards, the carbon dioxide and water molecules would not only need to absorb energy but also arrange themselves in a much more ordered pattern. Or to put this in formal terms, burning a match is both entropically and enthalpically favourable, whereas unburning it would be unfavourable by both measures, which is an outlawed combination. In this case, the boulder rolls down and stays firmly at the bottom of the hill.

PART 2: LIFE AND THERMODYNAMICS

Chapter 4 introduced the apparent conflict between life and thermodynamics. On first appearances it seems that the house does not win when life forms. Organisms are organised. Both words derive from the Latin *organ* for tool. Life is the coordination of tools like the liver, lungs or heart, just as a government or business organisation is subdivided into departments that cooperate in pursuit of a common goal. If there appears to be a conflict between an observable process and physical laws, one of two things is true: either the laws are wrong or we have not understood how they apply to the process in question.

Schrödinger accounted for the discrepancy by arguing that organisms suck orderliness from the environment. Starch granules are highly ordered when we harvest them from plants, but that order is absent once they have passed through out digestive system. The order we suck from these digested foodstuffs is used to drive life-sustaining biological processes. When biochemists investigate the chemical reactions that keep us alive, they consider not only their energy needs but also their entropy needs.

Accordingly, figures are quoted not in terms of energy but free energy, a concept we will explore in the next section. Viewed independently, every process that happens in our body meets the checks and balances of thermodynamic legislation. Moreover, viewed as a whole, on all but the hottest days our incessant metabolism provides a constant supply of heat to our surroundings, which in turn causes the entropy of the surrounding air molecules to increase. Although we are fantastically ordered entities, we can be sure that the house is still winning. Our existence causes the entropy of the universe to increase.

PART 3: FREE ENERGY

Free energy is calculated using eqn (A.1). The triangle is read "delta" and means change. G stands for Gibbs free energy, named after its discoverer Josiah Willard Gibbs. Delta H is the change in enthalpy, delta S is the change in entropy and T is the temperature, measured in kelvin. Exothermic processes have negative enthalpy because the products end up with less energy than the reactants. This seems weird because these reactions give off heat, but the point is that the surroundings *gain* heat at the expense of the molecules taking part in the reaction. In order for a process to occur spontaneously, delta G must be less than zero. An interesting thing this equation shows us is that processes may be more or less likely to happen depending on the temperature.

$$\Delta G = \Delta H - T\Delta S \qquad (A.1)$$

Freezing water is an example of how temperature dictates whether or not a process can happen spontaneously. At atmospheric pressure, water can only freeze below 0 °C. Let's see why we need to consider the enthalpy and then the entropy.

Freezing water is an exothermic process. This often confounds students because it happens at cold temperatures. As shown in Figure A3, the molecules in ice are held together by hydrogen bonds, just like the sugar molecules described above. Earlier, we saw that energy is released by the formation of bonds. In this case, it is not so much that new bonds get formed but rather that existing bonds are strengthened. Molecules of cold water are also

Ice

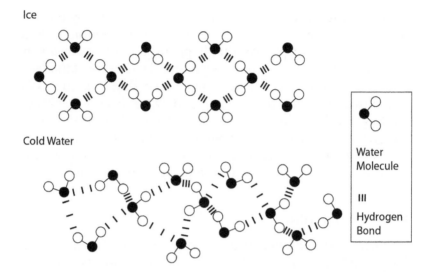

Cold Water

Water
Molecule

III

Hydrogen
Bond

Figure A3   Hydrogen bonds in water compared to ice. Individual hydrogen
bonds may be shorter, and therefore stronger, but the average
bond length is longer, and therefore weaker, in water than in ice.

held together by hydrogen bonds but, as they are all at different
distances from each other, the bonds vary in strength. When
cold water freezes, the molecules arrange themselves so that the
same distance separates all of them, leading to a greater average
bond strength. This is why freezing water releases heat.

Since exothermic processes are generally favourable, entropy
must be the thermodynamic barrier in this instance. Staying with
the model of entropy as a measure of disorder, we can see that the
formation of ice cubes will have a negative entropy change.
Crystals of ice are more ordered than cold water; therefore, the
process of freezing water is entropically unfavourable. This still
does not explain why water can freeze below 0 °C but not above.
From the Gibbs equation (eqn (A.1)), we know that changing the
temperature will affect the calculated value of the free energy. To
find out why, viewing the entropy as a disorder model is no longer
sufficient.

We can use the second law of thermodynamics to solve the
problem of the freezing water. The second law states that in a
spontaneous process the entropy of the universe increases.
Freezing water is spontaneous at temperatures below 0 °C but
not above it. What is so special about 0 °C? It is the temperature

above which freezing water no longer increases the entropy of the universe. But how can that be? How can the process be spontaneous at one temperature and not at another?

We need to consider how to calculate the entropy of the universe. This is achieved by adding the entropy change of the freezing water to the entropy change of the rest of the universe, as follows:

$$\Delta S_{\text{UNIVERSE}} = \Delta S_{\text{SYSTEM}} + \Delta S_{\text{SURROUNDINGS}} \qquad (A.2)$$

The system is the freezing water. In this case, water freezing into ice involves a decrease in disorder, so the entropy change is negative. Next, we consider the surroundings, the frontier of which is the chilly air in the freezer drawer. The entropy change is calculated as follows:

$$\Delta S_{\text{SURROUNDINGS}} = q/T \qquad (A.3)$$

In this equation, q is the heat change accompanying the process, measured in joules, while T is the temperature of the immediate surroundings. Here, the source of the heat, q, is the energy given off by the exothermic process of the ice freezing. Next, eqn (A.3) shows us how the temperature of the surroundings, the chilly freezer air, can affect the outcome. As the temperature increases, q/T will decrease, meaning that the entropy of the surroundings will decrease. Consequently, the higher the temperature of the surroundings, the lower the entropy change.

PART 4: A TRUER PICTURE OF ENTROPY

Why should the entropy change of the surroundings decrease as they get hotter? This question can be answered with the Boltzmann distribution, which represents the distribution of energy between particles.

In a container of gas, not all particles have the same amount of energy. Figure A4 shows a container full of gaseous particles. The differently sized arrows indicate not only their direction but also their relative speed. This is a snapshot showing the comparative speed and direction of each particle. Some particles are moving faster than others because the total energy of the particles is not distributed evenly between them. In a game of snooker, if a

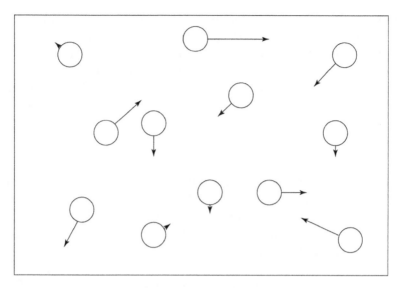

Figure A4 The particles in a gas all have different amounts of energy and hence move at different speeds. When particles move from one place to another, the process is called translation.

fast-moving ball hits a stationary one, the faster ball will slow down and the stationary ball will start moving, or, put another way, it will speed up. Particles behave in a similar way. They constantly collide with each other, emerging from collisions at different speeds than they entered them. Consequently, at any instant, some gas particles will have more kinetic energy than others.

The way that energy is distributed between the particles is shown in the form of a graph in Figure A5. The Boltzmann distribution curve is fairly simple and, like virtually everything in thermodynamics, it can be understood in terms of money. Imagine that the horizontal axis shows an amount of money and the vertical axis shows the number of people. The graph shows that no one has no money, most people have a bit of money and a very few people have loads of money. Atoms and molecules are the same. Like the diagram of gaseous particles shown in Figure A4, none of them have no energy, most of them have a bit of energy and a few of them have loads of energy.

That's one view of the Boltzmann distribution, but we can look more deeply. What the curve really shows is the population of particles in different energy levels. This is where we tumble down

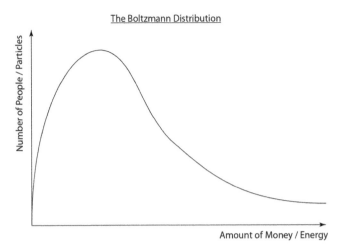

Figure A5 Boltzmann's distribution. The distribution of money between the members of society is analogous to the distribution of energy between the particles in a gas.

the rabbit hole of quantum physics because energy levels are the essence of this mind-bending field.

PART 5: ENERGY LEVELS

Energy levels represent the amount of energy that a particle has. There is more than one kind of energy level and there are strict rules about the amount of energy particles can have. The four different kinds of energy level are: vibrational, rotational, translational and electronic. The first three relate to different types of movement. If a particle vibrates, rotates or moves around, it does so at certain discrete levels of energy. If it rotates faster, it is elevated to a higher rotational energy level. If it vibrates more slowly, it falls to a lower vibrational energy level. Translational energy is when particles move from one place to another, as in Figure A4. Electronic levels are more familiar in the form of their alter-ego: electron shells, which are the large concentric circles decorated with dots or crosses when drawing atoms in chemistry lessons. What they really represent is levels of energy at which electrons can exist. For all of these different kinds of energy, it is not possible for a particle to exist *between* energy levels.

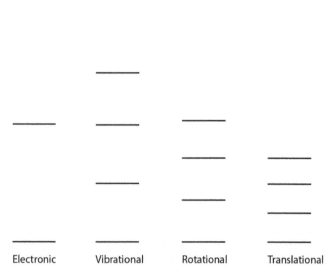

Figure A6   The comparative position of the energy levels representing the different degrees of freedom.

Figure A6 shows how the different sets of energy levels compare to each other. Collectively, the four sets are known as degrees of freedom. Note how the gaps between the levels get smaller going from the electronic levels to the translational. These gaps can be described as quanta and they represent the precise amounts of energy required to elevate molecules up the levels. The diagram shows that more energy is required to elevate particles up the electronic energy levels compared to the translational.

Energy levels can be considered in terms of gameshows. Some gameshows reward points with cash, so the prize fund differs with each game. Compare this with *Who Wants to be a Millionaire*, in which there are fixed values of money available. It is not possible to win £600 on *Millionaire*. You can win £500 or £1000, but £600 is just not possible. On a different gameshow, if each point was awarded £1 and the contestant got 600 points, then the £600 prize would be possible. Energy levels are like *Millionaire*; particles cannot exist at a level of energy between the prescribed levels.

Let's consider how this applies to the vibrational energy levels. Particles can only vibrate at certain frequencies. Consider a molecule formed from two atoms of an imaginary element. Let's say that at energy level 1, the atoms oscillate 10 times per second, that is, with a frequency of 10 hertz (Hz). At energy level 2, they oscillate at 15 Hz. It is not an option for the atoms to oscillate at 11, 12, 13 Hz or any other frequency in between. In order to go from oscillating at 10 Hz to 15 Hz, a precise quantum of energy is needed, let's say 5 arbitrary units (AU). If just 1 AU is supplied, it seems to make sense that the vibration will speed up by a fifth of the 5 Hz gap separating the energy levels, that is, the atoms would vibrate at 11 Hz. But they will not. They will continue oscillating at 10 Hz until all 5 AU have been supplied, at which point the frequency of their oscillation will jump up to 15 Hz. Similarly, if more than 2 AU is supplied, the atoms will not oscillate faster than 15 Hz unless the surplus energy is sufficient to elevate the particles to the third or subsequent energy levels. The same principles apply to the translational, rotational and electronic energy levels.

PART 6: THE RELATIONSHIP OF ENTROPY WITH ENERGY LEVELS

The different energy levels underpin nature's enthusiasm for chaos. They are used to calculate the entropies of substances at different temperatures, the values of which can then be used to calculate entropy changes. This can be used to explain why water does not freeze above 0 °C.

We will approach this complex subject with another analogy. Consider three people: Joshua, Jamila and Anna. How many different ways can three pound-coins be distributed between them? Anna could have three coins, while Joshua and Jamila get none; or Jamila could have two coins, Anna could have one and Joshua could have none; or each person could get one coin.

Imagine these characters are contestants in a gameshow. They are in some kind of pit and they have been directed to scrabble for coins as they are tossed inside. They must not use violence and we assume that each individual is equally motivated to maximise their share of coins. Demeaning as the format might sound, it can show us something useful about chemistry. If three pound-coins are tossed into the pit, who will end up with what? Statistics predict that one person will end up with two coins, another will get one and the last person will get nothing. Notice that I didn't say that Jamila would get two coins, even though she was the recipient of the two coins in the above example. It might turn out that Joshua or Anna get the two coins. Even if Jamila does get two coins, either Joshua or Anna could end up with the last coin. It turns out there are six different ways to distribute the coins this way, whereas there is only one way that everyone can end up with one coin. Put another way, three different arrangements are possible, but there are ten different ways of doing it, as shown in Table A1.

Ten different results are possible but six of them correspond to the same distribution. As such, the most likely outcome is that one of the contestants will get two coins, one will get one coin and one will get none because there are so many different ways for that distribution to play out.

How does that help us to understand entropy? This statistical analysis can tell us something useful about molecules. The three characters represent the molecules and the pound coins represent the energy. One strength of the analogy is the pound coins. It was not possible for one character to get £2 and for the others to get 50 pence each because the coins are not divisible. This is just like the quanta separating the different energy levels.

Next, we can apply the mathematics from the gameshow analogy to the molecules. How can three units of energy be distributed between three molecules? The outcome is exactly the

Table A1   The different ways three pound-coins can be distributed between three people.

Number of coins	3	2	1	0
Distribution A			Joshua, Anna, Jamila	
Distribution B	Joshua Anna Jamila			Anna, Jamila Joshua, Jamila Anna, Joshua
Distribution C		Joshua Joshua Anna Anna Jamila Jamila	Anna Jamila Joshua Jamila Joshua Anna	Jamila Anna Jamila Joshua Anna Joshua

same as in the gameshow, as shown in Figure A7. Just as before, there is only one way that all three molecules can have one unit of energy each, whereas there are six ways that they can fall into distribution Z. As such, we assume that, generally speaking, this is how the energy will be arranged between the molecules. This is called the Boltzmann distribution.[3]

Now let's consider ten molecules with three different values of total energy: 5, 8 and 11 units. In each case, it is possible to work out the distribution that can be achieved in the most different ways.

Let's return to the analogy briefly to get going. If ten contestants are ushered into the pit and told that they will compete for five pound-coins, what is the most likely distribution? Table A2 shows that there are seven possible distributions, which can be

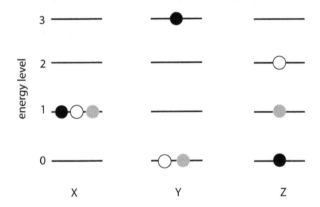

Figure A7 The different arrangements possible for three units of energy between three particles.

Table A2 The different ways five pound-coins can be distributed between ten people.

	Distribution and number of contestants						
Number of coins	A	B	C	D	E	F	G
5	1						
4		1					
3			1	1			
2			1		2	1	
1		1		2	1	3	5
0	9	8	8	7	7	6	5
Number of ways	10	90	90	360	360	840	252

achieved in 2002 different ways. The most probable is distribution F, which plays out like this:

One person gets two pound-coins
Three people get one pound-coin
Six people get nothing

This can be achieved in 840 different ways. For example, there are ten different people that can get the two pound-coins, which accounts for 10 of the different ways already. Once you factor in all the different combinations, it is easy to see how it can be done in 840 different ways.

Now let's imagine that 8 coins and then 11 coins are again distributed between the ten contestants. There are 22 different distributions possible for the 8 coins and 53 distributions possible for the 11 coins. But, in each case, there is a distribution that can be achieved in more different ways than any other. These distributions for the maximum number of ways are shown in Table A3.

Again, we can transfer the exact same mathematics to molecules. If we imagine that three sets of molecules, L, M and N,

Table A3 The different ways five, eight and eleven pound-coins can be distributed between ten contestants.

Money

Coins	£5	£8	£11	Energy level
4			1	4
3		1	1	3
2	1	1	1	2
1	3	3	2	1
0	6	5	5	0
No. ways	840	5040	15120	No. ways
	5 units	8 units	11 units	

Energy

have the total energy values of 5, 8 and 11 units, respectively, these same distributions are the ones that can be achieved in the most different ways. As such, we say that this is the most likely distribution in which the molecules will be found.

These calculations can be carried out for any number of molecules and used to find the most likely distribution. This does not mean that the molecules actually are arranged that way but that, *en masse*, they behave in accordance with predictions implied by the calculations. This mathematical treatment is called the Boltzmann distribution, described in its graphical format earlier.

Entropy can be equated to the number of different ways that the molecules can be arranged in the most probable distribution. The equation is as follows, where k_B is Boltzmann's constant:

$$S = k_B \ln W_{MAX} \qquad (A.4)$$

Although the relationship is not directly proportionate, the graph in Figure A8 shows that entropy increases with the number of ways (W_{MAX}) that molecules can be arranged in the most probable distribution.

Finally, we are in a position to understand how temperature affects the entropy change for a given amount of energy transferred to a substance. If the temperature of an object is increased, we can say

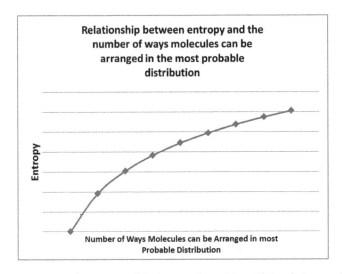

Figure A8   Entropy increases with the number of ways (W_{MAX}) that molecules can be arranged in the most probable distribution.

that the total energy of its molecules has increased; thus, increasing the temperature of an object is equivalent to increasing the number of pound coins distributed between the characters in the analogy.

What happens to the maximal value of W as energy is increased? If we look again at Table A3, we can see that W increases with energy and hence temperature. However, although there is an increase from $E = 5$ to $E = 11$, the rate of increase decelerates. That is, there is a much bigger increase in W going from $E = 5$ to $E = 8$ than there is from $E = 8$ to $E = 11$. The value of W for $E = 8$, let's call it W_8, is six times that of W_5, whereas W_{11} is only three times bigger than W_8. As such, although increasing the energy of an object does indeed increase its entropy, the rate at which entropy increases is in constant decline. Consequently, the entropy change resulting from the input of 10 joules of energy to a system will vary, depending on the temperature of that system.

This explains why water only freezes at or below 0 °C. When the water freezes, a certain amount of energy is released, as hydrogen bonds between the molecules are strengthened. The amount by which this energy increases the entropy of the surroundings will depend on the temperature of the surroundings. Thus, when we run the sums for temperatures above 0 °C, the increase in the entropy of the surroundings has a smaller value than the decrease of the entropy of the freezing water molecules. Entropy of the universe does not increase and so the process is outlawed. But when the temperature of the freezer drawer is below 0 °C, the balance shifts and the process becomes spontaneous.

We started by viewing entropy as a measure of disorder and we have moved on to see how the disorder is characteristic of the system that is mathematically more probable. It is fascinating that statistics can be used to predict the outcomes of chemical interactions. Even more remarkable is that many thermodynamic principles had already been established by experiments before anyone even know about energy levels and their separating quanta. Classical and quantum physics were merged to produce an even more reliable model of thermodynamics.

PART 7: THE HYDROPHOBIC EFFECT

Having addressed these thermodynamic issues, we can consider the question raised in Chapter 6. Why do oil and water not mix?

It is entropy rather than enthalpy that explains the phenomenon. It has often wrongly been said that water and oil do not mix because water molecules are more attracted to each other than they are to the oil molecules.[4] Actually, the statement is accurate on the strength of attraction, but it is not the reason for their immiscibility.

The hydrophobic effect is the name given to the phenomenon that oil and water do not mix. In the literal sense, hydrophobic means *water hating*, while hydrophilic means *water loving*. Cooking oils and hydrocarbons, like the butane that fuels disposable lighters, are categorised as hydrophobic because they are non-polar, while water and alcohol are hydrophilic because they are polar. Polar substances have an uneven distribution of electrons between the positive nuclei, leading to regions of partially positive or partially negative charge. Water is a polar substance because its electrons preferentially huddle around the oxygen atom, generating a partially negative charge around the central oxygen and partially positive charges around the electron-deficient hydrogen atoms. We say that a dipole exists along the axis of each bond because of the difference in electronegativity. This particular example of a dipole–dipole interaction is called a hydrogen bond and the term *polar* derives from *dipole*.

So much for the difference between hydrophobic and hydrophilic liquids, but what prevents them from mixing? Experiments have been conducted to measure the enthalpy change when these antagonistic solvents are made to mix in minute quantities.[5] The data show that the enthalpy change is negative, meaning that energy is released when immiscible liquids are made to mix. Since exothermic processes tend to be thermodynamically favourable, the problem here must be entropy.

Water becomes more ordered when it is mixed with oil. As we have considered, molecules of cold water are held together by hydrogen bonds. This lattice is literally more fluid than the lattice in solid ice, but it is there. When an oil molecule is thrust into water, something very interesting happens: hydrogen bonds still bind the water molecules together, but they have to rearrange themselves around the oil molecule. What actually happens is that the water forms a cage-like structure around the oil molecule, called a clathrate (from the Greek *klēthra*, for the

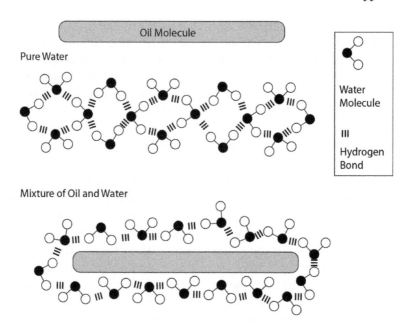

Figure A9   The formation of clathrates around molecules of oil in water.

bars of a cage), as shown in Figure A9. This word has also featured in media warnings that climate change may trigger its own acceleration. The greenhouse gas methane is "caged" into oceanic ice clathrates,[6] which jettison their Earth-cooking cargo more rapidly as the Earth warms. The clathrates formed when water accommodates oil are more ordered than pure water and it is this unfavourable decrease in entropy that prohibits their mixing.

The clathrate model explains two things. First, the entropic cost of forming the clathrates is thought to be the thermo-dynamic barrier behind the hydrophobic effect. Second, it explains the exothermic enthalpy change of mixing oil and water. Clathrates have often been described as "ice-like"[4] because of their resemblance to solid water. Earlier, we saw how the freezing of water is exothermic because the average strength of the hydrogen bonds between the water molecules is strengthened. The same is true for the formation of clathrates. In forming the more ordered lattice in the clathrate, the water molecules are forced into closer contact and the orientation between them becomes more uniform. This results in a greater average bond strength, which accounts for the exothermic change when hydrophobic liquids are forced to mix with water.

Oil and water can be made to mix in certain conditions. The Gibbs equation showed the role of temperature in gauging the feasibility of a process. As such, oil and water might be expected to mix above or below room temperature. In fact, the barrier to their mixing gets smaller as the temperature increases for the reason that the heat breaks down the lattice of hydrogen bonds between the water molecules.[7] If pressure is also increased, hydrophobic liquids can mix with water freely. Two Japanese scientists, named Shigeru Deguchi and Nao Ifuku, recently mixed the hydrophobic alkane dodecane with water at 374 °C and a pressure of 22.1 mega Pascals,[8] which is roughly the pressure you would experience if you dived 2200 metres below the surface of the ocean. As incoming heat erodes the lattice structure of water, clathrates stop forming and the entropic cost of mixing disappears.

Whether or not the clathrate model is accurate, entropy definitely seems to be the deciding factor in the miscibility of oil and water. Other models have been proposed, but they are variants on the theme that the entropy change must be negative. Experimental data confirm that the process of mixing water with hydrophobic liquids, such as cooking oil, is an exothermic process. From the Gibbs equation, this shows that the only explanation for their reluctance to mix is an unfavourable decrease in entropy when the liquids mix.

As a final word on the subject, the hydrophobic effect does not utterly preclude the mixing of oil and water. Thermodynamics are concerned with the behaviour of crowds of particles rather than individuals. A positive value of Gibbs free energy for a reaction does not mean that not a single pair of particles will react with one another, rather it predicts the *majority* of them will not react. In this case, if oil and water are shaken together in a flask, some of the water molecules will stray into the oil and *vice versa*, but the overwhelming majority of them will settle back into separate layers.

PART 8: PROTEINS, SPECIFIC HEAT CAPACITY AND THE HEAT RECEPTORS

Chapter 8 featured a simplified explanation of the mechanism of heat receptors. These are the ion channel proteins that signal to our brains when they detect changes in temperature. This section of the appendix will explore how the relationship between the hydrophobic effect and specific heat capacity affects protein folding, and hence underlies the emerging model for the mechanism of heat receptors.

Specific Heat Capacity and the Hydrophobic Effect

Another interesting feature of the hydrophobic effect is its effect on specific heat capacity, which is defined as the energy required to increase a given mass of a substance by a given temperature.[2] This is a school favourite with many fun demonstrations. For example, a water balloon will not burst when held in a flame (within limits—a flamethrower would burst it). The reason for this is that the water can absorb the heat from the flame rapidly enough to prevent the rubber from melting. By contrast, a balloon filled with air would burst immediately because air has a lower specific heat capacity than water. This means it cannot absorb heat at a sufficient rate to prevent the balloon's rupture.

Like entropy, specific heat capacity is closely related to energy levels. If we compare the water-filled and the air-filled balloons above, the water molecules have more energy levels among which to distribute themselves (and another obvious difference is the much greater density of water than air but even similar masses of air and water would differ in their heat capacity). In both the liquid and the gaseous phases, the molecules can move past each other; in other words, both can access translational energy levels. Molecules can also rotate in both states, meaning rotational energy levels are jointly accessible. But the situation is slightly more complicated with vibrational energy levels. Inside the molecules the atoms can vibrate relative to each other in both the liquid and gas phases. However, in liquids, vibration can also occur between whole molecules. In this phase, molecules are bound together by intermolecular forces, so not only can the atoms inside the molecule vibrate relative to each other, but molecules can also vibrate relative to their neighbours along the intermolecular bonds. Consequently, liquids have more degrees of freedom, meaning there are more accessible energy levels; therefore, equal masses of a liquid and the corresponding gas would need to absorb different amounts of energy to demonstrate the same temperature increase.

Experiments into the hydrophobic effect have found a relationship with specific heat capacity.[7] We have already seen that mixing hydrophobic and hydrophilic liquids is accompanied by a negative enthalpy change. Another interesting finding is that the specific heat capacity of the mixture is greater than the sum of its parts. This has also been attributed to the formation of the clathrates. Earlier, we saw that the average strength of the hydrogen bonds between water molecules is increased as they form clathrates. This, in turn, causes the spacing between the vibrational energy levels to increase. As a result, accelerating the vibration between water molecules guzzles more energy when they are arranged in clathrates, which accounts for the unexpectedly high specific heat capacity of mixtures containing clathrates.

Folding Proteins

The hydrophobic effect plays a central role in the behaviour of proteins. The hydrophobic effect has a huge impact on the way that proteins gain their characteristic shapes, a process known as folding. We will first consider this field and then borrow its findings to rationalise the conformation change of heat receptors.

To appreciate how proteins fold, we can start by pondering how they unravel, which is a much more familiar sight. When meat is cooked, it goes from having a chewy, gelatinous texture to a grainy and ideally tenderer one. Once again, we return to the actin–myosin complex, comprising two of the most abundant proteins in muscle. When they are heated, as happens when meat is cooked, the proteins lose their exquisitely precise shape and unravel into long chains. These chains clump together into fibres, which account for the grainy texture of cooked meat.[9] The proteins are said to have been denatured, meaning they can no longer carry out the function for which they are adapted, in this case, making muscles contract. In spite of this destruction, the protein maintains its primary structure. Every protein is characterised by a unique shape, which enables it to carry out a specific job. These shapes are rendered over several stages of production, a pivotal point in which is folding. Just as a strip of ribbon can be looped into an elaborate bow, so an unbranched chain of amino acids is folded into the protein's trademark shape. The first stage is to link the amino acids together in the sequence dictated by the corresponding gene. This is called the primary structure. The resulting chain is speckled with patches of charge that differ in sign and strength; they may be positive, partially positive, negative, partially negative or charge neutral. The attraction between the oppositely charged regions is one of the main factors that cause the protein to fold into shape. Hence, the sequence of amino acids, each of which is unique in its electrostatic character, dictates the final structure of the protein (ref. 10, p. 216). This suggests that a scientist should be able to predict the final structure from the primary structure, but no one has yet mastered this.

Electrostatic attraction between regions of opposing charge is not the only shaping factor. The hydrophobic effect also contributes to the final structure because amino acids can be hydrophilic or hydrophobic. Almost everything that reacts in our bodies is dissolved in water, so proteins must be adapted for an aquatic environment. Consequently, proteins fold themselves so as to keep their hydrophobic regions tucked inside and their hydrophilic regions on the outside, as shown in Figure A10.

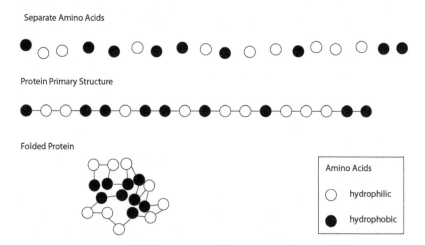

Figure A10   Protein formation. Amino acids link to form the primary structure of the protein, which then folds itself so that the hydrophobic regions are tucked inside.

Amino acids differ from oil and water in their degree of polarity. Technically speaking, all amino acids are polar because, by definition, they contain at least one amine group and at least one carboxylic acid group. As such, it is more accurate to describe an amino acid's *sidechain* as hydrophobic or hydrophilic, as shown in Figure A11.

Proteins fold themselves with their hydrophobic sections in their core and their hydrophilic sections on their surface. As we have seen, there is an entropic cost to mixing hydrophobic

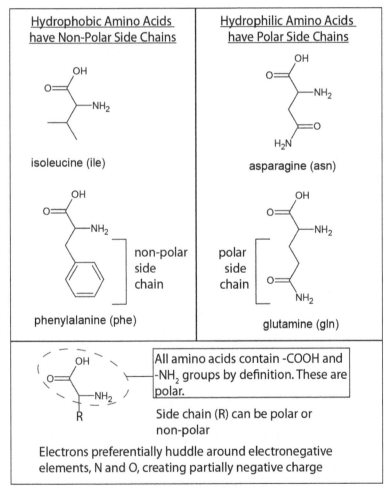

Figure A11 Amino acids with polar and non-polar sidechains.

substances with water, so it makes sense that proteins would draw their non-polar sections inside. Meanwhile, their hydrophilic exterior can form hydrogen bonds, which enables the entire protein to mix with water.

The Thermodynamics of Heat-detecting Ion Channels

At the time of writing, there was no definitive explanation for the phenomenon of how heat receptors work. However, a fascinating theory has emerged that blends all of the preceding thermodynamic concepts of Chapter 8 into one compelling explanation.

First of all, let's consider how heat receptors behave. They are ion channels, which are proteins that respond to specific stimuli by ferrying ions across cell membranes. Other ion channels we have met were triggered by changes in voltage or when thrown into contact with specific molecules, called ligands, but these are activated by changes in temperature. When it gets hotter, the ion channel opens up. This enables it to draw ions into the nerve cell, causing it to depolarise and hence signal the detection of heat to the brain.

A surprising feature of these ion channels has led some researchers to believe they might need help from supporting proteins because they play conflicting roles in different species. The ion channel named the transient receptor potential cation channel subfamily V member 1 (TRPV1) opens in response to heat in flies and snakes, but it detects cold in mice.[11] This has led to the suggestion that the receptor's ion channel component is subordinate to a temperature-detection module. Science always favours the simplest explanation and, while Clapham and Miller's proposal features extraordinarily complicated science, their theory is simpler because it shows how a single ion channel could signal both hot and cold temperatures, unaided by any auxiliary proteins.

Their theory might seem to fly against everything we have considered up to now. Clapham and Miller have suggested that the temperature change causes the proteins to thrust their hydrophobic sections into the water, as shown in Figure A12. We have already considered the fact that hydrophobic sections incline away from water because of the entropic cost of forging the water molecules into clathrates. But we have also considered that

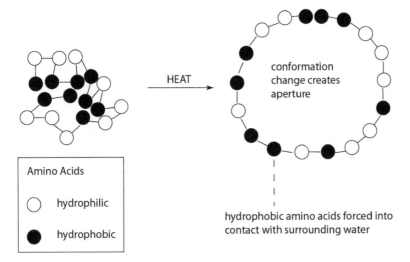

Figure A12 Heat receptor ion channels. Changes in temperature cause the protein to thrust its hydrophobic sections outwards, opening up the channel.

what cannot occur at one temperature may be possible at another.

Their theory hinges on the specific heat capacity of the proteins. Clapham and Miller have observed that these ion channels have a higher specific heat capacity in their open conformation. Earlier, we saw how the hydrophobic effect entails an increase in the specific heat capacity, so it makes sense that one way for the heat capacity of a protein to increase is for its hydrophobic innards to make contact with water. But why should an increase in specific heat capacity encourage a protein to unfold? As we deepen our understanding of thermodynamics, it emerges that specific heat capacity is closely linked to the two driving forces enthalpy and entropy.

Entropy is very important in the counter-intuitive behaviour of heat receptors. For any process to happen spontaneously, the entropy of the universe as a whole must decrease. As such, any process that involves a reduction in entropy, such as when bumbling liquid molecules align themselves into a regimented solid, must trigger an increase in entropy of greater magnitude in the surroundings. When a protein thrusts its hydrophobic sections into contact with water, clathrates will form just as when cooking oil is mixed with water. If we focus on the water

molecules alone, we should be satisfied that their entropy decreases as they form clathrates around the emerging hydrophobic sections of the heat receptor. But we need to think about
the entropy of the universe as a whole. Could it be that the entropy increase of the protein outweighs the entropy decrease
taking place as the clathrates form? In fact, the answer to this
question is very interesting. It is precisely because the clathrates
form that the process becomes thermodynamically favourable,
which is where specific heat capacity literally enters the equation.

Specific heat capacity and entropy are closely related. Although
their relationship is not linear, if one increases, so does the
other. Meanwhile, both of them increase with temperature.
(With one exception: as a liquid boils, its specific heat capacity
drops because the loss of bonds between particles reduces the
diversity of energy levels in which for incoming energy to be
distributed. However, as the temperature of the gas continues to
increase, specific heat capacity once again starts rising.) In fact,
if entropy is plotted against temperature, the gradient of the
graph gives the ratio of specific heat capacity and temperature.[12]
From this, aficionados of calculus will recognise that when
specific heat capacity is plotted against temperature, the area
under the curve gives the entropy.[2] It should not be too big a
surprise to find that entropy and specific heat capacity are linked
as each is closely related to the distribution of molecules between energy levels. The greater the temperature, and hence the
energy in the system, the more diverse is the number of ways the
molecules can be arranged between the energy levels, meaning
the greater will be the entropy. Meanwhile, the greater the
number of accessible energy levels, the higher the specific heat
capacity (Figure A13).

The formation of clathrates does two things: one, it decreases
the entropy of the water molecules that form them, and two, it
increases the specific heat capacity of the protein. Remember
that the average strength of the hydrogen bonds between the
water molecules increases as they form clathrates, making the
process exothermic. These toned bonds also increase the specific
heat capacity of the protein–water complex. This explains two
things. First of all, it explains how the entropy of the universe
can increase overall, even though it decreases in the forming
clathrates. Secondly, it explains how heat receptors can be activated

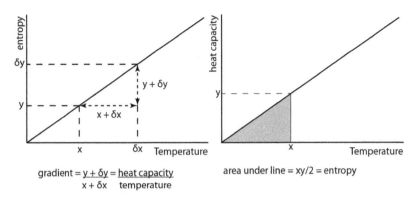

gradient = $\dfrac{y + \delta y}{x + \delta x}$ = $\dfrac{\text{heat capacity}}{\text{temperature}}$

area under line = xy/2 = entropy

Figure A13 Entropy and specific heat capacity are interrelated *via* temperature.

by rising, as well as falling, temperatures. Recall that Clapham and Miller's theoretical model accounts for the fact that the same proteins detect heat in snakes and flies, but cold in mice.

Earlier, we met the Gibbs equation for free energy (eqn (A.1)), which can be used to predict if a process will happen spontaneously. This is a very useful equation, but it does have limitations, the main one being that it only allows for a single value of temperature.

$$\text{Gibbs equation: } \Delta G = \Delta H - T\Delta S \qquad (\text{A.1})$$

This presents a problem for the heat receptor. The Gibbs equation only requires a single value for temperature, so which one should we use? We cannot use the temperature of the ion channel before it opens; the whole point is that it will not open at that temperature. So then, should we use the temperature of the ion channel after it has opened? This latter option seems like the best one, except for one problem. What if the entropy change changes? When we investigated freezing water earlier, we saw that the entropy change of the surroundings was different depending on the temperature of the surroundings. In other words, the entropy change for a process is different, depending on the temperature at which it takes place. This makes things very complicated for the situation with the heat receptor. What do you do if the thermodynamic changes accompanying a process themselves *change* as the process takes place? Answer: improve the equation.

Eqn (A.5) shows the relationship between entropy (S) and specific heat capacity (C_P) between two different temperatures. T_o is taken as a reference temperature, while T is the actual temperature at different stages during the conformational change of the heat receptor. ("ln" is a mathematical operator.)

$$\Delta S°(T) = \Delta S°(T_o) + \Delta C_P \ln(T/T_o) \tag{A.5}$$

We can see from the equation that an increase in specific heat capacity will increase the entropy change undergone by the heat receptor as it opens. Next, Clapham and Miller combine the above equations (eqn (A.1) and (A.5)) with others to produce the following behemoth:[11]

$$\ln K(T) = \ln K_o + (\Delta H°_o/T_o - \Delta C_P)(1 - T_o/T)/R - (\Delta C_P/R) \ln(T_o/T) \tag{A.6}$$

In this case, K is the equilibrium constant between the products and reactants (the proportion of open ion channels to the proportion of closed ion channels). K is related to K_o in the same way as T to T_o above; H is the enthalpy and R is the gas constant ($8.314 \ \mathrm{J\,K^{-1}\,mol^{-1}}$). It's a pretty forbidding equation until different values are plugged in and plotted as a graph, as shown in Figure A14.

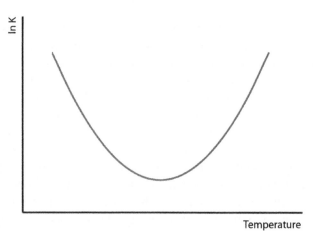

Figure A14   A graph of $\ln K$ against temperature for the heat receptor. K denotes the proportion of heat receptors that are open; "ln" is a mathematical operator. Whether temperature increases or decreases, the proportion of open ion channels increases.

What the graph shows is remarkable. Whether the temperature falls or rises, the effect on K is the same. K is the ratio of the number of open ion channels to the number of closed ion channels, which explains how the same receptor can detect both hot and cold temperatures. Whether the temperature rises or falls, the ratio of open ion channels increases, which explains how the same proteins are able to detect high temperatures in some species and low temperatures in others. But Clapham and Miller note that the range of temperatures between the ion channel's two open states would be too great to detect both hot and cold sensations in a single species of animal.

The reason why the process seemed counter-intuitive is the very reason why it works at all. The same forces that keep oil and water separate also account for the fact that this ion channel can detect temperature changes. The ion channel is finely tuned to exploit the hydrophobic effect over a very narrow temperature range. Compare this with a mixture of water and the hydrophobic alkane dodecane. At room temperature, they are immiscible, and they need to be heated to 374 °C and exposed to a pressure of 22.1 mega Pascals in order to blend.[8] But dodecane has a much greater surface area of hydrophobicity than any of the hydrophobic amino acids in the heat receptor. This makes the thermodynamic barrier much smaller for the unfolding of the heat receptor.

There is also the snowball effect. Remember that Clapham and Miller's equation (eqn (A.6)) only predicts that the *proportion* of open ion channels increases when the temperature changes. In Chapter 8 we learned that these ion channels respond not only to changes in temperature but also to changes in voltage.[13] As one ion channel opens, ions enter the cell, which changes the voltage across its membrane. This, in turn, inclines other heat receptors to open up, which ushers in more ions, which further alters the voltage. A few receptors lead the charge, but their compadres follow suit.

This is a remarkable accomplishment of evolution. A series of genetic mutations has perfectly engineered this protein for its function. At our normal temperature, the opening of the ion channel is thermodynamically forbidden because the entropic cost of forming clathrates cannot be met. Just as oil and water do not mix, so the protein's hydrophobic sections keep themselves tucked inside. But unlike oil and water, a very slight shift in temperature can totally transform the thermodynamic profile. The accompanying spike in specific heat capacity drives an overall increase in entropy sufficient to outweigh the thermodynamic cost of forming the clathrates. At this new temperature of just a few degrees more (or less in other species), the house wins. Now, as the protein unfurls, the entropy of the universe is increased overall, meaning that ions can pour inside the cell, triggering the depolarisation that will ripple along the nerve and signal the temperature change to the brain.

REFERENCES

1. J. Keeler and P. Wothers, *Why Chemical Reactions Happen*, Oxford University Press, Oxford, 2003.
2. T. Engel and P. Reid, *Thermodynamics, Statistical Thermodynamics, & Kinetics*, Prentice Hall, Tappan Hall, New Jersey, 3rd edn, 2012.
3. J. Keeler, in *Chemical Structure and Reactivity: An Integrated Approach*, ed. James Keeler and Peter Wothers, Oxford University Press, Oxford, 2008.
4. T. P. Silverstein, The Real Reason Why Oil and Water Don't Mix, *J. Chem. Educ.*, 1998, 75(1), 116.
5. A. Seidell, *Solubilities of Organic Compounds*, 3rd edn, van Nostrand, New York, vol. 2, 1941.
6. S. Cieślak and K. Gaj, Hazards of Uncontrolled Methane Release from Clathrates Analyse and Environmental Evaluation of Extraction Methods, *Environ. Prot. Eng.*, 2014, **40**(3), 99–111.
7. C. Tanford, *Hydrophobic Effect: Formation of Micelles and Biological Membranes*, John Wiley & Sons Inc, New York, 1973.
8. S. Deguchi and N. Ifuku, Bottom-Up Formation of Dodecane-in-Water Nanoemulsions from Hydrothermal Homogeneous Solutions, *Angew. Chem. Int. Ed.*, 2013, **52**(25), 6409–6412.
9. H. Mcgee, *McGee on Food and Cooking: An Encyclopedia of Kitchen Science, History and Culture*, Hodder & Stoughton, London, 2004.
10. H. Frauenfelder, *The Physics of Proteins*, Springer New York, New York, NY, 2010.

11. D. E. Clapham and C. Miller, A Thermodynamic Framework for Understanding Temperature Sensing by Transient Receptor Potential (TRP) Channels, *Proc. Natl. Acad. Sci.*, 2011, **108**(49), 19492–19497.
12. J. P. O'Connell and J. M. Haile, *Thermodynamics: Fundamentals for Applications*, Cambridge University Press, Cambridge, 2005.
13. B. Chanda, Personal Correspondence, 05 Aug 2015.

Acknowledgements

This book could not have been written without the support of a number of people who variously advised, informed and encouraged me. First and foremost, I would like to thank Matt Williams for his tremendous help in editing the manuscript both for technical accuracy and also for readability. His insights were of great value and his comments very reassuring. Thanks also to my sister Jess and brother-in-law Alex for bringing Matt on board and commenting on early drafts.

Also, I would like to thank the individuals at the Royal Society of Chemistry who agreed to publish the book. First of all, Karen Ogilvie gave my proposal a hearing and passed it on to her colleagues in the publishing department. I would like to thank Janet Freshwater not only for commissioning the book but also for giving me the freedom to let the project evolve. Thanks also to the anonymous reviewers who pointed out flaws in my initial proposal, which helped develop the book into a far superior product than initially conceived. Also thanks to Antonia Pass and Harriet Manning for their help with my various enquiries.

A number of researchers were kind enough to give up their time to discuss their work with me. Special thanks to Kim Sharp, Elaine Harris, Mark Post, Baron Chanda, Doug Stephan, Haig Kazazian, Michael Liebowitz and Galit Shohat-Ophir. Thanks also to my friends Simeon Dann and Tom Stafford for giving me the benefit of their professional expertise.

The Chemistry of Human Nature
By Tom Husband
© Tom Husband 2017
Published by the Royal Society of Chemistry, www.rsc.org

I am also indebted to the staff at ACD ChemSketch, whose software was used to render all of the molecular structures shown in this book.

Many of my friends were also good enough to read early drafts of the manuscript. In particular, Hannah Meese and Kim Li-Lakkappa gave me honest and invaluable advice on how to refine the book's central message. Paul Andrew gave similar advice but also greatly assisted me by patiently listening to and almost unwaveringly disagreeing with all my nuggets of newly acquired knowledge, helping me to assimilate them and weigh their respective value. Olly, Frank, Tim, Elliot, Andres, Charles, Rory and Jo also read early drafts and suggested improvements.

A number of other individuals were also supportive. Thanks to my family for their constant encouragement, Glyn for telling me to shut up and get on with it, other friends and colleagues for the progress updates they frequently requested, and finally the courteous staff at the British Library for their assistance and their inexhaustible amusement at my surname.

Subject Index

1984 303
2001: A Space Odyssey (film) 79

A Beautiful Mind (film) 273
A Midsummer Night's Dream
  (play) 179
A Treatise of Human Nature 353
Abbott Laboratories 145
actin 31, 39–40, 64
actin–myosin complex
    adenosine triphosphate 64
    calcium ions 163
    control 134
    muscle contraction
      39–40
    neuron stretching 289
    proteins 105
Action for Happiness
  Organisation 355
addiction
    dopamine 64–5
    drugs 65–6, 68
    gambling 69
    research 63–4
    smoking 64
    see also food addiction
adenine (nucleobase) 35, 86,
  107–8, 234

adenosine diphosphate (ADP)
  37–8, 190
adenosine triphosphate (ATP)
    actin–myosin complex 64
    gustatory nerve 55–6
    heart 190–3
    hydrolysis 37–8
    myosin 35–40, 47, 55
    neurons and taste cells 60
adenylyl cyclase 190, 220
adrenaline 186–7
"adrenaline rush" 189
adrenocorticotrophic hormone
  (ACTH) 278
African clawed frog (cellular
  cooperation) 236–7
agonists/antagonists 179–81
Alexander, Bruce 74
allele term 114
alleles 114–15
amino acids 213, 389
Aminoff, Michael 144
ammonia
    origins of life 90
    primordial atmosphere
      90
amphetamine 64–5, 68, 171,
  176

amygdala (brain) 60, 187–9, 223, 266, 273, 316, 320, 325, 358
Anderson, Kevin 10
Androgel 145
angiomotin (AMOT) 245–6
"angry eye contact" 315
Animal Farm 303
Anisman, Hymie 279
ankyrin repeat and kinase domain containing 1 (gene) 72
Anolis carolinensis (lizard) 321–2
Anslinger, Harry 74
anthocyanin (dye) 27
antibodies (*Wally finders*) 326–7
apomorphine 179–80
Aragona, Brandon 180
arginine 157
Arnold Rimmer (fictional character, *Red Dwarf*) 169–71
"aromatic" term 174
Aron, Arthur 181, 188
Asperger's Syndrome 199, 274
assisted self-mimicry 116
atomos (indivisible) 13
attachment
    cheating the sensors 209–18
    not a cuddle-hormone 218
    introduction 198–203
    neuromodulator 218–21
    oxytocin-cuddle hormone 206–8
    puppet masters of mammalian attachment 203–6

reward 221–4
    temperature 208–9
autism 273, 278
Aztecs 51

baby making
    conclusions 249
    epigenetic inheritance 248–9
    epigenetics 230–6
    foetal development 236–48
    introduction 229–30
baby making – foetal development
    description 236–9
    mechanism 242–4
    merging hypotheses 244–8
    three hypotheses 239–42
bacteria
    flagella 120–2
    receptors 121–2
Bad Science 266
Badcock, Christopher 273
Baron-Cohen, Simon 274
Barratt Simplified Measure of Social Status (BSMSS) 320
Barron, Andrew B. 130
Bayout, Abdelmalek 287–8, 292
Beatles 356
Beckerman, Stephen 113–14
bed nucleus of stria terminalis (BNST) 330
Beiderbeck, Daniela 298
Berridge, Kent 66, 73, 130
Beyond the Pleasure Principle 126
Big Bang 345–6

biochemistry and freedom of will 342
bipolar disorder 273
Birmingham, Wendy 206
Blair, Eric Arthur (George Orwell) 304
Boltzmann distribution 374–5
bonds (chemical) 19–20
Bonfire of the Vanities 305
Boyle, Danny 64
brain
    addiction research 63–4
    amygdala 60, 187–9
    dopamine 64
    functions 63
    neuroplasticity 358
    oxytocin 220
    prediction error 66–7
    reward network 181, 304, 315, 337
    see also nucleus accumbens; ventral tegmental area
Braun, Allen 280
Brave New World 169
Brown-Séquard, Charles-Édouard 142–4, 146
Brunner syndrome 292–4, 295, 297
Buckholtz, Joshua 298
Buddhism 352–4, 357–8
Butenandt, Adolf 144

Cadbury (chocolate) 26
cadherins (foetal development) 242–3
Caenorhabditis elegans (worm) 248
calcium and ion channels 41–2, 47

"camel's hump of misery" 275
Canis familiaris 41
capsaicin 211–12
Captain Corelli's Mandolin 199
Captain Kirk (fictional character, *Star Trek*) 171
Caspi, Avshalom 296
cell membranes 103–4
Chagnon, Napoleon 113, 299
Chanda, Baron 214–15, 217
character
    conclusions 336–7
    introduction 257–8
Charney, Evan 268–70, 276, 279
Chasing the Scream 74
chemiluminescence 326–7
Chinese-Restaurant syndrome 29
chlorofluorocarbons (CFCs) 328
chocoholism 51, 175
chocolate
    appeal 49
    bean roasting 48
    cocoa butter 48
    constituents 30
    eating 30, 48, 63, 70, 125, 134
    flavour sensation 48
    Fry, Joseph 26
    neurotransmitters 175
    phenylethylamine 174
    serotonin 174, 175
    tasting 60–8
    Valentine's day gift 26
chocolatl 25, 27
cholesterol and sex hormones 149–50
Chowdbury, Sandipan 214–15

Christian, Gary 74–5
chromosomes 117–18
"circling" term 14
citrulline 157
citrus fruits 42
Clapham, David 212–14, 217
Clark, Gregory 310–14, 331–2
clathrates 383–5
Clay Davis (fictional character, *The Wire*) 309
Cnemidophorus inornatus 202
Coatlicue 51
cocaine 29, 64–5, 68, 159, 223, 318–21
cognitive control network (CCN) 280
Comfort, Nathaniel 262
Conze, Edward 352–3
cooking oil and water 150–2
corticotrophin releasing hormone (CRH) 278
cortisol 277–8, 325
CpG islands (nucleotide pairings) 234–5, 247, 277, 289, 297
Creative Achievement Questionnaire 267
creative intelligence
    birth of a paradigm 261–6
    heritability studies 268–70
    how heritable is creativity? 266–8
    introduction 259–61
    nature–nurture duality 276–9
    price of creativity 270–6
    role for reward network? 280–1

Creative Personality Scale 268
creativity, genes 267
Creighton, Harriet 118
Crick, Francis 79, 82, 85
"cuddles" 209
cyclic adenosine monophosphate (cAMP) 189–94, 220
cyclic AMP response element binding protein (CREB) 68
cyclic guanosine monophosphate (cGMP) 162–3, 175, 194
cytosine (nucleobase) 86, 107–8, 234–5

Dalley, Jeffrey W. 319
Dancing Naked in the Mind Field 327–8
Darlow Smithson Productions 11
Darwin, Charles 79–80, 82, 98, 102, 113, 267
Darwin's Bulldog (Huxley, Thomas) 102
Davidson, Richard 358
Davis, Caroline 72
Dawkins, Richard 111, 272, 342, 350, 352
de Bernières, Louis 199
de Dreu, Carsten 278
default mode network (DMN) 275, 280–1
dehydroepiandrosterone (DHEA) 149–50
Democritus 13, 16, 353
Dennett, Daniel 135, 342, 343–7, 349
Dick, Philip K. 17
Dickson, Suzanne 68–71, 75

dihydrotestosterone (DHT)
    149–50
directed panspermia 79, 85
DNA
    acetyl/methyl
        modification 233–5
    adenine 35
    base pairings 86–7
    copying 83
    different types of cells
        230–1
    hormones 149
    length 154
    mutations 158
    nucleosomes 232
    order 96
    promoter region 155–6
    protein coding 108–9
    replication 7–8, 89, 119
    self-replicating
        molecules 6
cDNA (complementary DNA)
    330
dodecane 216–17, 385
dominance
    aggression 306–7
    cocaine 318–21
    conclusions 331–2
    introduction 303–5
    neuroscience 314–18
    reward network 332
    serotonin 321–3
    status 308–10, 332
    status heritability 310–14
    testosterone 316
    what is dominance? 305–8
    winner effect 323–31
dominance – neuroscience,
    processing information
    316–18

dominance – winner effect,
    Nobel legacies and reward
    mediation 325–31
dopamine
    addiction 64–5, 178
    agonists 180, 203
    antagonists 180, 203–4
    anti-depressants 178
    brain 64
    cocaine 318–19, 321
    dominance 319–21, 325
    Drosophila 129
    eating 70–1
    food addiction 71–3, 73–55
    G-protein coupled
        receptors 62
    gambling 69
    GPCRs 62
    learning 331
    ligands 179
    mating 224
    neurons 158
    neurotransmitters 62,
        173–4, 176, 316
    nitric oxide 159
    nucleus accumbens
        180, 224
    oxytocin 223–4
    prediction error 66–7
    rapoclide 185
    receptors 72, 178–80
    reward network 61, 298
    reward-seeking
        behaviour 316, 320
    signalling 66
    structure 61–2, 180
    tyrosine hydroxylase 327
    ventral tegmental area
        184
    wanting 224

Drews, Carlos 305, 310, 331
Drosophila 128–30
drug addiction 65–6, 68
dTRPA1 gene 128
Durrenberger, Stephen 275
Dutch famine, World War II 249
"dyadic sex" 147

Edmondson, Adrian 345
Einstein, Albert 14–16, 267, 274, 348
Elbow Room: The Variation of Free Will Worth Wanting 135
electrons
    atoms 14–15
    chemiluminescence 326
    electrolysis 21
    hydrogen chloride 22
    indeterminancy 341
    ion channel taste receptors 41
    oxygen 20–1, 33
    water 19, 87
"elixir of life" (testosterone) 141–3
Elton, Ben 345
embryonic development 238–9
Encyclopedia of Creativity 275
Endler, John 128
endogamous marriage 313
"endogenous amphetamine" 172
Epicurus 13, 341
epigenetics
    babymaking 230–6
    cell differentiation 231
    heritability 336
    inheritance 258–9

    modification 232, 332
    status of successive generations 331
Esch, Tobias 189
Escher, M C 309
ethanol, fruit flies 126–7
eugenics 261–3
"evitability" term 346
evolution
    developing choice 120–5
    introduction 102–3
    pleasure 125–30
    from replication to reproduction 110–20
    from self-replicators to organisms 103–10
    theory 79
evolution – from replication to reproduction
    multicellular organisms 111–13
    sexual reproduction 113–20
    single-celled organisms 111
evolution – from self-replicators to organisms
    cell membranes 103–4
    proteins 104–10
eye colour 114

Fabre, Jean-Henri 135–6
Facebook addiction 68
far-from-equilibrium state 6
FAT/CD36 (fatty acid translocase) receptor 48–9, 59
Father Durán (Franciscan missionary) 51
Father of Endocrinology (Brown-Séquard) 144

Federal Bureau of Narcotics,
US 74
Federation of American
Societies of Experimental
Biology (FASEB) 30
Feynman, Richard 82
fight-or-flight response
186, 188
Filipski, Patryk 328
Finkel, Michael 198
Finnerty, Gillian 23
Fiore, Palestina Guevara 128
Fischer, Stefan 36
Fisher, Helen 142, 164, 171,
177–8, 182, 184, 193, 199,
201, 252, 342
flagella
bacteria 120–2
rotation 120–1, 123–4
sperm 120
Flaherty, Alice 280
flavour components
nervous system 27
smell 27
taste 27
flour beetles 119
partial foetal alcohol
syndrome 288
foetal alcohol syndrome 289
food addiction
description 68–73
what next? 73–5
Fowler, Joanna 297
Franklin, Rosalind 79
free will
breaking the rules 351–5
Buddhism 357–8
healthy balance 355–7
indeterminancy 347–8
introduction 341

pleasure and the three
directives 348–51
unease about emerging
biology 341–3, 343–7
what should we
change? 359–60
free will – breaking the rules
Buddhism 352–4
can we break the rules or
not? 354–5
sex and drugs 351–2
free will – unease about
emerging biology, Dennett,
Daniel 343–7
Freud, Sigmund 126
frog's eye development 236–7
fruit flies
alcohol 126–7
feed and mate 130
neuropeptide F 128
neuropeptide Y 127
reward system 129
wasps 129
Fry, Joseph (chocolate) 26
fuel and building supplies
conclusions 134–6
pleasure 134–5
Fuxjager, Matthew J. 327, 330

G polymorphism and
oxytocin-related
behaviours 279
G-protein coupled receptors
(GPCRs)
dopamine 62
mechanism 45–6
noradrenaline 189–90
taste 40–1, 43–5, 47–8, 53
Gage, Matthew J. G. 119
Galton, Francis 260–4, 311

gambling
    addiction 69
    dopamine 69
Gates, Bill 314
genes
    alleles 114–15
    androgen receptors 330
    capture hypothesis 119
    competence 314, 332
    creativity 267
    definition 290
    environment 332
    heritability 268
    mutation 291–2
    nitric oxide synthase 235
    traits 265
    transcription 105–7
    translation 107–8
genius
    autism 273
    bipolar disorder 273
    insanity 274–5
    schizophrenia 273
Gershwin, George 348
Gerstein, Mark B. 291
Gibbons, Jennifer 242, 259–61
Gibbons, June 242, 259–61
Gibbs equation 385
Glenn, Andrea 299
glossopharyngeal nerves 60
glucocorticoid receptors
    277, 279
glycoproteins 112
Goldacre, Ben 266
Golden Fleece Award 342
Good Samaritan parable 218
Gregory, Simon 278
Grosberg, Richard K. 112
guanine (nucleotide) 86, 107,
    162

guanosine diphosphate
    (GDP) 45–7, 190–2
guanosine (nucleobase)
    234–5
guanosine triphosphate
    (GTP) 45–7, 162, 190–3
gustatory system 27, 30, 60

Haidt, Jonathan 275
Haldane, John 270, 272
haloperidol 179–80
Hamilton, William 11,
    270, 272
Hari, Johann 74–5, 252, 355
Harlow, Harry 207–8
Harrhuis, Paul (tennis player)
    323
Harris, John 74
heat-detecting ion
    channels 210–11
Hebebrand, Johannes 68–71,
    75
Heberlein, Ulrike 126
Heisenberg (uncertainty
    principle) 15
Henman, Tim (tennis player)
    323–4
HERC2 gene 114
heroine 207
Herrnstein, Richard 263
Higgs boson 14
HIPPO pathway (foetal
    development) 244–7
hippocampus, oxytocin 223
histamine 209
histidine 108
histones (protein) 154–5, 330
Hodgson, David 347–8
holograms 170
Holt-Lunstaad, Julianne 206

Homer (fictional character,
    The Simpsons) 209, 305
Hommer, Daniel 318
Homo sapiens 4, 194, 200–2,
    281
homosexuality and genes 265
Hooker, J. D. 82
hormone replacement
    therapy 144–5
Hu, Hailan 317
Hulihan, Terence 206
Hull, Elaine 154, 156, 158
Hume, David 347, 353
Hur, Yoon-Mi 267
Huxley, Aldous 169, 263, 267
Huxley, Andrew 267
Huxley, Julian 263, 267
Huxley, Thomas 102, 267
hydrochloric acid 21
hydrogen
    ionic bonds 21–2
    origins of life 90
hydrogen chloride 21–2
hypogonadism 146
hypothalamus 60, 188

inclusive fitness theory 270–1
Inner Life of a Cell 344
inositol triphosphate (IP$_3$)
    47, 221
Insel, Thomas 205
inside-outside hypothesis
    (early development) 242
intermolecular bonding 20–1
"interpersonal warmth and
    connection" (addiction) 75
ion channels 42, 47
IQ 264
isotocin 202
Izzard, Eddie 345–6

Jarecki, Brian 214–15, 217
Johnson, Martin 240
"jumping genes" 268–9
Jupp, Bianca 319

Kacsoh, Balint Z. 129
karaka (Buddhism) 353
karma (Buddhism) 353
Kazazian, Haig Jr 269
Keijzer, Fred 135
Kelley, Ann 51, 70
Kennedy, James 72
kidney beans 28, 212
King Midas 83
Kirschvink, Joseph 99
Kitcher, Philip 306
Knight, Christopher
    Thomas 198–9
Knuston, Brian 318
Kobilka, Brian 43
Kossel, Albrecht 35
Kringelbach, Morten 73
Kummer, Hans 306
Kusama, Yayoi 273–4
Kwok, Ho Man 29–30

Lancet 142–3
Laplace 13, 15, 343, 347
Larson, Earl 321
Laugerette, Fabienne 48
Lefkowitz, Robert 43
Leigh, Vivien 273
Leptopilina heterotoma
    (wasps) 129
leucine enkephalin 62
Liebowitz, Michael 171–4, 176,
    177, 181, 186, 193, 199, 223,
    252, 293, 342
Lieu, Maggie 9
life, definition 84

life's origins
    chemical evolution 95–8
    cosmic origins 98–100
    introduction 79–89
    RNA 90–5
Light, Kathleen 206
Lindt (conching process) 26
Linnaeus, Carolus 25–6
Liu, Siyuan 280
"lock and key" (enzymes) 44
Long, Kathryn 113–14
Lord of Misrule
    (documentary) 344
loser effect 323, 332
love and relationships
    conclusions 252–3
    introduction 139–40
"low status" 314
Lucas, Christopher 273
Lumley, Alyson J. 119
lust
    androgen and
        women 147–8
    hypothesis 164
    introduction 141–7
    priming genitals 159–63
    radical transmitter 157–9
    testosterone 148–57
Lutz, Antoine 358

McClintock, Barbara 117–18,
    268
MacDonald, Ryan 10
McDonnel, Michael 9
McGowan, Patrick 278
Macmillan's (magazine) 262
McQuaid, Robyn 279
"mad genius controversy" 280
Maggie Simpson (fictional
    character) 209

functional magnetic resonance
    imaging (fMRI) 182–4, 186,
    189
magnetic resonance imaging
    (MRI) 223
functional magnetic resonance
    imaging (fMRI) 275, 318
Malison, Robert T. 320
Marler, Catherine A. 327, 330
Marlo (fictional character,
    The Wire) 309
Mars
    life 3, 23, 99
    methane and water
        18, 22
    return trip 23
    suicide mission 4
    Total Recall 17
    trip 8, 9–10, 10–12
Mars One mission
    eating 8
    free will 341
    freedom of will 361
    human nature 4
    organisational
        strategy 11
    origins of life 23
    reproduction 9
    return trip 16, 22, 100
    sanity of volunteers 360
    self-replicators 12
    status 10
Martinez, Ruben 287
Matsuke, David 320
Mayall, Rik 344–7
Maynard Smith, John 115,
    117, 270
MDMA (3, 4-methylenedioxy-
    methamphetamine) 321–2
mechanoreceptors 209

medial prefrontal cortex
(mPFC) 316–18
methane
    combustion 17–18
    ice 18
    origins of life 90
    oxygen 17
micelles 103–4
Miller, Christopher 212–14,
    217
Miller, Stanley 90, 92
Milner, Peter 60–1, 70, 73
Minnesota Center for Twin
    and Family Research, US
    265
Money Can't Buy Me Love
    (song) 356
monoamine oxidase (MAO)
    172–4, 175–6, 178, 186, 293
monoamine oxidase A (MAO A)
    288, 290, 297
monoamine oxidase A (MAO A)
    gene 294–6, 342
monoamines 172–5, 175–7
monosodium glutamate (MSG)
    29–30
monozygotic twins 260, 270
Moreno, Alvaro 83
Morgan, Drake 319–20
morphine 207
morula (16-cell stage of foetal
    development) 239, 241–2
Motecuhzoma Ilhuicamina
    (Aztec ruler) 51
moths
    electrons 20
    light bulbs 14–15
Mullis, Kary
    autobiography 328
    PCR 326–8

Murchison meteorite (Victoria,
    Australia) 79–80, 82, 90, 92,
    99, 103
Murdock, George Peter 4–5
Murray, Charles 263
muscles – contraction
    calcium ions 160–1, 164
    mechanism of action
        30–40
    myosin 31–2
    see also actin-myosin
        complex
"mutation" term 110
myosin
    ATP 36–9, 45, 55, 160
    mechanism of action
        31–2
    muscle contraction
        31–2
    vesicle transport 343–4

Nabhan, Gary 211–12
Nader, Michael A. 319–20
Nahmias, Eddy 342
names status 310–14
Nash, John 273
natural gas (methane) 18
"natural pain killers" 61
"nature dominates
    nurture" 312
nature and nurture 336
"nature–nurture" term 261
"necklace bead"
    (replication) 6
Nestle (powdered milk) 26
Neumann, Inga 298
neurons
    ATP 60
    axon terminals 60
    connections 52

dopamine 158
gustatory nerve 54–60
myosin 59
receptor density 336
sexual behaviour 146
SNARE complex 59
structure 52–3
synapse 58
voltage-gated ion channels 56–7
neurons in gustatory nerve signal transmission
1. neuron activation 55
2. signal propagation 56–7
3. next neuron signalling 57–9
4. machinery resetting 59–60
description 52–3
neuropeptide F (NPF) 127–9
neuropeptide Y (NPY) 129
neuropeptide term 127
Newton, Isaac 267, 274
Nirvāna 354
Nirvannah Crane (fictional character, *Red Dwarf*) 170
nitric oxide (NO) 157–9, 161–3
nitric oxide synthase (NOS) 153–6, 159–60, 164, 231, 316
nitric oxide synthase (NOS) gene 235
nitrogen, oxides 157
nitrous oxide (N₂O, laughing gas) 158
noradrenaline 173–6, 187, 189–94, 327
North Pond Hermit, Maine, US 198, 356

nucleosomes 232
nucleotides
adenine 35, 86, 107–8, 234
cytosine 86, 107–8, 234
guanine 86, 107, 162
thymine 234
uracil 86, 107–8
nucleus accumbens (brain) 60, 65, 70, 180, 224, 299, 304, 318, 323, 326–7, 330

obsessive-compulsive disorder (OCD) 177, 182, 316
OCA2 gene 114
oestradiol 149
oil/water mixing 215–16
Olds, James 60–1, 70, 73
olfactory system 27
On the Origin of Species 82
opioids
agonists 72
attachment 223
enkephalins 70
feelings 224
liking 224
neurotransmitters 62
nucleus accumbens 224
parents/offspring 224
receptors 65
rewards 66
sensitivity 71
skin contact 207
social behaviour 224
oral contraceptives 147
orbitofrontal cortex (OFC) 182
Orgel, Leslie 79, 82
Orwell, George 303–4, 310
oxygen, electrons 20–1, 33

oxytocin
  amino acids 204
  amygdala 223
  antagonists 206
  autism 199, 278–9
  behaviours 279
  brain 220
  cuddle-hormone 206–8
  dopamine 223–4
  endocytosis of
    receptors 220
  functions 203
  hippocampus 223
  mammals 202
  neuromodulation
    pathway 221–2
  neuron function 220
  nucleus accumbens 224
  production 202–3
  social behaviour 203,
    218, 223–4
  structure 205
  synthesis 202–3
  TRPV3 217
  ventral tegmental area
    223

Page, Lionel 324
"palatable" term 69
panspermia 81
Pasadena Independent 51
Pasteur, Louis 102, 143
Pavlov 66
PCR (polymerase chain
  reaction) 326–9
quantitative (qPCR) 327–30
Pepsi Cola Addict 259
Peretó, Juli 83
Perez, Walter 288
Pericak-Vance, Margaret 278

Perry, Clint J. 130
phenylalanine 173
phenylethylamine (PEA)
  172–6, 177
photosynthesis 18–19, 96–7
Piffer, Davide 267
Pinker, Steven 63
pituitary gland 186
pleasure
  evolution 7–8
  food addiction 68–75
  introduction 51–2
  pleasure circuit 52–68
pleasure – circuit
  chocolate tasting 60–8
  description 52–3
  ions/potential 53–4
  neurons in gustatory
    nerve signal
    transmission 54–60
PNA (peptide nucleic acid) 95
polar compounds 33
polar/non-polar amino
  acids 213
Pollyanna (film) 34–5
polymerase chain reaction
  (PCR) 278
"pop sociobiology" 306
Portugal drug policy 74
positron emission tomography
  (PET) 184–6, 194, 296–7,
  318, 319–20
prediction error (brain) 66–7
Price, George 270–4, 342
primordial atmosphere 90
Pross, Addy 95–6, 97–8, 113,
  119
protein kinase A (PKA) 193,
  220
protein kinase G (PKG) 162–3

protein-coding 109–10
proteins
    amino acids 213
    changing shape 34–5
    DNA coding 109
    epigenetics 230
    evolution – from
        self-replicators to
        organisms 104–10
    histones 154–5
    myosin–actin complex
        105
    temperature sensitivity
        212, 214
    three-dimensional shape
        34
    transcription 106
    see also G-protein coupled
        receptors
protons 14, 41
Proxmire William 342
Prozac (SSRI) 60, 174, 322
pseudopanspermia 99
psychopaths 298–9
Pulitzer Prize (media) 308

quantum indeterminancy
    16, 348

Raleigh, Michael 322
rapoclide 185–6
Red Dwarf (TV programme)
    169–70
Redford, Donald 83
rhizoids 112
Rhodochorton purpureum 112
RNA
    composition 106–7
    copying 89, 97
    genetic information 85

nucleotides 86
polymerase 106
self-replication 88–9, 98
where did it come
    from? 90–5
Robocop (film) 349–51
romantic love
    dopamine receptors
        178–80
    Fisher, Helen 201
    humans 181–3
    hypothesis 177–8
    introduction 169–72
    lust 142
    mating drive 200
    monoamines 172–7
    neurotransmitters 183–6
    noradrenaline 189–94
    self-replication 200
    why does my heart go
        boom? 186–9
romantic love – monoamines,
    removal 175–7
Ruiz-Mirazo, Kepa 83
Ružička, Leopold 144
Russell, Bertrand 353, 355

Sachs, Greg 16
Santiago, Martina 22
Sapolsky, Robert 360
schizophrenia 273
Schlenke, Todd A. 129
Schrödinger, Erwin 82, 96–7,
    97–8, 116, 260
Schultz, Wolfram 66
Schwartz, Lisa 145
Schwarzenegger, Arnold 146
science, technology,
    engineering and Maths
    (STEM) 274

Scientific American 206
scientific determinism
  theory 13
Sciurus vulgaris (red squirrels)
  271
sea turtles 200–2
Search for Extraterrestrial
  Intelligence (SETI) 81
selective serotonin reuptake
  inhibitors (SSRIs) 60, 174,
  177–8, 321–2
serotonin 174–5, 175, 177,
  321–3
sex hormones and
  cholesterol 149–50
sexual reproduction 117, 120
Shamay-Tsoory, Simone 278
Share, Lisa 360
Sheldon Cooper (fictional
  character, *The Big Bang
  Theory*) 9
Sherman McCoy (fictional
  character, *Bonfire of the
  Vanities*) 305
Shohat-Ophir, Galit 126–8,
  130, 135
Shumay, Elena 297
sildenafil (Viagra) 163
Silicon Valley 257
"single-celled organism" 126
skandhas (Buddhism)
    acts of consciousness 353
    feelings 353
    impulse 353
    material 353
    perception 353
Smith, Jeremy 36
SNARE complex (neurons) 59
Snow, Jon 182
Snyder, Michael 291

social hierarchies 315
Sociobiology 263, 306
solar system 3
Soliz, Mark 287, 288–90
soluble guanylyl cyclase 161–2
somatosensory component 27
space travel
    chemistry fundamentals –
      way back 16–23
    introduction 3–4
    is it in our nature? 4–5
    reaction of self
      replication 5–8
    risky strategy 8–16
    universal human traits
      4–5
space travel – risky strategy
    eating 8
    free will 12–16
    reproduction 10–11
    status 10–12
"species" term 41
Spector, Tim 265–6
sperm, flagella 120
Sphex icheumoneus (wasp) 135
Stefano, George 189
steroid hormones, cell
  interaction 153
"steroid" term 149
"sticky" term 39
Stock, Solveig 207
Strathman, Richard R. 112
strychnine 27
Südhof, Thomas 59
Summers, Cliff 321
Sun Tzu 257
Sweden, government 314
synaptic plasticity 316–17
synaptotagmin 59
systems chemistry 95

Tarkowski, Andrzej 242
taste
    buds 40–7
    commonality 47–9
    introduction 25–30
    muscles 30–40
    receptors 40–1
taste – buds
    G-protein coupled taste
      receptors 40–1, 43–5,
      47–8
    ion channel taste
      receptors 41–2
    taste receptors 40–1
    what happens next? 45–7
tastes
    bitter 27, 47
    salty 27
    sour 27
    sweet 27, 47
    umami 27, 29–30, 45, 48
temperature detecting
  ion-channels 209
Terburg, David 315–16
Terman, Lewis 263
testosterone
    androgen receptors 235
    chemiluminescence 327
    competitive aggression
    325
    dominance 316, 325
    female supplements 148
    gene expression 330
    learning 331
    mechanism of action
    148–57
    social hierarchies
    315–16
    status 170
    supplements 141, 145–6

    transportation into cells
    150
    vasopressin 316
    winner effect 325
    see also lust
testosterone replacement
    therapy (TRT) 142
Tetris (computer game) 325
Tewkesbury, Joshua 211–12
thalamus 60
The Art of War 257
The Bell Curve 263
The Big Bang Theory (TV
    programme) 9
The Chemistry of Love 171
The Conquest of Happiness 355
The Diagnostic and Statistical
    Manual of Disorders
    (DSM-V) 69, 71
The Guardian 74
The Happiness Hypothesis 275
"the Imp" (inner voice)
    275–6
The Imprinted Brain: How Genes
    Set the Balance of the Mind
    between Autism and
    Psychosis 273
The Man I Love (song) 348
The Matrix (film) 54
The Selfish Gene 342, 350
The Simpsons (TV programme)
    28, 209, 265, 305
The Son Also Rises 310
The Wire (TV programme) 308
Theobrama cacao 25
thermodynamics
    crowd behaviour 215
    description 214
    laws 6, 96–7, 152–3, 157,
    214–15

thermodynamics, immiscible
liquids and heat receptors
(appendix)
energy levels 375–7
entropy 373–5
entropy/energy levels
relationship 377–82
folding proteins 387–90
free energy 371–5
heat-detecting ion
channels 390–6
hydrophobic effect 382–5
life and
thermodynamics 370–1
proteins, specific heat
capacity and heat
receptors 385–96
reactions 365–70
specific heat capacity and
hydrophobic effect
386–7
Thomas, J. Anderson Jr 177
thymine (nucleobase) 87, 234
TIR1/TIR3 taste receptors 45
Total Recall (film) 17
Trainspotting (film) 64
transcription
genes 105–7
proteins 106
*transient receptor potential
cation channel subfamily V
member 1 (TRPV1)* 128, 212,
217, 390
translation (genes) 107–8
Trifonov, Edward 83, 97, 116
tropomyosin 160–1
tryptophan 173–4
twins
dyzygotic 264–5
monozygotic 260, 264–5,
270

tyramine 174
tyrosine 173–4
Tzu, Sun 49

umami (delicious)
receptors 29–30, 37, 45, 48
uncertainty principle
(Heisenberg) 15
uracil (nucleotide) 86, 107–8
Uvnas Moberg, Kerstin 207–8

van Honk, Jack 315–16
van Houten, Coenraad
Johannes 26
vanillin 293
vanillyl mandelic acid (VMA)
293
vasopressin
amino acids 204
antagonists 205
mammals 202
neurotransmitters 316
production 203
social behaviour 203, 224
structure 205
synthesis 202–3
vasotocin 202
Vaulting Ambition 306
ventral tegmental area (VTA,
brain) 61, 66, 69, 181–2,
223, 280, 318, 320, 323, 326,
330
Verhoeven, Paul 349
Viagra (sildenafil) 163
Vigen, Tyler 146
vinegar 42
violence
Brunner syndrome 292–4
conclusions 299–300
epigenome 296–7
introduction 287–8

MAO A gene 294–6
 reward network 297–9
 what is a gene anyway?
 290–2
Vipassana (meditation) 358
voltage-gated ion channels
 56–7
Vonnegut, Kurt 349

Wallace, B. Alan 353
Wallace, Marjorie 260
Waller, Niels 267–8
Wang, Fei 317
Waorani people, Ecuador
 113–14
warm little pond (Darwin)
 99, 102
"warm touch" term 206, 208
wasps 128–9, 135
water
 origins of life 90
 structure 19–20
Watson, James 79
Watson, Neil V. 325
Weatherley, Nancy 287
"Wee for Wii" 53
Wegner, Daniel 348
Where's Wally (game)
 326

Who wants to be a Millionaire
 (TV programme) 376
Wilberforce, Samuel 102
Williams, George 115
Wilson, Edward 263, 306
winner effect 323–31
Winslow, James 205
Woese, Carl 85
Wolfe, Tom 305
Woloshin, Steven 145
work–life balance 355
Wróblewska, Joanna 242

Yalow, Rosalyn Sussman 326
Yang, Yaling 299
Yanomamö people,
 Venezuela 113–14, 299
yes-associated protein (YAP) 246
Younger, Jarred 188
Yuwiler, Arthur 322

Zaid, David 298
Zedong, Mao 312
Zilioli, Samuel 325
Ziomek, Carol 240
Zoloft (SSRI) 321
Zucker, Leila 8, 12
Zuckerberg, Mark 314
zygote development 237–40